BRADY

Fire Service Law

Lawrence T. Bennett

PEARSON

Prentice
Hall

Upper Saddle River, New Jersey 07458

Library of Congress Cataloging-in-Publication Data

Bennett, Lawrence T.
  Fire service law / Lawrence T. Bennett.
    p. cm.
  At head of title: Brady.
  Includes index.
  ISBN 0-13-155288-0 (alk. paper)
    1. Fire fighters—Legal status, laws, etc.—United States.   2. Fire fighters—Legal status,
laws, etc.—United States—Cases.   I. Title.   II. Title: Brady fire service law.

KF3976.Z9B56 2008
344.7305'37–dc22   2006053055

Acquisitions Editor: Stephen Smith
Developmental Editor: Alexis Breen Ferraro, Triple SSS Press Media Development
Marketing Manager: Katrin Beacom
Managing Editor (Production): Patrick Walsh
Production Editor: Julie Li
Manufacturing Buyer: Pat Brown
Cover Design:
Cover Illustration/Photo:
Director, Image Resource Center: Melinda Reo
Manager, Rights and Permissions: Zina Arabia
Manager: Visual Research: Beth Brenzel
Manager, Cover Visual Research & Permissions: Karen Sanatar
Composition/Full-Service Project Management: Karen Fortgang, Bookworks Publishing Services
Printer/Binder: Courier/Westford
Typeface: 10/12, TimesTen Roman

This book was set in TimesTen Roman by *Techbooks*. It was printed and bound
by Courier-Westford. The cover was printed by Phoenix Color Corp.

Pearson Education LTD.   Pearson Education Australia PTY, Limited
Pearson Education Singapore, Pte. Ltd   Pearson Education North Asia Ltd
Pearson Education, Canada, Ltd   Pearson Educación de Mexico, S.A. de C.V.
Pearson Education—Japan   Pearson Education Malaysia, Pte. Ltd

10 9 8 7 6 5 4 3 2 1
ISBN 0-13-155288-0

BRADY

Fire Service
Law

This textbook is dedicated to my wife, Jane, my three sons, Rob, John, and Bill, and to my three grandsons, Jasper, Warren, and Conrad, for all the love and joy they have given me.

Author and grandsons Jasper and Conrad Bennett in Piedmont (CA) Fire Department.

Author, grandson Warren Bennett, and author's pup, "Bruce," in Cincinnati (OH).

It is also dedicated to my brother, Bob Bennett, who served our nation as a helicopter and fixed wing pilot with the U.S. Army and Army National Guard for 26 years, including two tours in Vietnam, and again went to war in Desert Storm I. He fought his final battle, with cancer, in March 2006.

Author's brother, Bob Bennett.

Contents

Foreword

Chief William Kramer, Ph.D., Director, Fire Science Education, University of Cincinnati, and retired Fire Chief of Deerfield Township (OH) Fire and Rescue Department, addressing 9/11 Memorial Service for Warren County.

It has been my pleasure to work with author Larry Bennett for nearly ten years, first as a fellow member of the active fire service and then in an academic setting. It is rare that a person with detailed legal expertise has also maintained an active role in the fire service, as has Larry. This combination, however, has enabled Larry to present legal information that not only is quite current and relevant but is also directly applicable in the daily operations of a modern fire service organization.

This textbook covers the key areas in which a firefighter or fire department is likely to get into trouble or encounter litigation. The information presented explains how to avoid problems by learning from the experiences of others and how case law has shaped the rules of proper behavior for the future.

Just as fire prevention is considered superior to fire suppression, so, too, is litigation prevention superior to legal damage control. This textbook not only shows how the best practices drawn from the brightest minds in the national fire service can be used to prevent legal problems, but also how fire department officials should proceed once they are faced with a legal problem.

Our society today has been called "litigation happy." Indeed, the frequency of lawsuits is proliferating in both the private and public sectors. The fire service is not immune. Over the years, the protections afforded by "Sovereign Immunity" are eroding, and fire departments and their members are rightfully being held to quality standards. As the number of responsibilities of the fire departments have increased, their exposures to litigation have also increased.

This textbook contains easily understood legal principles, case studies, and relevant examples that will make the study of fire and EMS law a rewarding experience. The legal principles contained herein can serve as an excellent foundation of proper decision-making and protocol in a fire service organization.

Laws at the national, state, and local levels remain fluid and dynamic, even as this textbook goes to print. New case law replaces old, and the fire service adapts to a changing society. This textbook gives readers an opportunity to remain abreast of the very latest information available by directing them to pertinent websites, which are regularly updated with the latest court decisions affecting fire department operations.

Larry Bennett has been at my side to help during a few difficult moments, some of which could have been avoided if this textbook were available earlier. When key legal advice is necessary to protect the integrity of my fire service organization, I value the knowledge provided by Larry Bennett. It is satisfying to know that fire service members everywhere will have the benefit of his sage advice.

William M. Kramer, Ph.D.,
Director, Fire Science Education
University of Cincinnati

Preface

ORGANIZATION

This textbook is organized into 5 modules and 18 chapters. The author has selected these 18 chapters to make the course comprehensive and easy to follow.

The five modules are from the January, 2000 manual for the National Fire Academy's course, "Political and Legal Foundations of Fire Protection." The author teaches this course for the University of Cincinnati both as a "distance learning" online course and as a residency course. In spring 2006, Ed Kaplan of the National Fire Academy and Memphis University kindly appointed the author as "subject matter expert" to coordinate a team that would update the NFA's course manual for use by distant learning students. We wish to thank the National Fire Academy and Ed Kaplan, in particular, for allowing us to use these five modules.

MODULE I: OVERVIEW OF THE LAW

MODULE II: EMPLOYMENT AND PERSONNEL ISSUES

MODULE III: OPERATIONAL AND MANAGERIAL ISSUES

MODULE IV: THE FIRE OFFICIAL AS RULE-MAKER AND ENFORCER

MODULE V: LEGISLATIVE AND POLITICAL FOUNDATIONS

FEATURES

Each chapter has the following sections:

- **From the Headlines.** Each chapter begins with a current events news story, which illustrates some of the "real world" legal issues to be addressed in the chapter.
- **Key Terms.** A key terms list appears at the beginning of each chapter, and key terms are defined in the margins where first introduced. Full definitions are provided in the comprehensive glossary.
- **Key Statutes.** The most important federal statutes are discussed, along with reference to corresponding state statutes.
- **Introduction/NFPA.** Each chapter begins with an Introduction. At the end of the Introduction are references to the relevant National Fire Protection Association standards.
- **Case Studies.** Three main cases are in each chapter to reinforce key chapter concepts.
- **Chapter Review Questions.** These questions correspond to the three main cases to help the reader better understand the issues addressed in the main cases.
- **Expand Your Learning—Student and Group Activities.** Additional current cases are included for student and group assignments.

About the Author

The author during a "ride along" with the University of Cincinnati Air Care.

I have been a part-time firefighter/ emergency medical technician (EMT) for the past 26 years and a full-time attorney for over 35 years. This text has been written to share the legal lessons I have learned from current cases throughout the nation.

In my law practice as a partner in a Cincinnati law firm (where I write a monthly fire and EMS law newsletter, *www.katzmanlaw.com*), I have had the honor of representing fire chiefs, firefighters, and emergency medical service (EMS) personnel in federal and state courts, before the Ohio EMS Board, and before township trustees and other municipalities throughout Ohio. I have also been an adjunct professor at the University of Cincinnati, Fire Science Department and at Cincinnati State Technical and Community College, as well as a guest lecturer at the National Fire Academy, Executive Fire Officer class, and at the Ohio Fire Academy.

I am honored to be general counsel for the Warren County (OH) Fire Chiefs' Association and now, thanks to Dudley Smith, manager of air care at the University of Cincinnati Hospital, I have become legal counsel for the Commission on Accreditation of Air Transport Systems, which certifies air care operations nationwide.

This textbook was written not only to educate firefighters and EMS personnel but also to help them stay out of court and away from negative headlines in the media. By sharing "legal lessons learned" with current cases, this will help these dedicated public servants enjoy a rewarding career without the pain of litigation.

Legal Caution: This textbook does not seek to give legal advice, and the author specifically represents that no legal advice is provided. Readers must consult with their *own legal counsel,* who is knowledgeable of state and local laws and your department's practices, when faced with particular legal issues. In addition, cases cited in this textbook may be reversed or modified on appeal.

The author is contributing 50% of his royalties to the SW Ohio CISM team.

Acknowledgments

Many people have assisted the author in writing this textbook, but special thanks go to Deputy Chief Mike Cardwell, Deerfield Township (OH) Fire & Rescue Department, for his assistance on the initial draft of NFPA standards and many other resources. Recognition is also due to Richard L. Johnson, former training supervisor at the Ohio Fire Academy and retired firefighter from the Upper Arlington Fire Department, for his detailed review of the initial draft. Appreciation is also given to those who kindly reviewed my final draft: Fire Chief Trish Brooks, Forest Park (OH) Fire & Rescue Department; Fire Chief Nathan Bromen, Deerfield Township (OH) Fire Department; Professor John Glass, School of Fire Protection Engineering Technology, Seneca College, Toronto, Canada; and Dr. William Kramer, director of the Fire Science Department, University of Cincinnati.

I also wish to thank David Whelan, Director, Cincinnati Law Library Association, for his joyful research assistance. Finally, a special thank you to Alexis Ferraro, developmental editor at Triple SSS Press, who patiently and tirelessly helped me organize the chapters and content of this textbook.

REVIEWER LIST

Bill Benjamin, Professor
Johnson County Community College
Overland Park, KS

John M. Buckman III
Fire Chief
German Township Volunteer Fire Department
Evansville, IN

Debra J. Jarvis
Consultant, Ret. Chief, NFA Adjunct Instructor
Indianapolis, IN

John Eric Pearce, MPA, EMT-P
Instructor
Rio Hondo College
Whittier, CA

Brian Skipworth, FF/EMT-P
Public Safety Services Instructor
Morrow County EMS
Washington Township Fire Department
Mt. Gilead, OH

Jeff Travers
Program Coordinator
Great Oaks Institute
Cincinnati, OH

Introduction

This textbook will use the "case study" method to educate the reader about very current court opinions from throughout the nation that identify "real world" legal issues in the fire service. The goal is for the reader to identify the "legal lessons learned" in these cases and to share these lessons at their fire departments to avoid the pain and expense of litigation whenever possible. The author has been using this same case study approach in writing a fire law newsletter online for the past five years (*www.katzmanlaw.com*), and the feedback has been very positive.

Students can read entire case decisions by going to free web services such as *www.findlaw.com* or *www.lexisone.com* (provides cases for past five years), or to paid services such as *www.lexis.com* or *www.westlaw.com*. In this text, we delete footnotes and case citations when quoting from cases.

U.S. CONSTITUTION

The Founding Fathers drafted the Constitution to establish a federal government with limited powers, with three branches of government (executive, legislative, and judicial), each with "checks and balances" on the other branches. They also drafted the first ten amendments to the Constitution (known as the *Bill of Rights*) to protect citizens from their government. Throughout this textbook, you will read cases that involve disputes between individuals and governments.

It is therefore important that every student start by reading the U.S. Constitution, including the Bill of Rights (*www.usconstitution.com*).

LEGISLATIVE HISTORY

As you read these documents and want to understand better the intent of the Founding Fathers, you should turn to their other writings and speeches at the Continental Congress to learn the legislative history. Documents from the Continental Congress and the Constitutional Convention (1774–89) can be read at the Library of Congress website, *www.thomas.loc.gov*.

Courts involved in current litigation review the legislative history when determining the intent of Congress or the intent of state legislatures in enacting statutes. When the words of the statute are not clear, courts and the attorneys representing the disputing parties then carefully review the legislative history. For federal statutes, they look at the Conference Report in which the Senate and House members worked out final differences in the bill. They also turn to the Committee Reports of the Senate and House Committees that drafted the bills and to floor debates in the Senate and House. They also may look at the testimony of key witnesses.

14-Year-Old Firefighter Killed in Line of Duty

A detailed review of the legislative history is being used to help determine if a 14-year-old junior firefighter from Pennsylvania, killed riding his bike to the station, is included in the definition of "public safety officer" so his family can have the federal death benefits. An administrative law judge from the U.S. Department of Justice ruled that he was not covered, but a federal Court of Claims judge has found legislative history to support the family's contention that he is covered. The Department of Justice has filed an appeal. (See Chapter 2 and the March 27, 2006, decision *Julie Amber Messick, Administratrix, Estate of Christopher Nicholas Kangas, deceased v. United States,* 2006 U.S. Claim LEXIS 76.) To avoid similar disputes, a congressman from Pennsylvania has introduced a bill that would amend the definition of "public safety officer" to clearly include junior firefighters.

U.S. CODE

The federal statutes enacted by the U.S. Congress are published in the United States Code, and they can be read at *http://uscode.house.gov/search/criteria.shtml.* These statutes and the "legislative history," including U.S. Senate and U.S. House of Representatives committee reports, prepared remarks, and floor debates of senators and representatives, can be read at the Library of Congress website, *www.thomas.gov.*

CODE OF FEDERAL REGULATIONS

Federal agencies must enforce the statutes enacted by Congress and signed into law by the president. These agencies publish detailed regulations in the Code of Federal Regulations; for example, 29 CFR 553.212(a) is published by the U.S. Department of Labor, Wage and Hour Division and concerns firefighters earning overtime pay under the Fair Labor Standards Act. The regulations can be read at *www.gpoaccess. gov/cfr/index.html.* We suggest that you add this to your computer as a favorite site.

CASE PRECEDENTS—U.S. SUPREME COURT

In Chapter 1, we will review the American legal system and the role of the U.S. Supreme Court (see their excellent website, *www.supremecourtus.gov*). Decisions of this nine-member court not only finally decide who won and who lost the case but may also establish precedents that help shape the law for future cases in all federal and state courts. Only a very small percentage of cases ever reach the U.S. Supreme Court; therefore, case decisions by the various federal circuit courts of appeal set precedent in their circuit. Likewise, state supreme court decisions interpreting state statutes and law set precedent in state and local courts.

FIRST AMENDMENT "LANDMARK" DECISION

On May 30, 2006, the U.S. Supreme Court decided *Garcetti et al. v. Ceballos,* 126 S. Ct. 1951 (2006), *www.supremecourtus.gov,* a case that may have great impact on all public employees, including firefighters, and their First Amendment rights.

Richard Ceballos, a supervising district attorney, had filed a federal lawsuit against his public employer, the Los Angeles County District Attorney's Office, alleging he suffered "retaliation" (he had been transferred, his responsibilities reduced, and he was passed over for promotion) when he investigated a complaint by a defense attorney that a deputy sheriff had lied in an affidavit for a search warrant. Ceballos says that a thorough investigation, two memos to his supervisors, and a meeting with the county sheriff's office and the deputy, which became "heated," were all part of his official duties. Therefore, he is protected by "free speech" under the First Amendment to the U.S. Constitution. The U.S. Supreme Court majority (five to four) disagreed and ruled against him. Justice Kennedy wrote the majority decision (joined by newly appointed Chief Justice Roberts, newly appointed Justice Alito, and Justices Scalia and Thomas), stating, "We hold that when public employees make statements pursuant to their official duties, the employees are not speaking as citizens for First Amendment purposes, and the Constitution does not insulate their communications from employer discipline." The opinion stresses, "Our holding likewise is supported by the emphasis of our precedents on affording government employers sufficient discretion to manage their operations." (Read also the dissenting opinion of Justice Stevens, the dissenting opinions by Justice Souter, joined by Justice Ginsburg, and the dissenting opinion of Justice Breyer.)

APPEALS TO THE U.S. SUPREME COURT

To have an appeal decided by the U.S. Supreme Court, four of the nine justices must vote for a *writ of certiorari* (an order to the lower court to send the records). In the vast majority of the appeals filed with the Court, the *writ of certiorari* is denied (see "Orders" at *www.supremecourtus.gov*).

World Trade Center—Firefighter Families Can't Sue City of New York or Motorola

In January 2006, the U.S. Supreme Court refused to hear the appeal of spouses of New York City FDNY firefighters killed at the World Trade Center in *Lucy Virgilio et al. v. City of New York and Motorola, Inc.* They were appealing from the decision of the U.S. Court of Appeals for the Second Circuit, 407 F.3d 105 (2d Cir. April 29, 2005), which upheld the dismissal of their lawsuit by a U.S. District Court judge.

Cases are often decided on the interpretation of federal or state statutes, enacted by Congress or the state legislature. In the FDNY survivors lawsuit, *Virgilio v. City of New York and Motorola, Inc.,* the Court of Appeals wrote,

> In a series of tragic and terrifying attacks on September 11, 2001, terrorists killed thousands in Pennsylvania, Virginia, and New York, caused extensive damage to the Pentagon, and brought about the collapse of the North and South Towers of the World Trade Center ("WTC"). As with other catastrophes, true heroes responded, not the least among them the brave firefighters, police, and first-response units of the City of New York. Plaintiffs are the personal representatives of firefighters who lost their lives in responding to the WTC following the attacks. Plaintiffs' complaint focuses on the failure of radio-transmission equipment in the North and South Towers that prevented firefighters from receiving evacuation orders before the Towers' collapse. Plaintiffs commenced this action for wrongful death against New York City (the "City") on December 22, 2003, and filed an amended complaint as of right on January 20, 2004, that added Motorola, Inc. ("Motorola") as a defendant.
>
> Plaintiffs claim that Motorola negligently and intentionally provided the City with radio-transmission communication equipment for firefighters that Motorola knew to be

ineffective in high-rise structures like the Towers of the WTC, that Motorola made fraudulent material misrepresentations to secure contracts with the City, and that those acts and representations caused decedents' deaths. Plaintiffs also press a series of wrongful death claims against the City based upon its alleged failure to meet duties imposed on the City under New York law to provide adequate and safe radio-transmission equipment. Finally, in Count 8 of the Amended Complaint, plaintiffs allege that the City and Motorola engaged in concerted action in an attempt to deprive firefighters of adequate protection and to "engage in fraudulent misrepresentations and deceitful conduct."

Shortly after the disaster, Congress passed the Air Transportation Safety and System Stabilization Act (the "Air Stabilization Act" or the "Act"). Pub. L. No. 107-42, 115 Stat. 230 (2001). The statute limited liability for the air carriers involved in the tragedy to their insurance coverage, *see* Air Stabilization Act § 408(a); created the Victim Compensation Fund (the "Fund") to provide no-fault compensation to victims who were injured in the attacks and to personal representatives of victims killed in the attacks, *see id.* §§ 402(3), 405(a)(1), (b), (c); and provided an election of remedies—all claimants who filed with the Fund waived the right to sue for injuries resulting from the attacks except for collateral benefits, *see id.* § 405(c)(3)(B)(i). On November 19, 2001, the Act was amended by the Aviation and Transportation Security Act (the "Aviation Security Act"). Pub.L. No. 107-71, 115 Stat. 597 (2001). Significantly, the amendments extended liability limits to aircraft manufacturers, those with a proprietary interest in the WTC, and the City of New York, *see id.* § 201(b), while allowing Fund claimants to sue individuals responsible for the attacks notwithstanding the waiver, *see id.* § 201(a).

A number of September 11-related cases were consolidated before Judge Hellerstein. Judge Hellerstein dismissed the complaint in an unpublished decision. *See Virgilio v. Motorola, Inc.,* No. 03 Civ. 10156(AKH), 2004 WL 433789 (S.D.N.Y. Mar. 10, 2004). The district court adopted Judge Haight's decision noting that "the waiver provision applies to all of the claims against Motorola and the City of New York.... As plaintiffs have elected their remedy, they have also waived the right to bring a civil action for damages sustained as a result of the terrorist-related aircraft crashes of September 11, 2001."

We agree with the district court that under the plain language of the statute, claimants who have filed claims with the Fund have waived "the right to file a civil action ... for damages sustained as a result of the terrorist-related aircraft crashes of September 11, 2001" and that the waiver bars claims for "damages sustained" against non-airline defendants. We affirm the district court's determination and find plaintiffs' claim barred by their election of remedies.

FINDING CASE DECISIONS ON THE INTERNET

The cases in this textbook are from court decisions nationwide and are public records obtainable on several websites, including the free services, *www.findlaw.com* or *www.lexisone.com,* or paid services such as *www.lexis.com* or *www.westlaw.com.* We suggest you mark one of these sites as a "favorite" on your computer.

Federal court decisions, including U.S. Court of Appeals and U.S. District Court opinions and U.S. Supreme Court decisions, can be found at *www.uscourts.gov.*

CASE CITATIONS

Each case in the textbook will include a citation showing where it has been published in an official court reporter. For example, the famous "Miranda warnings" used by police officers came from the U.S. Supreme Court decision in *Miranda v. Arizona,* 384 U.S. 436 (1996). The case was decided in 1996 and is published in the United States Reports, volume 384 beginning on page 436.

U.S. Court of Appeals decisions are published in the Federal Reports, such as *Minch v. City of Chicago,* 363 F.3d 615 (7th Cir. 2004), concerning mandatory retirement age for firefighters. The case was decided in 2004 and is in the Federal Reporter, 3rd Volume, volume 363 beginning on page 615.

U.S. District Court (trial judge) decisions are published in the Federal Supplement, such as *Patterson v. Tortolano,* 359 F.Supp.2d 13 (D. Mass. 2005), concerning termination of a firefighter for missing work without notifying the department.

State court cases are published in regional reporters, such as *Ogden City Corporation v. Harmon,* 116 P.3d 973 (Utah Ct. App. 2005), concerning a fire captain who organized a fundraising activity using strippers.

Very recent court decisions are first published with reference to LEXIS, such as *Christian Reed Smith v. State,* 2005 Ida App. LEXIS 109 (Idaho Court of Appeals, 2006). This court decision is discussed in Chapter I, Case Study 1-1. LEXIS is a paid service of Lexis/Nexis; opinions on LEXIS are available through law libraries or other paid subscribers. There are also other subscriber services, including WESTLAW.

UNPUBLISHED DECISIONS

This textbook includes both published and unpublished decisions that relate to the fire service. "Unpublished" decisions refer to written judicial decisions that dispose of the case, but the court or trial judge who wrote the opinion concluded that it is important to the parties in the case but not worthy of formal, nationwide publication in the law journals. In the age of the Internet, however, these decisions are still distributed widely by LEXIS.

For example, on May 30, 2006, the California Court of Appeals issued an unpublished decision in *Harry K. Vernon v. City of Berkeley,* 2006 Cal. App. Unpub. LEXIS 4620. The court upheld the dismissal of a lawsuit by an African-American firefighter with the skin condition, *pseudofolliculitis barbae* (PFB). The Court held he must comply with the "no beard" rule for firefighters, based on prior California precedents. The court agreed with the City of Berkeley that the city was required by state law to comply with the Occupational Safety and Health Act (OSHA) Respiratory Protection Regulations adopted by Cal–OSHA in 1997 (see Chapter 6 for more cases on beards).

In May 2006, the U.S. Supreme Court voted to amend a longstanding rule prohibiting attorneys in federal courts from citing unpublished decisions in their briefs or in oral arguments. The new rule will allow these cases to be cited for all unpublished opinions written after January 1, 2007.

American Legal System
Search Warrants in Arson Investigations, Fire Code Enforcement, and Civil Litigation

1 CHAPTER

National Fallen Firefighters Memorial Service, National Fire Academy, Emmitsburg (MD). *Photo by author.*

Key Terms

qualified immunity, p. 5
governmental immunity, p. 5
summary judgment, p. 5

unreasonable searches and
 seizures, p. 8
exigent circumstances, p. 8

voluntary consent, p. 9

From the Headlines

Rock Band Manager Gets Four Years in Prison

The Station nightclub fire in Warwick (RI) killed 100 people on February 20, 2003. Daniel Biechele, former tour manager for the Great White band, pleaded guilty to 100 counts of misdemeanor involuntary manslaughter for the pyrotechnics display that he set up. Superior Court Judge Francis Darigan sentenced him to 15 years in prison but suspended all but 4 years based on the remorse of the defendant. After his imprisonment, he sent handwritten letters of apology to the families of victims, including the family of Nick O'Neill, age 18. His letter began, "To Nick O'Neill's family and friends, Please allow me to start by apologizing for the part I played in Nick's tragic death and for taking so long to convey this apology to you."

Fire Inspections/Local Fire Marshal Remains Defendant

A federal judge has dismissed as defendants the State of Rhode Island and the state fire marshal from civil lawsuits stemming from the February 20, 2003, fire at The Station nightclub but has refused to dismiss the town's fire marshal, who had inspected the nightclub and deemed it safe. Other remaining defendants include the club owners; members of the Great White band; the band's management company, Manic Music, Inc.; and the American Foam Corporation, the company that sold the club highly flammable foam to use for soundproofing. State law caps civil liability of municipalities to $100,000 per family. The U.S. Department of Commerce's National Institute of Standards and Technology (NIST) has issued a report on the fire at The Station nightclub in West Warwick (RI) that killed 100 people (*http://fire.nist.gov*).

U.S. Supreme Court Holds That Consent Search of Home Is Illegal If Other Spouse Tells Police to Get Search Warrant

In *State of Georgia v. Scott Fritz Randolph,* No. 04–1067 (S. Ct. 3/22/2006) (S. Ct. 2006), in a five to three vote, the U.S. Supreme Court held on March 22, 2006, that cocaine found in Mr. Randolph's bedroom cannot be used in evidence against him, because he objected when his estranged wife told the police they could search their home. This decision, which can be read at *www.supremecourtus.gov,* has implications in arson cases in which multiple people live in the same residence and one refuses to consent to a search.

Tribute to the fallen at the World Trade Center in New York City. *Photo by author.*

World Trade Center, New
York City. *Photo by author.*

◆ INTRODUCTION

Our nation's Founding Fathers established three branches of federal government to protect both citizens and state governments from the federal government. They created three branches of power in the federal government: the executive branch, with a president and cabinet officials who head various federal agencies; the legislative branch, with the U.S. Congress, with two senators from each state and multiple representatives from each state based on population; and the judicial branch with the Supreme Court (nine justices) and lower federal courts (U.S. Courts of Appeal and U.S. District Courts).

Each branch of the federal government has powers meant to "check and balance" the other. The judicial branch, for example, includes lifelong appointments for Article III judges (from Article III of the U.S. Constitution) who are appointed by the president, and must be confirmed by the U.S. Senate. These judges cannot be removed from the bench except by impeachment.

◆ VIOLATIONS OF THE FOURTH AMENDMENT

The judicial branch can "control" the action of FBI agents, fire marshals, and other executive branch law enforcement officers by declaring confessions obtained in violation of Miranda rights and evidence seized in violation of the Fourth Amendment to the Constitution to be inadmissible in evidence in the criminal trial. The Fourth Amendment provides that no search warrants shall be issued "but upon probable cause, supported by Oath or affirmation, and particularly describing the place to be searched, and the persons or things to be seized."

◆ ARSON SEARCHES WITHOUT WARRANT

One of the leading U.S. Supreme Court decisions on arson investigations and search warrants is *Michigan v. Tyler*, 436 U.S. 499 (1978) (*www.supremecourtus.gov* or *www.findlaw.com*). Loren Tyler and Robert Tompkins were convicted in a Michigan trial court of conspiracy to burn real property; their convictions were reversed by the Supreme Court. The Supreme Court held, "an entry to fight a fire requires no warrant, and that, once in the building, officials may remain there for a reasonable time to investigate the cause of the blaze. Thereafter, additional entries to investigate the cause of the fire must be made pursuant to the warrant procedures governing administrative searches. Evidence of arson discovered in the course of such investigations is admissible at trial, but if the investigating officials find probable cause to believe that arson has occurred and require further access to gather evidence for a possible prosecution, they may obtain a warrant only upon a traditional showing of probable cause applicable to searches for evidence of crime. These principles require that we affirm the judgment of the Michigan Supreme Court ordering a new trial." Please carefully read this important decision and focus on the facts, the various searches that occurred, and the impact of the decision on current arson investigations.

◆ CONVICTION REVERSED—EVIDENCE SEIZED ILLEGALLY FROM A METH LAB

Some appellate decisions can appear to be "splitting hairs" and can be very frustrating to law enforcement. For example, on May 17, 2006, the Court of Appeals of Idaho in *State of Idaho v. Bryan L. Bunting,* 2006 Ida. App. Lexis 49, vacated the conviction of Bryan Bunting for possession of methamphetamine and "trafficking in methamphetamine by manufacturing," because of an illegal search of his garage. There had been an explosion in the garage, and a neighbor called 9-1-1. Police and fire were dispatched, and police arrived first. With Bunting's consent, three police officers entered the house and the garage. They observed a scorched table with a butane torch, a burner, a small propane tank, a roll of aluminum foil, and an empty pill bottle. Bunting told the police that a neighbor was using his garage to clean car parts. Firefighters arrived and entered the garage looking for hot spots. They notice a closed blue cooler under the scorched table. At the request of a police officer, the firefighters opened the cooler and found jars, tubing, and funnels—all used to manufacture methamphetamine. The police officers then called a drug enforcement officer to the scene, who entered the garage and observed the cooler and other items. After securing these items, the drug officer filed an affidavit and obtained a search warrant for both Bunting's home and his neighbor's home. The warrant was served the next day, and more drug manufacturing equipment was seized. The trial judge denied his defense attorney's motion to suppress the evidence, and the jury convicted Bunting (unanimous vote of 12 jurors required in most states). He was sentenced by the trial judge to seven years in prison: five years for trafficking and two years for possession. The Idaho Court of Appeals (three to zero vote) vacated the conviction, holding, "The initial entry into Bunting's garage and house was lawful. Neither traditional law enforcement exigency nor the following-in-the-footsteps of emergency personnel existed (since there was no fire when the fire department arrived) and the subsequent police entries into Bunting's garage and house were unlawful. The lawfully obtained evidence presented to the magistrate was insufficient to show probable cause to issue a search warrant."

◆ FIRE CODE ENFORCEMENT—QUALIFIED IMMUNITY OF FIRE MARSHALS AND FIRE INSPECTORS

In this chapter, we will also review civil litigation against arson investigators and other fire service officials, in which arson charges were subsequently dropped or the jury acquitted the defendant. Fortunately, most lawsuits are quickly dismissed against fire service officials on the basis of **qualified immunity**. Qualified immunity applies to lawsuits against individual public officials. Lawsuits against cities and other political subdivisions are also generally dismissed on the basis of **governmental immunity**.

> **qualified immunity** Public officials, such as fire inspectors, generally will not be personally liable if the officials can prove they were performing their official duties and their conduct did not constitute willful or wanton misconduct.

◆ LAWSUITS BY SURVIVORS ALLEGING FAILURE TO ENFORCE

If your community suffers a terrible loss of life at a nightclub, such as the 2003 fire at The Station nightclub in Rhode Island, the families of the deceased will likely retain legal counsel to file lawsuits. Your fire department's inspection records ("public records" in most states) will be copied and carefully scrutinized by the expert witnesses retained by plaintiff's counsel—perhaps retired fire inspectors from elsewhere in the county. If a lawsuit is filed, the complaint may name the city or other political subdivision, the fire inspector, and other members of the fire department. Hopefully, the trial judge will grant the defense motion to dismiss the lawsuit or grant **summary judgment** for the defendants, including the city (governmental immunity) and the individual fire inspectors and other fire officials (qualified immunity) prior to trial.

> **governmental immunity** A political subdivision, such as a city, township, or village, may enjoy immunity from liability under state statutes as long as its employee was adequately trained and not acting with willful or wanton misconduct.

◆ FAILURE TO CORRECT FIRE AND MUNICIPAL CODE VIOLATIONS

In *City of Quincy v. Weinberg,* 363 Ill. App.3d 654, 844 N.E.2d 59 (January 26, 2006), the city filed a lawsuit in September 1999 seeking an injunction against Donald L. Weinberg, a licensed attorney, for repeated failure to correct numerous fire and municipal code violations. In August 1999, they had obtained a search warrant and found privies, a casket, old signs, and garbage on his property that would preclude access by fire vehicles and presented a danger to neighbors. The court proceedings included years of court monitoring. Finally, the trial judge found the defendant in direct civil and criminal contempt in October 2004 and fined him $33,141.34. In April 2005, the court ordered the defendant to divest himself of title to the real estate as a sanction for his contempt. The property owner appealed, and the Illinois Appellate Court held that a forced sale of the property was too drastic. "[I]n selecting sanctions, a court is obliged to use the least possible power to the end purpose," citing the U.S. Supreme Court decision in *Spallone v. U.S.,* 493 U.S. 265, 276 (1990).

> **summary judgment** A trial judge dismisses the lawsuit prior to trial, based on his or her conclusion that, as a matter of law, the facts in the case support one party's position.

◆ FAILURE TO ENFORCE BUILDING OR FIRE CODES

Numerous lawsuits have been filed for alleged negligent failure to enforce building or fire codes. For example, the Supreme Court of Florida held in *Trianon Park Condominium Association, Inc. v. City of Hialeah,* 468 So. 2d 912 (1985); 1985 Fla. LEXIS

University of Cincinnati Air Care lands at Children's Hospital with a young driver injured in a motor vehicle accident. *Photo by author.*

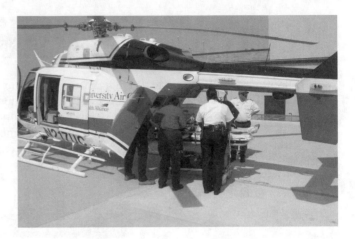

2962, that even though 49 of the 65 new condos had leaky roofs, they could not sue the City of Hialeah. "To give effect to Trianon's position would make the taxpayers of each governmental entity liable to individual property owners for the failure of governmental inspectors to use due care in enforcing the construction requirements of the building code." The Florida Supreme Court majority (two dissenting judges) held, "Governments must be able to enact and enforce laws without creating new duties of care and corresponding tort liability that would, in effect, make the governments and the taxpayers virtual insurers of the activities regulated." In Florida, however, a lien filed against a home from a fire code enforcement action will not prevent the homeowner from selling the property, under the Florida "homestead" property protection laws. *Aristides Pelecanos, et al. v. City of Hallandale Beach,* Court of Appeals of Florida, Fourth District (November 23, 2005), 914 So. 2d 1055, 2005 App. LEXIS 18638.

◆ KEY STATUTES

FOURTH AMENDMENT TO THE U.S. CONSTITUTION

The Fourth Amendment provides, "No Warrants shall issue, but upon probable cause, supported by Oath or affirmation, and particularly describing the place to be searched, and the persons or things to be seized." See *www.usconstitution.com.*

Arson investigators and firefighters at scenes of arson fires and other crimes must understand the rules concerning "search and seizure." When in doubt, consult your local or federal prosecutor and obtain a court-ordered search warrant. Fire personnel who obtain evidence in violation of the Fourth Amendment may learn later that the trial judge has granted a defense motion to "suppress the evidence," and the defendant avoids conviction and punishment.

STATE CONSTITUTIONS

Many states have their own constitutional provisions prohibiting the use of illegally seized evidence in state criminal prosecutions. All state and federal courts are bound by U.S. Supreme Court decisions requiring trial judges from allowing illegally seized evidence to be shown to a jury.

Greater Cincinnati/
Northern Kentucky
Airport crash truck. *Photo
by author.*

◆ **CASE STUDIES**

CASE STUDY 1-1   Search of Arson Site without a Warrant

Christian Reed Smith v. State, 2005 Ida App. LEXIS 109 (Idaho Court of Appeals, 2006); substituted opinion for 2006 Ida App. LEXIS 13, Docket No. 31315 (Idaho Court of Appeals, December 1, 2005).

FACTS

Christian Smith's couch caught fire inside his apartment located in an old, Victorian-style house. He and a friend dragged the couch downstairs and out to the driveway where they tried to extinguish the fire with a garden hose. They left the smoldering couch in the driveway. When dragging out the couch, they damaged the wall in the hallway. He left a note for the landlord, stating that he was sorry for the damage and that it would be repaired. Smith went to bed a few hours later.

Early in the afternoon the following day, the fire department responded to a 9-1-1 call from a neighbor. Firefighter Jason Blubaum responded in his own vehicle and found a small amount of smoke coming from a corner of the couch. At the landlord's request, he went to the Smith apartment to confirm there was no further fire danger and to investigate the cause of the fire. The landlord let him in the apartment, and he determined there was no further fire danger and that the couch apparently had been placed too close to a baseboard heater.

When the firefighter left the apartment in about five minutes, he saw Battalion Chief Aaron Watson and an engine company arrive non-emergency. He briefed the battalion chief, who sent him home and told the engine company to return to the station.

The landlord asked the battalion chief to look at the apartment. Chief Watson, accompanied by Moscow Police Officer Keith Cox, re-entered the apartment, and found evidence that may have disproved that the couch fire was caused by being too close to the baseboard heater. They also saw a small, white ivory pipe, some marijuana seeds, and a packet of rolling paper sitting on the coffee table.

The battalion chief needed to talk with Smith, who was at work. He and the police officer drove to Smith's workplace, and he agreed to ride back with them in the police car. Once back at the apartment building, the police officer informed Smith for the first time that they had already been in the apartment and had seen the marijuana pipe and seeds. He said they intended to search his apartment and asked him to sign a consent form. The police officer and another urged him to sign, telling him they could easily get a search warrant based on what they saw on the first entry. After first balking, Smith eventually signed the consent form. The police officers and the battalion chief re-entered the apartment, and the officers seized one growing marijuana plant and several items of drug paraphernalia.

Smith was charged with manufacture of a controlled substance. Prior to trial, his attorney filed a motion to suppress on the grounds that the first investigative entry by the battalion chief and the police officer violated the Fourth Amendment and that Smith's subsequent consent to search was involuntary.

YOU BE THE JUDGE

Discuss whether the trial judge should suppress the evidence by answering the following questions:

1. Was the initial entry by the battalion chief and the police officer lawful?
2. Was Smith's consent given involuntarily?

HOLDING

The trial judge ordered the evidence suppressed. The State of Idaho appealed to the Idaho Court of Appeals, and a three-judge panel reversed the decision and ordered the trial judge to hold a hearing on the consent issue. The Court of Appeals decision states, in part:

unreasonable searches and seizures   Under the Fourth Amendment to the U.S. Constitution, evidence seized without a court-ordered search warrant may be determined to be inadmissible in evidence, unless the seizure is deemed by a judge to be reasonable under the circumstances.

exigent circumstances   Emergency circumstances, including extinguishing fire, will authorize warrantless entry for cause and origin investigation.

> The Fourth Amendment to the United States Constitution and article I, § 17 of the Idaho Constitution protect individuals against **unreasonable searches and seizures**. Although a warrantless search is presumptively unreasonable, it may still be permissible if it falls within a recognized exception to the warrant requirement or is otherwise reasonable under the circumstances. Under the **exigent circumstances** exception, agents of the state may conduct a warrantless search when the facts known at the time of the entry, along with reasonable inferences drawn thereupon, demonstrate a "compelling need for official action and no time to secure a warrant." *Michigan v. Tyler,* 436 U.S. 499, 509 (1978); *State v. Barrett, 138 Idaho 290,* 293, 62 P.3d 214, 217 (Ct. App. 2003). The burden is on the government to show the applicability of this exception to the warrant requirement.
>
> Turning to the present case, we find Chief Watson and officer Cox's entry to investigate the fire in Smith's apartment analogous to those entries found to be constitutional in *Tyler*. Specifically, Chief Watson's entry to investigate the fire was no more than an actual continuation of firefighter Blubaum's initial entry. The record shows that Chief Watson and officer Cox entered Smith's apartment within minutes after Blubaum's exit. This is a substantially shorter time frame than that allowed between entries in *Tyler* and strongly suggests that the entries were part of a single fire investigation. Furthermore, the singular nature of this investigation is demonstrated by noting the sheer fortuity of firefighter Blubaum looking out the window to see Chief Watson and then exiting the apartment to meet him. Had Blubaum not peered out the window and simply awaited the fire chief's imminent arrival, there could be no legitimate claim that two separate investigations occurred. This demonstrates, to paraphrase the *Tyler* Court, that the only purpose to be served by requiring firefighter Blubaum to remain in Smith's apartment would have been to remove any doubt about the legality of Chief Watson's later entry. *Id.* at 511. We agree with other courts that constitutional distinctions should not turn on such occasions of happenstance.... Accordingly, the entry by Chief

Watson and officer Cox to investigate the cause and extent of the couch fire in Smith's apartment must be considered no more than an actual continuation of firefighter Blubaum's initial entry and, therefore, proper under the Fourth Amendment as interpreted in *Tyler.*

Our holding comports not only with the letter, but also the spirit of *Tyler.* That decision did not view firefighting as a task delineated by clear and unchanging boundaries, but rather acknowledged that the circumstances of particular fires and the role of firemen and investigating officials will vary widely. *See Tyler,* 436 U.S. at 510 n.6. The Supreme Court predicted that, in some circumstances, it may be necessary for officials—pursuing their duty both to extinguish the fire and to ascertain its origin—to remain on the scene for an extended period of time repeatedly entering or re-entering the building or buildings. Thus, *Tyler* admonished that appropriate recognition must be given to the exigencies that confront officials serving under these conditions, to the public safety, as well as to individuals' reasonable expectations of privacy. *Id.* Here, the exigencies confronting Chief Watson—a recently-doused couch in a driveway, a trail of soot leading to an apartment in an old building housing other tenants, and a request to investigate by the building's landlord—were of sufficient proportion to render his warrantless but timely reentry "reasonable," regardless of whether a subordinate firefighter had made some preliminary determinations. These principles require that we reverse the decision of the district court finding no exigency for the second entry.

The U.S. Supreme Court set forth the standard by which the **voluntariness of consent** is measured in *Schneckloth v. Bustamonte,* 412 U.S. 218 (1973). The Supreme Court held that consent must be "free from any aspect of official coercion," *Id.* at 229, and further explained:

> [T]he Fourth and Fourteenth Amendments require that a consent not be coerced, by explicit or implicit means, by implied threat or covert force. For, no matter how subtly the coercion were applied, the resulting "consent" would be no more than a pretext for the unjustified police intrusion against which the Fourth Amendment is directed.

In the present case, officer Cox's statement that a warrant would be secured if Smith refused consent was based on Cox's belief that he had sufficient probable cause to secure a warrant, i.e., his personal observation of the marijuana pipe and seeds when he was in Smith's apartment to investigate the fire with Chief Watson. As a result, officer Cox did not falsely or erroneously claim to have the ability to secure a warrant and, under *Abeyta,* this claim did not *ipso facto* render Smith's consent involuntary. Our analysis does not end here, however, since the threat of obtaining a search warrant is only one factor in determining whether consent was voluntary. Rather, it must still be determined whether this, together with all of the other circumstances surrounding Smith's consent, amounted to the type of coercion prohibited by the U.S. Supreme Court in *Schneckloth.*

Smith claims that the officers' coercive actions in this case include: the police taking Smith from work to his apartment in the backseat of a police car; the officers' precluding Smith from walking away; one or more uniformed police officers accompanying Smith at all times; the police telling Smith they had already been inside his apartment and had already seen his contraband; the officers' suggestion that a warrant could be procured; the police telling Smith he would be detained, restrained from entering his apartment, and that the scene would be frozen until a warrant was obtained; the police telling Smith he needed to hurry to make his decision regarding consent, and that there would be lenient treatment if Smith consented; and the police implying that the judge would be angry if a warrant was sought on a Sunday. The district court, however, made no detailed findings of fact regarding Smith's consent nor related conclusions of law under the standards set forth in *Schneckloth* and *Abeyta.* We are, therefore, unable to make a final determination as to the voluntariness of Smith's consent as a matter of law. Consequently, we must remand this issue back to the district court for further proceedings.

voluntary consent Individuals may normally consent to a search of their property without a search warrant, as long as they know that they can refuse.

LEGAL LESSONS LEARNED

To avoid motions to suppress evidence and obtain more pleas of guilty, consider keeping a fire officer at the property until an inspector can be present to conduct a thorough search of the premises. Arson investigators and fire inspectors should periodically conduct training programs and invite their local prosecutor or other attorney knowledgeable of search and seizure cases to review recent court decisions and discuss "best practices."

CASE STUDY 1-2   Cause and Origin Search of Detached Storage Shed

Kristy R. Garrison v. Texas, 2005 Tex. App. LEXIS 5323; Case No. LWC-4752 (Texas Court of Appeals, Second District, Fort Worth, July 7, 2005).

FACTS

A one-alarm fire occurred at Ms. Garrison's residence on March 25, 2003. The fire department responded and eventually extinguished the blaze. Investigator Hardeman was dispatched to determine the cause and origin. When he arrived, the incident commander (IC) told him that the main body of the fire had been extinguished but the fire department was still eliminating hot spots. Hardeman told Garrison that he would be entering the residence to investigate.

The residence is a one-story, wood-frame house with a covered patio and carport in the back. The investigator discovered several one-gallon cans of acetone in the patio/carport. He also noticed several extension cords running through the house and learned that a gas generator was being used to provide electricity to the residence.

He followed the extension cords to the rear of the house and followed them to a "lean-to" shed attached to the house. The rear wall of the shed had burned away, and the investigator ultimately determined that the generator was the heat source of the fire.

Additional extension cords extended into the yard. An orange extension cord was draped over a van parked near the house and extended into a detached metal shed. Investigator Hardeman followed the cord to the shed. The shed doors were open about four inches. He opened the doors to the shed and observed a methamphetamine lab. He notified the IC, and police and narcotics officers were called to the scene and seized the evidence. Detectives later obtained a search warrant to search the residence. No further evidence was discovered.

Garrison was charged with manufacture and possession with intent to manufacture. Her attorney filed a motion to suppress, arguing that there had been no fire in the shed.

During the hearing on the motion, the following questions were asked of the investigator:

> **"[Prosecutor:]** And, Investigator Hardeman, why were you trying to search and find the—where the power cords snaked to or trying to follow the power cords from what you determined to be the gas generator, the origin of the heat source?
> **[Hardeman:]** It was—as part of the investigation, my fire investigation, I'm trying to determine what caused this fire. With an electric generator, it's required to provide electricity. I was trying to determine what was being required of this generator. Was it being overloaded? What caused it to malfunction, if it did malfunction, to cause this fire?

[Prosecutor:] And is that normal protocol for you when you determine—in this type of situation?
[Hardeman:] Yes, it is."

YOU BE THE JUDGE

1. If you were defense counsel, what arguments would you make to urge the trial judge to suppress the evidence?
2. If you were the prosecutor, what arguments would you make to allow the evidence to be admitted and shown to the jury?

HOLDING

Garrison pleaded guilty to manufacture of a controlled substance of 400 grams or more of methamphetamine and to possession of methamphetamines with intent to manufacture. She was sentenced to five years' confinement but reserved the right to appeal the trial judge's decision not to suppress the evidence.

The Court of Appeals wrote:

> Because Investigator Hardeman saw the extension cord running from the generator to the shed and entered the shed in furtherance of his investigation into the cause of the fire, his entry did not require a warrant. *See Michigan v. Tyler,* 436 U.S. at 510, 98 S. Ct. at 1950. If officers are permitted to conduct a warrantless cause and origin search of a home in the immediate aftermath of a fire, they are equally entitled to conduct a warrantless cause and origin search of a detached storage shed when it reasonably appears related to the cause or origin of the fire.
>
> Garrison also contends that Investigator Hardeman could not enter the shed without a warrant because there was never any fire in the shed. Again, this argument would be more persuasive if Investigator Hardeman had not observed the orange extension cord running from the generator to the shed. It was this extension cord, linked to the point of the origin of the fire, that made Investigator Hardeman suspect that something in the shed may have caused the generator to overload or malfunction. Because Investigator Hardeman was authorized to search areas where the cause of the fire would likely be found, he was—in light of the orange extension cord—authorized to search the shed. *See Michigan v. Clifford,* 464 U.S. at 295 n.6, 104 S. Ct. at 647 n.6 ("In searching solely to ascertain the cause [of the fire], firemen customarily must . . . search other areas where the causes of fires are likely to be found. An object that comes into view during such a search may be preserved without a warrant.") Because Investigator Hardeman discovered the methamphetamine lab and contraband in plain view inside the shed, this evidence was properly used to establish probable cause to obtain the search warrant.
>
> We hold that the trial court did not abuse its discretion by determining that Investigator Hardeman's cause and origin investigation fell within the exigent circumstances exception to the warrant requirement or by denying Garrison's motion to suppress.

LEGAL LESSONS LEARNED

When a fire inspector is investigating the cause and origin of a fire and the fire department is still at the scene, courts will generally recognize broad authority of the inspector to search all possible sources of the fire. A fire department Standard Operating Procedure (SOP) concerning this subject could be very helpful in court if an arson defendant files a motion to suppress evidence.

CASE STUDY 1-3   Hotel Fire/Governmental Immunity

Loren Hage et al. v. Stade, 304 N.W.2d 283 (Minn.), 1981 Minn. LEXIS 1211.

FACTS

A suspected arson fire at the Stratford Hotel in Breckenridge (MN) on January 28, 1977, killed 13 people. Their heirs sued the hotel owner, the State of Minnesota, and the state fire marshal. The only smoke alarm that sounded was a manual alarm in the lobby, about ten feet from the hotel check-in desk. The hotel clerk that night was 81 years of age and apparently was overcome by smoke before she could get to the fire alarm. There were no smoke alarms or sprinkler in the hotel.

The hotel was last inspected on June 11, 1975, by Deputy State Fire Marshal C. C. Tallman. This was several months prior to the state adopting the Uniform Fire Code. There were no follow-up inspections by the state fire marshal's office. In June, 1976, the hotel was inspected by the fire chief of Breckenridge, but apparently there was no discussion of the need for smoke detectors. The trial judge granted summary judgment to the State of Minnesota and the fire marshal.

YOU BE THE JUDGE

Was this lawsuit properly dismissed prior to trial, on the basis that the State enjoys governmental immunity and the state fire marshal enjoys qualified immunity?

HOLDING

Yes.

The majority opinion (there were four dissenting judges) stated,

> The thrust of plaintiffs' claims is that the State of Minnesota and its agents negligently failed to enforce proper fire safety measures at the Stratford Hotel and that this failure directly contributed to the death of plaintiffs' decedents. Plaintiffs argue that state agents failed to perform a required annual inspection of the hotel, negligently conducted the inspections that did take place, and failed to require correction of safety hazards known to them. The trial court did not reach the issue of whether respondents had been negligent. Instead, the trial court concluded that the state had undertaken to enact fire safety regulations and perform inspections for the benefit of the public as a whole and not to protect a particular class of persons. Therefore, negligent inspection by a state agency did not breach a duty to third parties, and the third parties (here, plaintiffs) did not have a cause of action against the state. We agree with the trial court that the state can only be liable for negligent omissions if it owes a special duty to a particular class of persons.
>
> The distinction between a public duty and a special duty was recently discussed and reaffirmed by this court in *Cracraft v. City of St. Louis Park,* 279 N.W.2d 801 (Minn. 1979). In *Cracraft,* the City of St. Louis Park had enacted a fire prevention ordinance which provided that all places of public assembly must be inspected at least once every month. Pursuant to the ordinance, a city inspector inspected a high school. He found no major violations, and sent a letter to school officials outlining the minor violations that he had discovered. Forty-four days after his inspection, a large drum of duplicating fluid ignited on the loading dock of the school, which was adjacent to the football field. Two students were killed and a third was severely injured. The presence of the drum on the dock was a violation of the city fire code.

This court held in *Cracraft* that although the common-law doctrine of sovereign immunity from tort liability had been abolished by the Minnesota Legislature, Minn. Stat. § 3.736 (1978), in response to our decision in *Nieting v. Blondell,* 306 Minn. 122, 235 N.W.2d 597 (1975), the distinction between public duties and special duties is still valid. A duty owed to the public in general cannot be the basis of a negligence action, but a special duty owed to individual members of the public or to a special class can be the basis of such a claim. As stated in *Cracraft*:

[A] municipality does not owe any individual a duty of care merely by the fact that it enacts a general ordinance requiring fire code inspection or by the fact that it undertakes an inspection for fire code violations. A duty of care arises only when there are additional indicia that the municipality has undertaken the responsibility of not only protecting itself, but also undertaken the responsibility of protecting a particular class of persons from the risks associated with the fire code violations. 279 N.W.2d at 806.

We hold as a matter of law that there are no facts indicating actual knowledge in this case. The record indicates that, although Deputy Fire Marshal Tallman had inspected the Stratford Hotel in the past, a state inspector had not been to the hotel since the enactment of the Uniform Fire Code. No claim is made by plaintiffs that the Stratford failed to meet the former fire safety requirements of Minn. Stat § § 157.01-.14 (1974). Thus, there is no evidence that the state had actual knowledge of any dangerous conditions which were violations of any fire code and which would serve to impose a special duty on the state under the first *Cracraft* factor.

LEGAL LESSONS LEARNED

While the majority of the court upheld the trial court's dismissal of this lawsuit, four dissenting judges would have permitted the lawsuit to go to a jury trial. One dissenting judge wrote, "I would permit plaintiffs to use the alleged breach of Minnesota [fire code] as a basis for proving that the state breached its duty of care." One could only imagine how jurors would react with the courtroom packed with the survivors of the 13 people who died in the fire. Likewise, in The Station nightclub fire in Rhode Island, where 100 died, a civil suit against a fire marshal and his city could be very emotional.

■■■

Chapter Review Questions

1. In Case Study 1-1, *Christian Reed Smith v. State,* the court addressed the issue of arson investigators entering a premises, after firefighters have left the scene, to conduct a search for evidence of arson. In Ohio and many other states, the state fire marshal's office will often send their arson investigator to assist at a major fire. Sometimes it will take hours for the state investigator to arrive at the scene, particularly if the state's local investigator is already at another scene or is not otherwise available. Discuss what practical steps a fire department can take to ensure that they have maintained "control" of the fire scene so that state inspectors will normally not need to obtain a search warrant.

2. In Case Study 1-2, *Kristy R. Garrison v. Texas,* the court addressed the issue of how far an arson investigator may search when the fire was in the house but evidence used to convict her was found in the detached garage. Discuss what instructions you would give to fire department arson investigators about

how far they can search without a search warrant and when it would be wise to obtain a search warrant.

3. In Case Study 1-3 *Loren Hage, et al. v. Stade,* the state and the fire marshal were dismissed from the lawsuit by the families of 13 who died in the hotel fire. Describe what steps a local fire department can take to ensure that all hotels in the fire district are regularly inspected.

■ ■

Expand Your Learning

Read the following cases and complete the individual student or group assignment, as directed by your instructor.

1. **100 Dead at Nightclub Fire.** Please read the NIST report on the tragic Rhode Island fire at *http://fire.nist.gov.* Your fire chief has asked you to prepare a memo of the "lessons learned" for the fire inspectors on your fire department. Please summarize those findings in the report that are most damaging to the town's fire marshal.

2. **Consent to Search by Wife.** In *State of Georgia v. Scott Fritz Randolph,* No. 04-1067, the U.S. Supreme Court on March 22, 2006, held that law enforcement may not rely on the consent of a wife to search a home when her husband tells police they cannot search without a search warrant. The decision can be read at the U.S. Supreme Court's website, *www.supremecourtus.gov.*

   Mrs. Randolph had separated from her husband and was living in Canada. On July 6, 2001, she was at her husband's house to pick up clothes, and she called the police when her husband took their son to a neighbor's house to prevent her from taking the child to Canada. Police arrived, and she told them of her husband's cocaine upstairs in his bedroom. The police asked him for consent to search the house. He refused (he is an attorney). They then called his parents, who owned the house—they refused. They then asked his wife for consent—she agreed, and took them upstairs to his bedroom and found a "piece of cut straw" and cocaine. Later, they obtained a search warrant and seized more cocaine. The trial judge refused to suppress the evidence, finding she still co-owned the house and could consent to the search. The Georgia Court of Appeals reversed, and the Georgia Supreme Court agreed the consent was illegal. The State of Georgia appealed to the U.S. Supreme Court, which held that none of the cocaine may be introduced in evidence.

   Assuming there is an arson investigation of a fire in a college dormitory, and suspicion focused on three college roommates, discuss whether arson investigators may search their dormitory room for evidence if two of the roommates consent, but one declines.

See Appendix A for additional Expand Your Learning activities related to this chapter.

Line of Duty Death (LODD) and Safety
Litigation and Fireman's Rule

Cincinnati (OH) firefighters salute outside the church at the funeral of firefighter Oscar Armstrong, killed in a flashover in a residential structure fire. *Photo by author.*

Key Terms

Fireman's Rule, p. 19   **rescue doctrine, p. 25**

From the Headlines

$2.8 Million/Turnout Gear

Family of a volunteer firefighter, killed January 1, 2001, in basement fires, settles lawsuit filed in federal court in New Jersey. The lawsuit alleged the fibers in his turnout gear had broken down in the heat. A study at North Carolina State

University's College of Textiles reportedly found that the fibers would fail under certain conditions.

Cincinnati Firefighter Grateful to Be Alive

A Cincinnati FD firefighter was crawling along the floor and looking for the seat of a fire in a house when his air hose got caught on something, and his mask was ripped off his face. He expressed thanks to two other firefighters who were able to get him out of the building and to the minimum manning of four firefighters per engine.

National Program to Reduce LODD

National Fallen Firefighters Foundation announces "Everyone Goes Home Program" aimed at reducing fatalities by 25 percent over the next five years and by 50 percent within 10 years. One immediate action that fire departments can take is to require a self-contained breathing apparatus (SCBA) to be worn during overhaul when the air quality is poor; another action is to install fire station exhaust systems for all engines and trucks. Another action for fire departments is to include in their training the video by Olympia, Washington, firefighter Mark Nobel, who later died of cancer in 2005. In Mark Noble's video, he discusses in graphic detail his brain tumor, its possible connection to his not wearing a SCBA during overhaul, and lack of ventilation devices on engines in the station. The video can be obtained through *www.ergonomics.org*.

Cemetery service for Cincinnati firefighter Oscar Armstrong. *Photo by author.*

House where flashover killed Cincinnati firefighter Oscar Armstrong. *Photo by author.*

◆ INTRODUCTION

This chapter will review cases involving litigation concerning firefighter line of duty deaths (LODD) and other safety issues. Approximately 100 firefighters die in the line of duty each year. There is a national effort to reduce these tragic losses. The National Fallen Firefighters Foundation, located in Emmitsburg (MD) has issued 16 major initiatives to reduce line of duty deaths, following the Firefighter Life Safety Summit in Tampa (FL) in March 2004. Additional information can be obtained by calling (301) 447-1365 or visiting *www.firehero.org*.

◆ LITIGATION—SHOCK THE CONSCIENCE

There is increased litigation against fire departments and senior fire officers, filed by the surviving families of firefighters killed in the line of duty, seeking compensatory and punitive damages from a jury. Many of these suits allege deprivation of constitutional rights (42 U.S.C. 1983), claiming that the fire department failed to implement improvements identified in prior incidents. If the plaintiff can establish sufficient facts to "shock the conscience" of the court, the trial judge may order the case to trial.

◆ NIOSH FIRE FIGHTER INVESTIGATION AND PREVENTION REPORTS

Fire departments should also identify "lessons learned" from LODDs at other departments. It is strongly recommended that fire departments assign a specific officer (such as a safety officer or training officer) to read every NIOSH Fire Fighter Fatality Investigation and Prevention Report (*www.cdc.gov/niosh/fire/*) and identify training and other action items for your department. (You can register online to receive these reports automatically.)

For example, on May 22, 2006, NIOSH issued three reports: Florida career firefighter/EMT who died in an ambulance crash (F2005-12), New York captain who died of a pulmonary embolism (F2005-33), and a New Jersey firefighter who died of sudden cardiac arrest (F2005-30).

◆ FIREFIGHTERS WITH CARDIAC ARRESTS

Cardiac arrest is a common cause of death of firefighters. The New Jersey firefighter in Fire Fighter Fatality Investigation Report F2005-30 was 58 years of age and was found slumped over a rowing machine at his station. The autopsy revealed a 75 percent blockage in his left anterior descending coronary artery. The 252-member department did not have annual physicals, nor did it provide medical questionnaires and clearance for employees using SCBAs. The NIOSH report states, "The Occupational Safety and Health Administration (OSHA)'s Revised Respiratory Protection Standard requires employers to provide medical evaluations and clearance for employees using respiratory protection. Such employees include fire fighters who utilize SCBA in the performance of their duties. These clearance evaluations are required for private industry employees and public sector employees in states operating OSHA-approved state plans. New Jersey is a state-plan state for public sector employees only; therefore, public sector employers are required to comply with OSHA standards."

◆ FEDERAL DEATH BENEFITS—14-YEAR-OLD JUNIOR FIREFIGHTER'S DEATH

Applications for LODD benefits are submitted to the U.S. Department of Justice, and occasionally there is a legal dispute. For example, the family of the 14-year-old firefighter, killed by a motorist as he rode his bike to the fire station after alarms were sounded for a fire, has been on appeal. On March 27, 2006, the U.S. Court of Federal Claims in *Julie Amber Messick, Administratrix, Estate of Christopher Nicholas Kangas, deceased v. United States,* 2006 U.S. Claim LEXIS 76 held that Christopher was a "public safety officer." On May 4, 2002, Christopher was riding his bicycle to the Brookhaven (PA) Volunteer Fire Department. A hearing officer with the U.S. Department of Justice, Bureau of Justice Assistance, had refused line of duty death benefits to the estate of this volunteer apprentice firefighter. The hearing officer concluded that he was not a "public safety officer" because Pennsylvania Child Labor Law 48.3 limited junior firefighters from "suppression of fires." The Court of Federal Claims disagreed. The court stated that Christopher had been issued official firefighter equipment, had completed 58.5 hours of in-house training in 22 different areas of firefighting, including rescue operations, SCBA, electronics, carbon monoxide detection, and hose rolling. He was also certified in CPR and had responded to 24 house drills. The Brookhaven Fire Department had authorized Christopher to "be part of the firefighting team by participating at the scene of a fire, including bringing out portable equipment and fire hoses, providing food, drink and first aid to other firefighters, and cleaning up after fires that were under control." The court therefore concluded that he met the definition of "public safety officer" under the Public Safety Officers Death Benefits Act, Omnibus Crime Control and Safe Streets Act of 1968, as amended, 42 U.S.C. 3796-3796c. The U.S. Department of Justice has filed a notice of appeal, so this case continues. Legislation has also been introduced by Congressman Curt Weldon (a former PA fire chief) to further broaden the definition of "public safety officer."

◆ FIREMAN'S RULE

This chapter addresses the **Fireman's Rule** and litigation by families of deceased firefighters against business and homeowners in which building code violations may have caused the fire. Chapter 18, Legislative Process, discusses instances in which several states have enacted laws modifying the Fireman's Rule. The Fireman's Rule prohibits public safety officers, including firefighters, EMS, and police officers, or their survivors from filing suit against a property owner for their death or injury unless there has been an extraordinary hazard created by the property owner.

Fireman's Rule A rule based on judicial decisions in many states prohibiting public safety officers, including firefighters, EMS, and police officers, from filing suit against a property owner for their death or injury, unless there has been an extraordinary hazard created by the property owner.

◆ OCCUPATIONAL SAFETY AND HEALTH ACT (OSHA)

This chapter reviews the role of OSHA and states that have their own OSHA-approved state plans. Congress enacted the Occupational Safety and Health Act in 1970, 29 U.S.C. 651, et seq., and the department issued enforcement regulations, 29 CFR Parts 1900 to 2400. OSHA generally does not apply to employees of state and local governments, including fire departments, unless their state has an OSHA-approved state plan.

Some states that are federal OSHA states have also issued regulations that require all fire departments to follow the OSHA respiratory protection standard. Other OSHA standards may also be adopted by states, such as Personal Protective Equipment, Employee Access to Medical and Exposure Records, and Hazard Communication.

In states with an OSHA-approved state plan, inspections by state safety compliance officials often occur after firefighter fatalities and serious injuries. For example, on November 25, 2002, three firefighters died when the roof of an auto parts store collapsed in Coos Bay, Oregon (see NIOSH Fire Fighter Fatality Investigation F2002-50 *http://www.cdc.gov/niosh/fire/*), the State investigators issued 16 citations against the fire department concerning the lack of an incident command system meeting NFPA standards, violation of two-in, two-out, inadequate personnel accountability, and other charges (*http://www.iaff.org;* search "coos"). The fire department strongly disputed these citations but agreed to settle for $8,000 without admitting wrongdoing.

OSHA has published its own schedule of fines for willful violations, penalties of not less than $5,000 for each violation and not more than $70,000; for serious violations, a discretionary penalty of up to $7,000 for each violation, with downward adjustment for the employer's good-faith improvements; and for other than serious violations, a discretionary penalty of up to $7,000 for each violation. Under federal OSHA, willful violations involving a death can result in criminal prosecution, punishable by a court-imposed fine of up to $250,000 for an individual, $500,000 for an organization, and up to six months in prison. On May 22, 2006, the U.S. Department of Labor issued its revised *Employment Law Guide* (*www.dol.gov/compliance/guide/index/htm*), including a section on occupational safety and health and a section on whistleblower protection.

◆ WHISTLEBLOWER PROTECTION

Whistleblower protection also exists under OSHA and many state statutes. For employees covered by OSHA, they may file a complaint with OSHA if they believe their employer has retaliated or discriminated against them for raising a safety concern internally or reporting a possible violation to OSHA, 29 U.S.C 660(c). After the bankruptcy of Enron Corporation and widespread accounting fraud by other companies, Congress passed the Sarbanes-Oxley Act and directed OSHA to investigate any retaliation against corporate fraud whistleblowers, 18 U.S.C. 1514A.

◆ NFPA STANDARDS

- *NFPA 1500, Standard on Fire Fighter Occupational Health and Safety* (2002 edition). This standard identifies best practices for creating an environment for firefighters that is as safe as possible in all facets of emergency and non-emergency operations, for the purpose of reducing line of duty deaths.
- *NFPA 1584, Recommended Practice on the Rehabilitation of Members Operating at Incident Scene Operations and Training Exercises* (2003 edition). This standard requires that an ALS-level emergency medical crew be present at all working incidents and that all personnel undergo an evaluation after strenuous activity, in an effort to identify firefighters who may be having a cardiac arrest or other medical emergency.
- *NFPA 1710, Standard for the Organization and Deployment of Fire Suppression Operations, Emergency Medical Operations, and Special Operations to the Public by Career Fire Departments* (2004 edition). This standard identifies benchmarks for career organizations for the arrival of emergency service personnel in a timely manner and with adequate resources to mitigate the emergency situation. It was written in an effort to create an overall staffing level that was safer for firefighting operations.
- *NFPA 1720, Standard for the Organization and Deployment of Fire Suppression Operations, Emergency Medical Operations, and Special Operations to the Public by Volunteer Fire Departments* (2004 edition). This standard identifies benchmarks for volunteer organizations.
- *NFPA 1981, Standard on Open-Circuit Self-Contained Breathing Apparatus (SCBA) for Emergency Services* (2002 edition). This standard identifies issues of safety and interoperability regarding SCBAs and integrated Personal Alert Safety Systems (PASS), including necessary inspection, care, and maintenance issues.

  [Note: In response to new requirements for SCBA cylinder interchangeability, the National Fire Protection Association has issued a Report On Proposals *concerning a 2006 revision. See* www.nfpa.org/Assests/Files/PDF/ROP/1981-07-ROP.pdf. *See also International Association of Fire Chiefs Association information,* www.iafc.org, *December 28, 2005.]*

- *NFPA 1982, Standard on Personal Alert Safety Systems* (1998 edition). In December 2005, NIOSH reported that exposure to high heat may cause the loudness of the PASS alarm signals to be reduced. NIOSH tests showed this reduction may begin in temperatures as low as 300°F (*www.cdc.gov/niosh/homepage.html*). This standard addresses issues of design and operation of firefighter PASS systems and devices, including their mandatory use when operating in immediate danger to life and health (IDLH) environments. This standard should be reviewed in conjunction with NFPA 1981.
- *NFPA 1901, Standard for Automotive Fire Apparatus* (2003 edition). This standard identifies criteria for designing and specifying safe fire apparatus. Appendix D of the standard concerns the use and refurbishment of older apparatus.

  [Note: Task Force Chairman Robert Tutterow advised Fire Department Safety Officers' Association, Apparatus Specification and Maintenance Symposium, January 2006, that NFPA should include the following: remove SCBAs from cabs; hold officers and operators accountable for unsafe operations; prohibit the use of personal vehicles for emergency response.]

Phoenix (AZ) Fire Department's Command Training Center, includes a replica of the grocery store where firefighter Brent Tarver died in the line of duty March 14, 2001. Read NIOSH Report F2001-13 (*www.cdc.gov/niosh/fire/*). *Photo by author.*

◆ **KEY STATUTES**

42 U.S. CODE 1983

This statute has been used frequently in lawsuits against police officers and their municipal employers after police shootings and other serious incidents. These so-called "1983 suits" often allege that the officer was inadequately trained; therefore, the municipality should pay damages for the deprivation of the federal constitutional rights of the injured or killed individual. In recent years, lawsuits under the statute are being filed in federal and state courts against fire department officials and their municipalities, such as the suit filed against the former fire chief of Washington (DC) in the Cherry Road fire (see Case Study 2-1).

PUBLIC SAFETY OFFICERS' BENEFITS (PSOB) ACT OF 1976 (42 U.S.C. SEC. 3796)

The Public Safety Officers' Benefits (PSOB) Act established a $250,000 federal death benefit (this is adjusted with increases in Consumer Price Index; $283,385 as of October 1, 2005) and educational assistance for children of public safety officers who have died as the direct and proximate result of personal injury suffered in the line of duty. The program is administered by the U.S. Department of Justice, Bureau of Justice Affairs. Go to *www.ojp.gov/BJA/grant/psob/psob_def.html* or *www.usdoj.gov* and search for "Public Safety Officers' Benefit Program."

PUBLIC SAFETY OFFICERS' BENEFITS ACT AMENDMENT, OCTOBER 2000 (PUBLIC LAW 106-390, SEC. 305)

The PSOB Act was amended to include Federal Emergency Management Agency (FEMA) employees as "public safety officers" under the PSOB Act if they are performing official, hazardous duties related to a declared major disaster or emergency. The legislation also indicates that state, local, and tribal emergency management or civil defense agency employees working in cooperation with FEMA are, under the same circumstances, considered public safety officers under the PSOB Act.

THE MYCHAL JUDGE POLICE AND FIRE CHAPLAINS PUBLIC SAFETY OFFICERS' BENEFITS ACT OF 2002 (42 U.S.C. SEC. 3796)

The statute is retroactive to September 11, 2001, and amends the Public Safety Officers' Benefits Act of 1976 in the following ways:

- Includes chaplains in the definition of public safety officers.
- Defines a chaplain as "including any individual serving as an officially recognized or designated member of a legally organized volunteer fire department or legally organized fire or police department who was responding to a fire, rescue, or police emergency."
- Adds a new category of beneficiary. If the public safety officer had no surviving spouse or eligible children, the individual designated as the beneficiary on the officer's most recent life insurance policy is eligible for benefits. An eligible child is defined as any natural, illegitimate, adopted, or posthumous child or stepchild of the public safety officer who, at the time of the officer's death, was 18 or under, or between 19 and 22 (inclusive) and a full-time student at an eligible educational institution, or age 18 or older and incapable of self-support due to mental or physical disabilities.

HOMETOWN HEROES SURVIVORS BENEFITS ACT OF 2003 (PUBLIC LAW 108-32)

On December 15, 2003, President Bush signed Senate Bill S.459 into law, extending the PSOB to public safety officers who die from heart attacks or strokes within 24 hours of "non-routine stressful or strenuous activity." These activities include fire suppression, rescue, hazardous material response, and emergency medical services. (See *www.ojp.gov/BJA/grant/psob/psob_death.html.*)

◆ CASE STUDIES

CASE STUDY 2-1   Line of Duty Death

Estate of Anthony Phillips et al. v. D.C., August 1, 2006, 2006 U.S. App. LEXIS 19280, U.S. Court of Appeals for the District of Columbia Circuit.

The three-judge appellate court reversed the federal district judge's two prior decisions: March 2, 2005, U.S. District Court for the District of Columbia, 355 F. Supp.2d 212 (D.D.C. 2005); 2005 U.S. Dist. LEXIS 232; reconfirming March 31, 2003 decision at 257 F. Supp. 2d 69, 2003 U.S. Dist. LEXIS 5029.

FACTS

A lawsuit was filed by the families of the two deceased firefighters, Anthony Sean Phillips, Sr., and Louis J. Matthews, and by the two injured firefighters after a townhouse fire on Cherry Road in 1999. The defendants include the District of Columbia and the following officers sued in their official and personal capacities at the date of the fire: former Fire Chief Donald Edwards, former Deputy Fire Chief Thomas Tippett, Battalion Chief Damian A. Wilk, and Lt. Frederick C. Cooper, Jr.

Read the NIOSH Fire Fighter Fatality Investigation Report, FACE 99-F21, *www.cdc.gov/niosh/firehome.html,* FACE-99-F21.

The suit alleged a violation of "constitutional rights" under 42 U.S.C. 1983 (deprivation of rights by a governmental entity) and "intentional torts" (willful misconduct or gross negligence), claiming senior fire officials had a "policy and custom" of not improving fire ground operations, ignoring shortcomings identified after another firefighter, John Carter, died in the line of duty two years earlier at a grocery store fire. The District of Columbia filed a motion for summary judgment, claiming governmental immunity, and the fire officers claimed "qualified immunity." The report on the LODD in 1997 called for improved incident command, improved radios, and improved structure fire procedures. This lack of follow-up, the suit claimed, resulted in the deaths and injuries at the Cherry Road fire.

The NIOSH report documents the importance of coordination between the primary search teams and the truck companies performing ventilation, particularly in a basement fire. That coordination apparently did not happen in this fire, with tragic results.

The U.S. District judge, in his two lengthy written opinions, described improper ventilation of the walk-out basement that killed the two firefighters on the primary search team on the ground floor. The battalion chief/incident commander, using his portable radio, had twice asked Lt. Redding on the primary search team for their location. The IC received no response. The IC refused permission for another engine company to spray water in the basement, because he didn't know the location of the team on the first floor. A truck company arrived at the scene and began ventilating the first floor by a positive pressure fan in the window. Another truck company went to the rear of the townhouse to ventilate from the basement.

There was an immediate increase in fire and temperature on the first floor. Lt. Redding ran outside with his face and back burning. He called the IC on the radio to warn him that other firefighters were still in the building. Phillips and Matthews were ultimately located by an RIT team—neither survived their extensive burns.

The defendants asked the trial judge to dismiss the lawsuit, arguing that the courts have long recognized "governmental immunity" for municipalities and "qualified immunity" for government officials who must make policy decisions, balancing limited budgets against the need for improved IC training and better radios.

YOU BE THE JUDGE

1. Should the city be dismissed from the lawsuit?
2. Should the former fire chief be dismissed from the lawsuit?

HOLDING

The U.S. Court of Appeals ordered that the former fire chief be dismissed from the lawsuit since he enjoyed "qualified immunity." This decision should also lead to the dismissal of the deputy fire chief and all other individual defendants from the lawsuit.

The appellate judges disagreed with the federal district judge's written decision. The district judge wrote,

> The Court is ever-cognizant of the need for deference to government policy decisions— decisions that require a weighing and balancing of numerous, often competing, factors and interests amid inevitable constraints—and the slippery slope that would result if

(Sec.) 1983 actions against government officials were allowed to proceed every time a policy choice negatively impacted an individual or group of individuals. In this case, however, plaintiffs' allegations do not amount to a case wherein the Fire Chief and Deputy Chief simply made a policy decision in light of budgetary, manpower or other outside pressures that it was not feasible for the Fire Department to enact mandatory operating procedures, adequately train firefighters, or heed clear warnings that continued inaction would almost certainly lead to death or serious injury. Rather, plaintiffs allege that the City and the three individual defendants did nothing because they simply don't care. If true—and again, at this stage of the proceedings this Court must assume the allegations as pled—this deliberate indifference continues to shock the Court's conscience and nothing in the FOP decision [2004 decision of U.S. Court of Appeals for the D.C. Circuit] persuades this Court that its previous conclusion is flawed.

The former fire chief took an immediate appeal. The U.S. Court of Appeals reversed the district court judge and ordered former Fire Chief Donald Edwards dismissed from the lawsuit, finding that he enjoyed "qualified immunity." The court based its ruling on the U.S. Supreme Court decision in *DeShaney v. Winnebago County Department of Social Services,* 489 U.S. 189 (1989), where a mother filed suit against social workers and their state employer for failing to act after observing a father had inflicted some injuries on a child. The father later inflicted permanent injuries. The Supreme Court held that the social workers and the state could not be sued under 42 U.S.C. 1983 since they never had actual custody of the child. The Supreme Court stated that "if the Due Process Clause does not require the State to provide its citizens with particular protective services, it follows that the State cannot be liable under the clause for injuries that could have been averted had it chosen to provide them." (489 U.S. at 196-97) Applying this standard to the Cherry Road fire case, the Court of Appeals dismissed Chief Edwards from the case. The court wrote, "The facts here, like those in *DeShaney,* are indeed tragic. Joseph Morgan and Charles Redding suffered severe injuries and Anthony Phillips and Louis Matthews died attempting to save lives and property of others. But the Constitution does not provide a basis for holding Edwards individually responsible. The Firefighters have not alleged the deprivation of a clearly established constitutional right and Edwards is therefore entitled to qualified immunity from suit in his individual capacity."

LEGAL LESSONS LEARNED

Fire chiefs and other senior fire officials must address and document their follow-up actions carefully in response to reports of deficiencies in safety-related issues. Ventilation is a valuable tool in fighting a structure fire, but when conducted prematurely, it can have tragic consequences. Fire departments that are reluctant to detail their mistakes and corrective action plans should understand that failure to do so can lead to additional fire ground deaths. See the impressive report by the Cincinnati Fire Department after the LODD of Oscar Armstrong (*www.cincyfire.com*).

CASE STUDY 2-2   Highway Safety/Fireman's Rule

Espinoza v. Schulenburg, Arizona Supreme Court, March 15, 2006, 129 P. 3d 937, 2006 Ariz. LEXIS 27.

FACTS

Elizabeth Espinoza, a Phoenix firefighter/EMT, was driving home with her nine-year-old daughter on February 10, 2002, when she came upon an automobile accident. The car was being driven by an underage driver, Carrington Schulenburg. Her parents were also in the car. Firefighter Espinoza stopped and offered assistance to a Department of Public Safety police officer who was already at the scene. The accident vehicle was partly on the left shoulder and partly in the left through lane, so the firefighter leaned into the vehicle to activate its emergency flashers. As she did so, another vehicle driven by Casey Barnett struck the rear of the vehicle. Firefighter Espinoza suffered a broken hip, broken wrist, torn knee ligaments, a broken finger, and other injuries. Espinoza sued the operator of the other vehicle, the owner of the accident vehicle for allowing her daughter to drive without a license, and the Arizona Department of Public Safety. The Schulenburgs filed a motion for summary judgment, requesting that the lawsuit against them be dismissed based on the Fireman's Rule, and the trial judge agreed. The firefighter appealed, claiming she was off-duty and therefore the Fireman's Rule did not apply.

YOU BE THE JUDGE

Should the off-duty firefighter be permitted to sue the Schulenburgs, the owners of the vehicle?

HOLDING

The Arizona Supreme Court held that the Fireman's Rule does not apply to an off-duty firefighter rendering assistance as a Good Samaritan.

The court wrote:

> (W)e find that to the extent the fireman's rule bars recovery because the firefighter or police officer expects to encounter hazards while on the job, or that he renders aid not from a humanitarian impulse to help but because he is being paid, these justifications can not support extending the rule to an off-duty public safety professional who makes a voluntary effort to assist. This type of effort is precisely what the **rescue doctrine** was designed to protect, and we can conceive of no public policy that would be advanced by precluding such a volunteer from the benefit of the rescue doctrine."

The Arizona Supreme Court held:

> Off-duty professionals who risk injury to volunteer aid in emergency situations fall outside the policy rationale for the firefighter's rule because they are under no obligation to act. In Volunteering, they are thus going beyond the scope of their employment. They are also not at the time paid and may not be equipped to confront the situation. They generally lack identification, safety equipment, or the support of trained colleagues. They are, instead, acting just like those whom the rescue doctrine is intended to protect. Application of the firefighter's rule to preclude suit by such off-duty professionals is therefore inappropriate.

rescue doctrine Protects individuals acting as "Good Samaritans" when trying to rescue or render aid to others in an emergency from liability for their negligence, as long as they acted in good faith and were not grossly negligent.

LEGAL LESSONS LEARNED

The firefighter in this case was seriously injured, and she will now "get her day in court" and hopefully recover substantial damages. Fortunately, her nine-year-old daughter was not hurt. The Fireman's Rule has been abolished or amended in many states, either by statute or case decisions (see Case Study 2-3 and also Chapter 18).

[Note of caution: Rendering aid on a highway is extremely dangerous. On January 17, 2006, the U.S. Fire Administration sent out notice of the death of an Ann Arbor (MI) female firefighter who was struck and killed by an out-of-control vehicle at the scene of several weather-related motor vehicle accidents on January 7, 2006.]

CASE STUDY 2-3   Building Code Violation/Fireman's Rule

Foiles v. V.J.L. Construction Corp., 794 N.Y.S. 2d 27 (N.Y. App. Div. 2005).

FACTS

Firefighter Foiles responded to a basement fire in a three-family house and received severe injuries, leaving him permanently disabled. The fire was caused by flammable liquids being stored too close to a water heater. The firefighter sued both the building owner and the tenant, and they filed motions to dismiss based on the Fireman's Rule.

YOU BE THE JUDGE

Should the firefighter's lawsuit be dismissed?

HOLDING

No. New York has amended the Fireman's Rule by a statute (see Chapter 18 for states that have also amended the rule by statute or court decision).

The trial court refused to dismiss the lawsuits, and the appellate court agreed. Under New York law, liability is imposed for injury or death of a firefighter if there is a connection between violations of building codes and the injury. The New York statute is designed to encourage property owners to comply with building codes. In this case, the basement had been rented out, but the building owner never obtained a certificate of occupancy as required by the municipal code. A fire inspector, just three weeks earlier, had issued a citation and required removal of a partition and installation of a window.

LEGAL LESSONS LEARNED

The Fireman's Rule has been amended or abolished in many states. The New York statute allowing suits by firefighters when the fire was caused by a building code violation is very attractive middle ground. There is a strong public policy served when building owners can be held liable if firefighters can sue for injuries related to violation of building and safety codes.

■ ■

Chapter Review Questions

1. In Case Study 2-1, *Estate of Anthony Phillips et al. v. D.C.,* the U.S. Court of Appeals reversed the decision of the federal district judge, and ordered the former fire chief be dismissed from the lawsuit. That Court of Appeals opinion does not bind other federal Courts of Appeal around the nation. Discuss whether fire department personnel should purchase their own private liability insurance to supplement the insurance the city or other political subdivision may provide.

2. In Case Study 2-2, *Espinoza v. Schulenburg,* the Arizona court modified the Fireman's Rule and held that it did not apply to off-duty firefighters who stop at a motor vehicle accident (MVA) as a Good Samaritan to assist the motorists. Discuss the advantages of having your state legislature pass a law that addresses this issue (see various state statutes in Chapter 18).

3. In Case Study 2-3, *Foiles v. V.J.L. Construction Corp.,* the New York court authorized the lawsuit by the firefighter to

proceed, because it alleged that the property owner had violated the building code. Because most building owners and homeowners have insurance against injury by invited guests, it is highly likely that the annual insurance rates will go up if your state passes a statute similar to the New York statute. Discuss whether state elected officials in your state might be willing to "take the heat" from property owners for this rate increase.

▪▪

Expand Your Learning

Read and complete the individual student or group assignment, as directed by your instructor.

1. Read the line of duty death report of firefighter Bill Ellison, a full-time member of Anderson Township Fire Department, Hamilton County (OH), who was working on March 8, 2001, as a part-time firefighter with the Miami Township Fire Department, Hamilton County (OH); NIOSH Fire Fighter Fatality Investigation Report F-2001-16 (*www.cdc.gov/niosh/firehome.html*). The widow of firefighter/paramedic Bill Ellison filed a lawsuit, naming the heating and ventilation company that sold the hot water heater, the plumber who installed it, and the homeowner. It was installed in violation of building codes, because two water heaters were connected to one vent pipe, and this pipe is required to be a "pipe within a pipe." After a lengthy pre-trial "discovery" in which depositions were taken of many experts and many firefighters who were at the scene, the case settled for an "undisclosed amount" by the insurance companies for the defendants. (The complaint and other documents in this lawsuit can be read online at Hamilton County Court of Common Pleas website, *www.courtclerk.org;* Case No. A0106222, Victoria A. Ellison v. Anderson Automatic Heating and Cooling, filed September 21, 2001).

   Discuss whether the Fireman's Rule in your state should be modified to allow lawsuits by firefighters or their families when there is proof of a violation of a building code designed to prevent the safety hazard that injures or kills the firefighter.

2. Read the Cincinnati Fire Department's March 26, 2004, report on the death of Firefighter Oscar Armstrong (*www.cincyfire.com*) and the NIOSH Fire Fighter Fatality Investigation Report F2003-12 (*www.cdc.gov/niosh/firehome.html*).

   Describe the operational changes that the Cincinnati Fire Department has made in response to this tragedy, and whether you would recommend any of these changes for your fire department.

See Appendix A for additional Expand Your Learning activities related to this chapter.

CHAPTER 3

Homeland Security
National Incident Management System, USA Patriot Act, and War on Terrorism

Ground Zero, World Trade Center, New York: Taken from the rear of the church where George Washington once prayed. *Photo by author.*

Key Terms

FISA court, p. 32
waiver of right to sue, p. 34

petition for a writ of habeas
corpus, p. 35

From the Headlines

Canadian Terrorists

The Royal Canadian Police have arrested 17 suspects, ages 43 to 19, who had taken steps to acquire three tons of ammonium nitrate (three times the amount used to blow up the Oklahoma City federal building on April 19, 1995) and other bomb-making materials. According to Canadian authorities, the suspects were "inspired by al-Qaida" and planned to blow up numerous buildings, including the Canadian spy agency in Toronto.

Terrorism Funding Changed

Michael Chertoff, the Secretary of the U.S. Department of Homeland Security, announced changes in the criteria for distributing $765 million in counter-terrorism funds, so that the greatest money will go to the metropolitan areas with the greatest risk. New York City received $207 million in 2006 and was very pleased with the news; it was bad news for 11 cities facing complete elimination of funding, including Las Vegas, Buffalo, Tampa, and Louisville.

FEMA Sends Phoenix USAR Home

FEMA sent Phoenix's 28-member Urban Search and Rescue (USAR) team home from relief efforts in the Gulf Coast because the team included armed Phoenix police officers, who were observed embarking on a helicopter sortie with a loaded shotgun, working on Hurricane Rita relief efforts. Phoenix Mayor Phil Gordon is quoted, "This is crazy. This is a rule that was designed before the world changed, pre-9/11."

Tribute to the fallen, in the church across the street from the World Trade Center, New York City. *Photo by author.*

Ohio Task Force in St. Bernard Parish (LA), October 2005, helped a resident save handcrafted model boats: The resident has now gratefully donated one of the models for permanent display at the Ohio Fire Academy. *Photo supplied by Captain Tim Keene.*

◆ INTRODUCTION

This chapter discusses the legal issues involving Homeland Security and the fire service, including the response to Hurricane Katrina and Hurricane Rita. It also reviews legal issues facing our nation concerning the war on terrorism.

◆ NATIONAL INCIDENT MANAGEMENT SYSTEM (NIMS)

The National Incident Management System (NIMS) was created March 1, 2004, so that emergency responders at the federal, state, and local levels could work more effectively together concerning domestic incidents. On September 8, 2004, the U.S. Department of Homeland Security issued a letter to the governors of all 50 states outlining the requirements of NIMS to be eligible for federal grants in Fiscal Year (FY) 2005 and beyond. Homeland Security Presidential Directive 5 (HSPD-5) established tight deadlines for NIMS implementation in FY 2005, and grantees must certify they will meet all NIMS requirements in FY 2006.

Federal grant money requires NIMS compliance, including the Assistance to Firefighters Grant (AFG), Fire Prevention and Safety (FP&S), and Staffing for Adequate Fire and Emergency Response (SAFER). See the website at *www. firegrantsupport.com.* Grant applicants must certify that they have met or will meet the FY 2005 NIMS requirements. At the state level, each state must incorporate NIMS into existing training programs and emergency operation plans and also promote intrastate mutual aid agreements. At the local level, including fire departments, grant applicants must certify that personnel have taken the NIMS Awareness Course, IS 700, an independent study guide (*http://training.fema.gov/EMIWEB/ IS/is700.asp*).

◆ **THREATS TO DISPATCH CENTERS**

Other threats to homeland security include cyber threats, including terrorist attacks on 9-1-1 call center computers. On May 19, 2006, the Department of Homeland Security issued an INFOGRAM identifying these risks, and steps to "harden" these computer systems (see *http://www.cert.org*). To receive e-mail INFOGRAMS, call the U.S. Fire Administration, National Preparedness Network, Emmitsburg (MD) at (301) 447-1853.

◆ **NFPA STANDARDS**

- *NFPA 1561: Standard on Emergency Services Incident Management System* (2005 edition). This standard identifies the establishment, use, and training requirements for the Incident Management System (IMS), including the need for an integrated IMS system compatible with outside agencies. This standard is consistent with the U.S. Department of Homeland Security's NIMS requirements.
- *NFPA 1670: Standard on Operations and Training for Technical Search and Rescue Incidents* (2004 edition). This standard identifies the criteria for establishing and training operating teams that are compatible with the multi-level USAR system.

Residents of Louisiana, displaced by Hurricane Katrina, arrive in Cincinnati (OH) for temporary housing by the American Red Cross in an elementary school. *Photo taken by author while on duty with Cincinnati chapter of the Red Cross.*

AIR STABILIZATION AND SYSTEM STABILIZATION ACT

Soon after the September 11, 2001, attacks on the World Trade Center (WTC) and the Pentagon, Congress passed the Air Stabilization and System Stabilization Act, Pub. L. No. 107-42, 115 Stat. 230 (2001). The statute limited liability for the air carriers involved in the tragedy to their insurance coverage. It also created the Victim Compensation Fund (the "Fund") to provide no-fault compensation to victims who were injured in the attacks and to personal representatives of victims killed in the attacks. All claimants who filed with the Fund waived the right to sue for injuries resulting from the attacks except for collateral benefits. On November 19, 2001, the Act was amended by the Aviation and Transportation Security Act (the "Aviation Security Act"), Pub L. No. 107-71, 115 Stat. 597 (2001), and provided liability limits to aircraft manufacturers, those with a proprietary interest in the WTC, and the City of New York, while allowing Fund claimants to sue individuals responsible for the attacks notwithstanding the waiver. The U.S. Supreme Court in January 2006 refused to hear the appeals of several spouses of Fire Department of New York (FDNY) firefighters in *Lucy Virgilio et al. v. City of New York and Motorola, Inc.*, 407 F.3d 105 (2nd Cir. April 29, 2005), in which the Court of Appeals held that this statute prohibited the spouses from suing because they collected compensation from the federal Victim Compensation Fund set up under the statute.

ASSISTANCE TO FIREFIGHTERS GRANT PROGRAM (AFG) (2000)

Congress created the AFG grants program in 2000 to address the fact that too many fire departments lacked the most basic needs, including adequate turnout gear, training, radios, and public education programs. The U.S. Department of Homeland Security (DHS), Office of Grants and Training, awards the grant directly to local fire departments through a competitive and peer review process. Fire departments must "buy in" to the program by contributing part of the cost. Congress appropriated $545 million for FY 2006. (See *http://www.firegrantssupport.com.*)

EMERGENCY MANAGEMENT ASSISTANCE COMPACT (EMAC) (PUBLIC LAW 104-321)

This helpful federal statute authorizes governors to declare an emergency and request mutual aid from neighboring states. The responding personnel will have the same legal authority and certificates to practice as in their home state and will be covered by workers' compensation.

FOREIGN INTELLIGENCE SURVEILLANCE ACT OF 1978 (50 U.S. CODE 1808)

FISA court Congress established this court under the Foreign Intelligence Surveillance Act (FISA) of 1978, in which selected U.S. District Court judges review requests for federal electronic surveillance wiretaps.

In response to public concerns about FBI domestic surveillance of Vietnam War protestors and civil rights activists, Congress passed new restrictions on domestic wiretaps and electronic surveillance. The Foreign Intelligence Surveillance Act (FISA) established a new **FISA court**, in which designated U.S. District Court judges would review requests for electronic surveillance intercept warrants.

INTELLIGENCE REFORM AND TERRORISM PREVENTION ACT OF 2004

This statute, signed into law by President George Bush on December 17, 2004, expressed "the sense of Congress" that all emergency response agencies should adopt and follow the National Incident Management System (NIMS). The law also requires that fire departments train on and use NIMS in order to receive federal grant funds.

SAFER ACT OF 2003

On November 7, 2003, Congress passed the Staffing for Adequate Fire and Emergency Response (SAFER) Act, expanding the firefighter grant program and appropriating $650 million for direct grants to fire departments by the U.S. Fire Administration. House Bill H.R. 1588 amended the Federal Fire Prevention and Control Act of 1974.

USA PATRIOT ACT OF 2001

In response to the 9/11 terrorist attacks on New York City and Washington (DC), Congress amended FISA (Senate vote on October 25, 2001, was 98 to 1; House vote, 356 to 66) to grant additional authority for electronic surveillance of international communications, authority to delay notice to individuals that they have been intercepted, and the controversial Section 215 provision on seizure of an individual's library records. Section 215 is in 50 U.S. Code 1861(b)(2). For the complete text of this and other federal statutes, see the Library of Congress website, *http://thomas.loc.gov.* See also the U.S. Department of Justice website on the Patriot Act, "The USA Patriot Act: Preserving Life and Liberty," and also the April 27, 2005, Congressional testimony of U.S. Attorney General Alberto R. Gonzales and FBI Director Robert S. Mueller (*www.usdoj.gov*). See also the congressional testimony of U.S. Attorney General Alberto R. Gonzales on March 1, 2005, on the need to maintain and strengthen the Patriot Act (*www.usdoj.gov/ag/testimony.html*).

USA PATRIOT IMPROVEMENT AND REAUTHORIZATION ACT OF 2005

In December 2005, congressional opponents to Section 215 and to *The New York Times* report of domestic wiretapping with court orders reached a compromise with the White House and agreed to a limited extension until February 2006 of the Patriot Act in order to seek improved language.

Cincinnati (OH) school children drew messages to welcome Katrina victims staying in a Cincinnati elementary school, converted to a Red Cross shelter. *Photo by author.*

CASE STUDY 3-1   Failure of Radios at the World Trade Center

Virgilio v. City of New York and Motorola, 407 F.3d 105 (2nd Cir. 2005)

FACTS

Lucy Virgilio, the widow of deceased FDNY firefighter Lawrence Virgilio, who was killed in the collapse of the World Trade Center during the attacks on September 11, 2001, and the families of several other firefighters filed a lawsuit in federal court alleging that failure of radio transmissions in the North and South towers prevented firefighters from receiving evacuation orders. She alleged that Motorola negligently and intentionally provided the city with radio-transmission equipment for firefighters that Motorola knew to be ineffective in high-rise towers. She also alleged that the city is liable for wrongful death because it breached its duty to provide adequate radio-transmission equipment.

The defendants filed motions to dismiss the lawsuit, because the plaintiffs had all filed for compensation from the federal Victim Compensation Fund. In 2001, Congress enacted the Air Transportation Safety and System Stabilization Act (Pub. L. 107-42), which limited the liability for the airlines involved in the attack and created a Victim Compensation Fund to provide a no-fault compensation program for victims. Under this statute, all who file claims with the fund **waive their right to sue** for injuries. Congress amended the act on November 19, 2001 (Pub. L. 107-71) to extend the liability limits to protect aircraft manufacturers, the city of New York, and the owners of the World Trade Center but also allowed Compensation Fund claimants to sue individuals responsible for the attacks.

waiver of right to sue   After the 9/11 attacks, Congress established a no-fault Victim Compensation Fund, which includes a provision that all claimants waive their right to sue the airlines or others for damages arising out of the terrorist attacks.

YOU BE THE JUDGE

1. Should this lawsuit against Motorola be dismissed?
2. Should the lawsuit against the City of New York be dismissed?

HOLDING

The federal U.S. District Court judge dismissed the lawsuit against all defendants. The plaintiffs appealed, and on April 29, 2005, a three-judge panel of the U.S. Court of Appeals affirmed the dismissal.

The Court of Appeals wrote:

> We agree with the district court that under the plain language of the statute, claimants who have filed claims with the Fund waived the right to file a civil action . . . for damages sustained as a result of the terrorist-related aircraft crashes of September 11, 2001 and that the waiver bars claims for damages sustained against non-airline defendants.

The court also wrote:

> The overall structure of the Act highlights two predominant concerns: to insulate the airline industry from massive—virtually limitless—liability arising from the sudden and devastating acts of wanton cruelty on 9/11 and to provide an adequate no-fault system of compensation to victims.

The court concluded with the following:

> We close with a general observation. The events of September 11, 2001, changed this nation in ways that will not be fully understood for generations to come. However, the pain and sense of loss that the victims and their families feel need not wait the judgment of history—their anguish, we are sure, is a daily companion. As judges, we are not unmindful of the great sacrifice that many of New York's bravest men and women made on behalf of those who were trapped in the burning towers at Church and Vesey Streets. If Article III of the Constitution somehow gave us the power to turn back time and undo the disaster we would set to the task without reservation. Unfortunately, we have only the power to assess the law as it is given to us by Congress. Such is the nature of judging.

[Note: The plaintiffs filed an appeal to the U.S. Supreme Court: On January 16, 2006, the court declined to hear the case: It requires four Justices to vote for a writ of certiorari, *which is a court order to have records brought up from a lower court.]*

LEGAL LESSONS LEARNED

Test your radios throughout your fire district, and try to eliminate the weak spots by additional transmission relays on buildings and towers. Share your test results with the department personnel and the public, and spend the money necessary to fix the problem areas.

CASE STUDY 3-2 Al Qaeda Prisoner Held in Navy Brig, South Carolina

Padilla v. Hanft, U.S.N. Commander, Consolidated Naval Brig., 423 F.3d 386 (4th Cir. 2005); 2005 U.S. App. LEXIS 19465; Case No. 05-6396 (4th Cir. 2005).

FACTS

Jose Padilla is a U.S. citizen. While he was on a religious pilgrimage to Saudi Arabia, he was recruited by al Qaeda and brought to Afghanistan. He was trained as a terrorist in Afghanistan and took up arms against U.S. forces in that country in our war against al Qaeda. He escaped to Pakistan, armed with an assault rifle, where he met with a senior al Qaeda operations planner, Khalid Sheikh Mohammed, and was trained and directed to return to this country and blow up apartment buildings. He received cash, travel documents, and a communication device, and on May 8, 2002, he flew to Chicago's O'Hare Airport to begin carrying out his assignment. Fortunately, he was detained by the FBI and arrested on a material witness warrant issued by the federal district court in New York City, where a grand jury was investigating the 9/11 attacks on the World Trade Center.

Padilla was held in a civilian correctional center until, on June 9, 2002, the president of the United States signed a letter to the Secretary of Defense, designating Padilla as an "enemy combatant" and saying that he constitutes a "continuing, present and grave danger to the national security of the United States."

Padilla has been held in military custody since that date. On June 11, 2002, Padilla filed a **petition for a writ of habeas corpus** with the federal court in New York City. He claimed that the federal government was violating his rights under the U.S. Constitution, because he had not been charged with any federal crimes nor brought before a federal judge for arraignment and setting of bail. This petition ultimately was denied, and on appeal, the U.S. Supreme Court ruled that he should have filed the petition in the federal U.S. District Court in Charleston (SC) where he was being held in the Naval Brig. *Rumsfeld v. Padilla*, 124 S. Ct. 2711 (U.S. 2004).

petition for a writ of habeas corpus A petition filed by a person held in a jail or prison seeking a judge to review the lawfulness of his or her detention.

On July 2, 2004, Padilla filed the petition for a writ of habeas corpus in the federal court in South Carolina. The district judge ruled in his favor, holding that the president of the United States lacked the authority to detain him without charging him with a federal crime. The United States appealed to the U.S. Court of Appeals for the 4th Circuit in Richmond (VA), which scheduled the case for expedited oral argument on July 19, 2005.

The United States argued that Congress gave the President authority to hold enemy combatants, including U.S. citizens, when it passed the Resolution for Use of Military Force Joint Resolution on September 18, 2001. That Joint Resolution provides:

> [T]he President is authorized to use all necessary and appropriate force against those nations, organizations, or persons he determines planned, authorized, committed, or aided the terrorist attacks that occurred on September 11, 2001, or harbored such organizations or persons, in order to prevent any future acts of international terrorism against the United States by such nations, organizations or persons. [Pub. L. 107-40, Sec. 2 (a); 115 Stat. 224 (September 18, 2001).]

YOU BE THE JUDGE

If you were one of the three appellate judges hearing this oral argument, how would you decide?

Would it make a difference if Padilla were not a U.S. citizen?

HOLDING

The three-judge panel reversed the trial judge and held that the President of the United States has the authority to hold Padilla indefinitely without charging him with a crime. The court wrote in its decision of September 9, 2005:

> The Congress of the United States, in the Authorization for Use of Military Force Joint Resolution, provided the President all powers necessary and appropriate to protect American citizens from terrorist acts by those who attacked the United States on September 11, 2001. As would be expected, and as the Supreme Court has held, those powers include the power to detain identified and committed enemies such as Padilla, who associated with al Qaeda and the Taliban regime, who took up arms against this Nation in its war against these enemies, *and* who entered the United States for the avowed purpose of further prosecuting that war by attacking American citizens and targets on our own soil—a power without which, Congress understood, the President could well be unable to protect American citizens from the very kind of savage attack that occurred four years ago almost to the day.

LEGAL LESSONS LEARNED

The War on Terrorism is presenting new legal questions and will undoubtedly present new legal issues for the fire service. We are living in a fast changing world.

[Note: After this decision, Padilla filed an appeal to the U.S. Supreme Court, but the Supreme Court will probably never hear the case, because the U.S. Department of Justice decided to seek a criminal indictment against Padilla by a grand jury in the U.S. District Court in Miami. He has now been indicted for several felony offenses, and on January 4, 2006, the Supreme Court agreed he could be transferred from the Naval Brig in South Carolina to a federal jail in Miami. Like any federal criminal defendant, he had an arraignment before a federal District Court judge, pleaded not guilty, and is in federal prison awaiting criminal trial before a 12-person jury.]

CASE STUDY 3-3   Mutual Aid (Indiana, 2005)

FACTS

[Note: This is a "real world" event involving the author of this textbook.]

On July 8, 2005, fire departments from Indiana, Ohio, and Kentucky responded to the multi-alarm fire at the six-story plastic recycling plant of Alternative Plastic Services in Greendale (IN). Fire Chief Ed Noel had called for mutual aid assistance from the tri-state region; they quickly took a defensive attack on the fire. Fortunately, none of the responding firefighters were injured, nor were any of the 50 to 75 employees of the plant. The *Cincinnati Enquirer* (*www.cincinnatienquirer.com*) reported that Indiana State Fire Marshal Roger Johnson said a preliminary investigation indicated that the fire was caused by an electrical malfunction in a fuse box on the ground level of the building.

The night of the fire, I was returning from a meeting of the Warren County Fire Chiefs' Association when I received a cell phone call from Frank Young, director of the Warren County Communications Center and Warren County EMA director. I was suddenly on a teleconference with numerous other Ohio county EMA directors, and several fire chiefs. They wanted quick "legal advice" on the following issue. The governor of Indiana has not declared this fire a "disaster," nor has the governor of Ohio. Under the interstate mutual aid compact statute in Ohio, such a declaration is required to guarantee that all Ohio responding firefighters are covered under Ohio's workers' compensation if injured fighting a fire outside of Ohio.

Legislation has been proposed on a Local Emergency Management Assistance Compact (LEMAC) in which county EMA directors could make the "disaster" declaration, but it has never been enacted into law.

Many states have passed legislation covering interstate (responses across state lines) and intrastate (responses within a state). For example, Ohio has an *interstate* mutual aid compact (Ohio Rev. Code 5502.40) and also an *intrastate* mutual aid compact (Ohio Rev. Code 5502.41).

Interstate Mutual Aid Compact

Some key provisions of the Ohio Rev. Code 5502.40 Interstate Emergency Management Assistance Compact (EMAC) are as follows:

- Mutual assistance

  The purpose of this compact is to provide for mutual assistance between the states entering into this compact in managing any emergency or disaster that is duly declared by the governor of the affected state(s), whether arising from natural disaster, technological hazard, man-made disaster, civil emergency aspects of resources shortages, community disorders, insurgency, or enemy attack.

  (B) The authorized representative of a party state may request assistance of another party state by contacting the authorized representative of that state. The provisions of this agreement shall only apply to requests for assistance made by and to authorized representatives. Requests may be verbal or in writing. If verbal, the request shall be confirmed in writing within 30 days of the verbal request.

  Each party state shall afford to the emergency forces of any party state, while operating within its state limits under the terms and conditions of this compact, the same powers (except that of arrest unless specifically authorized by the receiving state), duties, rights, and privileges as are afforded forces of the state in which they are performing emergency services. Emergency forces will continue under the command and control of their regular leaders, but the organizational units will come under the operational control of the emergency services authorities of the state receiving assistance. These conditions may be activated, as needed, only subsequent to a declaration of a state of emergency or disaster by the governor of the party state that is to receive assistance or commencement of exercises or training for mutual aid and shall continue so long as the exercises or training for mutual aid are in progress, the state of emergency or disaster remains in effect or loaned resources remain in the receiving state(s), whichever is longer.

◆ Article VI—Liability

Officers or employees of a party state rendering aid in another state pursuant to this compact shall be considered agents of the requesting state for tort liability and immunity purposes; and no party state or its officers or employees rendering aid in another state pursuant to this compact shall be liable on account of any act or omission in good faith on the part of such forces while so engaged or on account of the maintenance or use of any equipment or supplies in connection therewith. Good faith in this article shall not include willful misconduct, gross negligence, or recklessness.

◆ Compensation

Each party state shall provide for the payment of compensation and death benefits to injured members of the emergency forces of that state and representatives of deceased members of such forces in case such members sustain injuries or are killed while rendering aid pursuant to this compact, in the same manner and on the same terms as if the injury or death were sustained within their own state.

[2002 H 319, effective 2-1-02. (The legislative history of House Bill 319 and the entire statute can be read at www.legislature.state.oh.us.*) Ohio also has a statute covering intrastate mutual aid responses, Ohio Rev. Code 5502.41.]*

YOU BE THE JUDGE

Are the Ohio firefighters protected under Ohio workers' compensation if injured in Indiana?

HOLDING

Because the governor of Indiana and the governor of Ohio have not declared an emergency need for mutual aid, the statute does not apply. The author therefore suggested that as soon as practical each responding Ohio fire department enter into a written agreement with the requesting Indiana fire department that confirms the mutual aid requests and provides that employee injuries will be covered by the responding fire department.

LEGAL LESSONS LEARNED

States should enact a Local Emergency Mutual Aid Compact (LEMAC), authorizing county EMA directors to declare a local emergency and send county fire departments.

Chapter Review Questions

1. In Case Study 3-1, *Virgilio v. City of New York and Motorola*, the court dismissed the lawsuit by New York City firefighter families concerning poor radio communications in the World Trade Center. If your fire department is going to enter into a contract to purchase radios, describe the "performance requirements" that you would include in your specifications for communications from inside large buildings in your fire district, and what type of penalty clauses that you might include in the contract for breach of the specifications.

2. In Case Study 3-2, *Padilla v. Hanft, U.S.N. Commander, Consolidated Naval Brig.*, the court addressed the issue of holding terrorists without criminal charges. Padilla is a U.S. citizen, and his appeal to the U.S. Supreme Court will probably never occur because the U.S. Department of Justice indicted him in Miami, and he will now get his "day in court."

   Discuss whether U.S. citizens should have a right to be indicted and charged for terrorist activities.

3. In Case Study 3-3, we discussed the Indiana plastics company fire. In 1966, Congress enacted the Emergency Management Assistance Compact, Public Law 104-321 (please read the statute at *http://thomas.loc.gov/*). Article V provides that persons with a license in one state (for example, a paramedic license) are deemed licensed in the state where they are rendering mutual aid. Likewise, in Article VI, for purposes of liability, they shall be considered "agents of the requesting state" and shall not be liable "for any act or omission in good faith." "Good faith in this article shall not include willful misconduct, gross negligence, or recklessness."

   Assume that your fire department decides to send one engine and a crew of four career firefighters/paramedics, and you are not selected. One of your best friends has been selected, urges you to drive to Louisiana in your own vehicle, and quietly hook up with the engine company when you get there.

   Describe what legal liability risks you are personally incurring when rendering medical assistance as an Ohio paramedic who is "freelancing." If the officer on the engine allows you to participate, is your Ohio municipality also facing potential liability? Discuss.

Expand Your Learning

Read and complete the individual student or group assignment, as directed by your instructor.

1. One of the most controversial sections of the Patriot Act concerns the authority to look at library cards and similar information. Please read Section 215 of the Patriot Act (Pub. Law 107-56) by going to the Library of Congress website (*http://thomas.loc.gov/*). Under "Find More Legislation," search "Public Laws," go to 107th Congress, and click on Law No. 56. The Attorney General has stated that this "library card" issue is not a real-world risk to the American public but is nothing but a political ploy by those opposed to President Bush.

   If you were a congressman who wanted to amend the Patriot Act, what changes would you offer to Section 215 regarding searches of library cards?

2. In July 2005, the *New York Daily News* reported that retired FDNY firefighter Brian Laine, age 47, has filed a lawsuit

to get his job back. He is one of nearly 2,000 New York City firefighters and police officers, out of 11,000, who decided to retire after the 9/11 attacks, with pensions that were enhanced because of all the overtime after the 9/11 attacks. Lane was with FDNY for 17 years and last assigned to Ladder 295 in Whitestone, Queens. Lane is quoted as stating: "I don't want to do anything else. I have a B.S. from Cornell and an M.B.A. from Baruch College. I could choose to do other things. I just want to be a helpful civil servant." One year after the stress of 9/11, Lane is feeling better, in great physical shape, and wanting to return "to the job I had always loved." The lawsuit alleges that Fire Commissioner Nicholas Scoppetta has discriminated against firefighters who risked their lives after the 9/11 terror attacks and that he has rejected every retiree's application to rejoin the department.

If you were the fire commissioner, what review process might you establish to determine whether to accept requests for retired firefighters to rejoin the fire department?

3. In *McNamara v. Hittner,* 767 N.Y.S.2d 800 (N.Y. Sup. Ct. App. Div. 2003), McNamara was a firefighter who was injured when a civilian driver drove a car over the firefighter's ankle as he was stepping out of a fire truck. The firefighter sued the civilian, and a jury awarded him $300,000 for past pain and suffering and $2 million for future pain and suffering. The trial judge commented to the jury that it had been only four months since the 9/11 attacks on New York City and that "we should never forget what happened four months ago. We can't bring back the dead, but we should never forget our losses." The trial judge then asked for a moment of silence in remembrance. The jury awarded the firefighter $2 million for future pain and suffering. The Court of Appeals held that the trial judge made an error in his comments but did not require a mistrial. However, the jury's award of future damages was unreasonable, and the court ordered it reduced from $2 million to $400,000, or the firefighter may have a new trial.

Discuss whether the injured firefighter, suing a civilian under an exception to the Fireman's Rule, should be able to appear in court wearing his Class A uniform, or whether this would unfairly influence the jury.

See Appendices A and B for additional Expand Your Learning activities related to this chapter.

Incident Command
Fire Scene Operations, Training, and Immunity

4 CHAPTER

Incident command at a trench incident. Hamilton County Urban Search and Rescue Team assists local fire departments. *Photo supplied by Deputy Chief Ray Mueller, Mason (OH) Fire Department.*

Key Terms

assumption of duty, p. 48   governmental immunity
deliberate interference, p. 50   statute, p. 52

From the Headlines

State Trooper Arrested Deputy Fire Chief Who Refused to Move Fire Engine

The New York *Daily Record* (and also in *www.firehouse.com*) reported that Rockaway Township Deputy Fire Chief Robert Jenkins was arrested at the scene of a rollover crash when he refused a state trooper's order to move a fire engine used to provide a safety zone for ambulance and other emergency personnel at the MVA. The incident raises the serious issue of who "controls" the scene of an accident. Who has the authority at an MVA scene in your state? In Ohio, the Ohio attorney general issued an opinion in 1994 (opinion 94-076, *http://www.ag.state.oh.us/legal/opinions/1994/94-076.htm*) that concludes the fire department has the authority.

NIMS/FEMA Online Training

FEMA has announced that 1,072,355 local, state, and federal emergency personnel have taken NIMS training online at FEMA's Independent Study Program (*http://training.fema.gov/emiweb/IS*) including Q-462, "Introduction to All-Hazards NIMS ICS for Operational First Responders," and Q-463, "Basic All-Hazards NIMS ICS for Operational First Responders.

FDNY: Hazardous Materials

"Fire Chief Challenges New York Emergency Plan," *The New York Times.* Story about the FDNY criticizing the mayor's decision to give police the initial control at hazardous materials emergencies. The fire chief is quoted as stating that the emergency response was "deeply flawed" and represented a "power grab" by the Police Department.

Phoenix Fire Department Command Training Center. *Photo by author.*

Phoenix Fire Department. Another view of the Command Training Center. *Photo by author.*

This chapter reviews litigation arising out of incident command and the tragic deaths of firefighters and civilians, in which lawsuits have been filed seeking compensation from the fire department or its municipality or other political subdivision. These suits may also include as defendants the incident commander, fire chief, and other personnel.

◆ GOVERNMENTAL IMMUNITY AND QUALIFIED IMMUNITY

Fortunately, most lawsuits against municipalities are dismissed prior to trial on the basis of "governmental immunity," and fire officials are also dismissed on the basis of "qualified immunity." This immunity is "qualified" rather than absolute, because in rare circumstances where there is some evidence of gross negligence, the trial judge may require that the case be tried by a jury.

In Chapter 2 we studied the lawsuit involving the Washington (DC) 1999 Cherry Road townhouse fire, in which two firefighters died and two were seriously injured (see NIOSH Fire Fighter Fatality Investigation report 99-F21, *www.cdc.gov/niosh*).

On August 1, 2006 the U.S. Court of Appeals held that the former fire chief enjoyed "qualified immunity" and must be dismissed from the lawsuit. Since other federal circuits are not bound to follow this decision, it is important that we understand the federal district judge's earlier decision ordering individual fire officers, including the incident commander, to stand trial.

A federal judge had ordered the lawsuit against the city, the former fire chief, and others, to go to a jury trial. *Estate of Anthony Phillips, et al. v. District of Columbia, et al.,* January 11, 2005, 355 F. Supp. 2d 212, 2005 (D.D.C.). The lawsuit was filed by the families of the two deceased firefighters and two of the injured firefighters and alleges that the incident commander allowed ventilation of the basement, even though he had lost radio contact with the primary search team on the first floor. The suit also alleges that a similar incident occurred two years earlier, when firefighter John Carter died in a grocery store.

The federal judge's chilling description of the events focused on the incident commander:

> While firefighters were in the house, the Incident Commander (IC) twice radioed Redding to locate his position. However, Redding did not receive this transmission. The IC had not established a fixed command post and was relying on a weaker portable radio device rather than the stronger radio mobile. The firefighters inside the house were unaware of each other's presence. Communications were impaired and visibility was poor. Redding did not even have a hand light with which to illuminate the inside of the townhouse. The improper and untimely ventilation of the house resulted in a sudden increase in temperature. Redding ran from the townhouse, with his face and back burning. He relayed to the IC that Matthews was still in the townhouse. Redding was unaware that Morgan and Phillips were also in the townhouse at that time. The IC did not order a rescue effort until approximately 90 seconds later, when firefighter Morgan exited the house critically injured. Firefighter Phillips was found unconscious and severely burned approximately eleven minutes after the rescue effort began. Phillips died of his injuries approximately 23 minutes after his removal from the townhouse, while Matthews died of his injuries on the following day.

◆ FIREFIGHTER DROWNED DURING SCUBA TRAINING

In *Deborah Frayne v. Dacor Corporation et al.,* 840 N.E.2d 294 (Ill. App. November 29, 2005), the Illinois Court of Appeals held that the trial court properly dismissed the lawsuit by the wife of deceased firefighter Kenneth Frayne, Channahon Fire Protection District, who drowned during SCUBA training. She had sued the numerous fire departments participating in the joint drill and several firefighters who were in command, alleging that someone supplied her husband with a defective regulator. The training occurred October 13, 2001, at a lake owned by the Coal City Area Club. The defendants all filed a joint motion for summary judgment, claiming statutory immunity from any liability under Section 3-110 of the Illinois Local Governmental and Governmental Employees Tort Immunity Act. The Illinois statute states, "Neither a local public entity nor a public employee is liable for any injury occurring on, in, or adjacent to any waterway, lake, pond, river or stream not owned, supervised, maintained, operated, managed or controlled by the local public agency." The Court of Appeals affirmed the dismissal of the lawsuit, writing:

> While plaintiff notes that defendants, or at least some of them, controlled and supervised almost every aspect of the dive itself, including everything from what type of equipment would be used to who the dive partners would be, the trial court correctly found that the defendants, as a matter of law, did not supervise, control, or maintain the lake and, therefore, are immune under the plain language of the Immunity Act.

◆ DEATH OF CHILDREN

Lawsuits by citizens claiming negligent conduct of fire departments are becoming more common, particularly involving deaths of children. For example, on January 10, 2006, in *Rhonda Davis v. City of Detroit,* 2006 Mich. App. LEXIS 11, the Court of Appeals of Michigan reversed a Wayne Circuit Court judge and ordered him to dismiss the lawsuit against the city. Ms. Davis had filed suit on behalf of her deceased children who died in a structure fire when the first two fire hydrants did not work. The Court of Appeals held that the city was "absolutely immune from suit" because it operated the hydrant strictly for governmental reasons and was not selling water as a profit-making enterprise.

In another Detroit case, on April 13, 2006, in *Scheherazde C. Love v. City of Detroit,* Court of Appeals of Michigan, 2006 Mich. App. LEXIS 1097, the trial judge dismissed the lawsuit alleging gross negligence in the death of four children. Although the plaintiff submitted affidavits from neighbors who alleged the fire department responded with a utility van and took "40 to 60 minutes" to respond with fire trucks with ladders, the Court of Appeals (two to one vote) upheld the dismissal of the lawsuit. The Court wrote, "Government employees are immune from liability for injuries they cause during the course of their employment if they are acting or reasonably believe they are acting within the scope of their authority, if they are engaged in the exercise or discharge of governmental function, and if their conduct does not amount to gross negligence that is the proximate cause of the injury or damage." The Court defines "gross negligence" as "conduct so reckless as to demonstrate a substantial lack of concern for whether an injury results."

◆ **NFPA STANDARDS**

- *NFPA 1021: Standard for Fire Officer Professional Qualifications* (2003 edition). This standard identifies the knowledge and skills needed by officers at all levels of command, in order to command emergency personnel effectively and safely. It is the basis for the National Fire Academy's Officer Development Model.
- *NFPA 1041: Standard for Fire Service Instructor Professional Qualifications* (2003 edition). This standard identifies the knowledge and skills that instructors must possess to deliver quality instruction safely. It includes specific criteria for developing and implementing safety training classes and programs.
- *NFPA 1061: Standard for Professional Qualifications for Public Safety Telecommunicator* (2002 edition). This standard identifies the knowledge and skills required of all personnel trained and employed to receive emergency calls and requests for service from the public, to provide emergency instructions to these callers, and pass those requests on to emergency response agencies.
- *NFPA 1403: Standard on Live Fire Training Evolutions* (2002 edition). This standard has a checklist by which a trained and certified instructor can safely provide quality instruction to firefighters in a live fire structure. It should be followed step by step prior to any live fire training.
- *NFPA 1521: Standard for Fire Department Safety Officer* (2002 edition). This standard identifies the knowledge and skills required to fill the role of safety officer for an emergency response organization, as well as the function of an incident safety officer (ISO) on emergency scenes. The standard calls for use of ISOs at all working incidents.
- *NFPA 1561: Standard on Emergency Services Incident Management System* (2005 edition). This standard identifies the development, training, and implementation required for an Incident Management System (IMS), including use of such a system at all emergency incidents to maintain accountability and control or firefighter safety.

Cincinnati Fire Department Rapid Assistance Team training. *Photo by author.*

EMERGENCY MANAGEMENT ASSISTANCE COMPACT OF 1996 (PUB. L. 104-321)

This statute establishes a national mutual aid response program. Article V provides that persons with a license in one state (for example, a paramedic certificate) are deemed licensed in the state where they are rendering mutual aid. Likewise, in Article VI, for purposes of liability they shall be considered "agents of the requesting state" and shall not be liable "for any act or omission in good faith." "Good faith in this article shall not include willful misconduct, gross negligence, or recklessness."

INTELLIGENCE REFORM AND TERRORIST PREVENTION ACT OF 2004 (S. 2845)

On December 17, 2004, President Bush signed the bill into law, which expresses the "sense of Congress" that all emergency response agencies should adopt and follow the NIMS. The bill also states that regular use of NIMS, and training in NIMS, should be a condition for receiving federal grant funding. It also requires FEMA to establish a mutual aid program for all hazards.

CONSTITUTIONAL CLAIMS (42 U.S.C. 1983)

So-called "1983 lawsuits" allege deprivation of federal constitutional rights by federal, state, or local government agencies, such as a city's alleged failure to adequately train police officers or firefighters. Suits can be filed in federal or state court. The statute provides a cause of action for "Every person who, under color of any statute, ordinance, regulations, custom, or usage of any State or Territory . . . subjects . . . a citizen of the United States . . . to the deprivation of any rights, privileges, or immunities secured by the Constitution. . . ."

◆ **CASE STUDIES**

CASE STUDY 4-1   Power Lines Down in Rear of Home: Electrocution of Child after Fire Department Leaves Scene

City of Muncie v. Thomas Weidner and Lauren Weidner, 832 N.E.2d 206 (Indiana Ct. App. 2005).

FACTS

The Weidners filed a wrongful death suit against the City of Muncie for the death of their son "A.W." [only the son's initials are used by the court]. On July 29, 2002, a storm caused more than 5,300 Muncie residents to lose electric power. That afternoon, Tony Gothard went home from work in response to his wife's call that the home's circuit breaker was sparking. He found that the house had no electric power and that the outside power line to the house was sagging. He called the Indiana Electric Company.

While Gothard was outside, he heard crackling and popping sounds from the backyard of his neighbor's house. He observed a power line hanging down into the middle of the bushes that separated the two homes. He immediately reported this to the Muncie Fire Department.

The fire department responded with an engine, with Mark Hill riding as the Officer-In-Charge (OIC). The power line had stopped crackling when the engine arrived. OIC Hill told Gothard to keep his family out of both backyards because electricity can travel through the bushes and also through the ground. The OIC also called the dispatcher, who sent a faxed report to the electric company of the downed power line.

The fire department left in about five minutes, prior to arrival of the electric company. The next day at about noon, Gothard's wife called him at work and advised that the electric company arrived, looked at the sagging wire in their front yard, and left without speaking to her. Gothard came home and called the electric company, which promised to send someone out.

About 2 p.m. that day, a person cutting another neighbor's lawn discovered the Weidners' son lying in the Weidners' backyard, unconscious, and on his back. An ambulance and police officer responded. As Paramedic Gary Gardner was assessing the boy, Police Officer Linda Cook stepped back, received an electric shock, and in turn, fell on the paramedic and shocked him. The paramedic and the police officer were not badly hurt, but the Weidners' son was dead.

The Weidners filed suit against the City of Muncie, alleging that the fire department was negligent in failing to protect their son from a downed, live, power line. The city filed a motion for summary judgment, asserting governmental immunity under the Indiana Emergency Management and Disaster Law, Indiana Code 10-14-3, because the firefighters were responding to a danger caused by a severe storm and could not stand by at this house waiting for the power company. The trial judge refused to dismiss the suit, and the city has filed an immediate appeal.

YOU BE THE JUDGE

Should the case be dismissed without any civil pre-trial discovery?

HOLDING

Trial court reversed; case dismissed.

The Muncie Fire Department has no duty with respect to downed power lines; that is the responsibility of the electric company. Indiana law does recognize that a duty may be created by "gratuitous or voluntary assumption."

The three-judge court of appeals wrote, "Our determination of whether the Muncie Fire Department is liable for the death of A.W. hinges on the type of duty assumed."

"We find that Muncie designated sufficient evidence, which was not rebutted by the Weidners, to establish that Muncie did not assume a duty for which it could be held liable."

The court said that the fire department's response did not increase the risk of harm to the child. The court further noted,

> The duty to maintain the lines was AEP's. Upon learning of the downed power line,
> Gothard's first call was to AEP to report the line was down. It was not until Gothard

heard popping noises emanating from the bushes near the downed line that he thought to call the fire department. A.W. would have been a victim of the same tragic accident regardless of whether the fire department failed to respond, had never been called, or, as in this case, responded and warned Gothard and his family to stay out of their backyard.

The Weidners contend that there was an increased risk of harm because if the fire department had done its job and stayed on the scene, someone would have been there to warn A.W. about the live wire. This contention misstates the standard for **assumption of duty**. It says only that the risk of harm to A.W. would have decreased had the fire department assumed the duty and carried it out reasonably. The standard by which we judge the fire department's action, however, is not whether the risk of harm would have *decreased* had the fire department acted with reasonable care. Rather, it is whether the fire department's failure to exercise reasonable care *increased* the risk of such harm. There is no designated evidence that the risk of harm to A.W. was increased by the fire department's actions.

assumption of duty A fire department assumes a duty when it responds to assist a resident, such as when power lines are down, but is not liable for injury or death, unless its conduct increases the danger to the resident.

LEGAL LESSONS LEARNED

Unfortunately, when fire departments are responding to numerous calls for power lines down, it is often not possible to take an engine out of service to cover one backyard. Consider putting up yellow scene tape near the downed power line prior to leaving for the next run.

CASE STUDY 4-2 Titanium Fire/Fire Department IC Orders Water Sprayed on Fire

State Automobile Mutual Insurance Company v. Titanium Metals Corporation and Oakwood Village Fire Department, Ohio Supreme Court, April 19, 2006; 108 Ohio St. 3d 540, 2006 Ohio 1713; 844 N.E. 2d 1199; 2006 Ohio LEXIS 981.

FACTS

The plant of an Ohio company, Ohio Briquetting, Inc., was seriously damaged by explosion and fire on November 12, 2002. The explosion was allegedly caused by contaminated scrap titanium supplied by Titanium Metals Corporation, and the fire was spread when the fire chief ordered water to be sprayed on the burning titanium.

The owner of the building, M. Alan Properties, Inc., and another tenant of the building, Bencin Trucking, Inc., were insured by the plaintiff, State Automobile Mutual Insurance Company, which paid approximately $850,000 in claims for property damage. State Auto thereafter brought a suit "in subrogation" against Ohio Briquetting and TIMET, alleging that their combined negligence was responsible for the loss sustained by its insured. Ohio Briquetting answered the lawsuit denying liability, filed a cross-claim against TIMET for contribution and indemnification, and filed suit against the Oakwood Village Fire Department ("the village") and its then fire chief.

Ohio Briquetting alleged that the fire department had been reckless in the manner in which it suppressed the ensuing chemical fire, thereby increasing the resulting property damage not only to the insured but to Ohio Briquetting as well. Ohio Briquetting alleged that the fire chief was aware of the various chemicals stored at its facility and knew that water-suppressant methods would exacerbate any explosions and fire already in progress.

The manager of Ohio Briquetting alleges he approached the village's fire officials, *at the scene of the fire,* and reminded them that metallic chemical elements were stored inside its facility and that they were not amenable to water suppression methods. He informed the fire chief of the type of metallic chemical elements stored at its facility, warned against the use of water as a fire suppressant, and thereafter relied on the fire department to suppress the fire and not cause further damage by its actions.

The lawsuit alleges that the fire department's use of water also violated "local, state and national fire codes." According to Ohio Briquetting's allegations, the fire chief ignored its warning and violated its own fire code when it used water to suppress the fire, which thereafter allegedly increased the damage not only to its building, but also those of surrounding businesses.

The fire department filed a motion with the trial judge to dismiss the lawsuit, claiming "governmental immunity" under Ohio law. The motion was denied without any written opinion, and the fire department appealed.

The Ohio Court of Appeals heard the appeal and refused to dismiss the fire department or the fire chief from the lawsuit. The fire department and the fire chief then filed an appeal to the Ohio Supreme Court.

YOU BE THE JUDGE

1. Should the lawsuit be dismissed?
2. Should it be remanded to the trial court for further discovery?

HOLDING

The Supreme Court ordered the case back to the trial judge for further discovery.

The Ohio Supreme Court wrote that since the trial judge had not issued a written opinion and order on the immunity issue, there was "no final, appealable order" and the Ohio Court of Appeals lacked jurisdiction to consider this matter. "The court of appeals considered the issue prematurely. The record below must be developed in order to reach this issue."

The Ohio Supreme Court also declined to issue any opinion on the constitutionality of an important 2003 amendment to the Ohio Revised Code that permits Ohio political subdivisions and employees to take an immediate appeal when a trial judge refuses to dismiss them from a lawsuit. Ohio Revised Code 2744.02 (C), which became effective on April 9, 2003, provides: "An order that denies a political subdivision or an employee of a political subdivision the benefit of an alleged immunity from liability as provided in this chapter or any other provision of the law is a final order.

In 1989 the Ohio Supreme Court in *Commerce & Industry Ins. Co. v. Toledo,* 45 Ohio St. 3d 96, 543 N.E.2d 1188, held that an insurance company's lawsuit against the City of Toledo could proceed to trial. In that case, the Toledo Fire Department responded to a fire at a commercial building. A fire officer allegedly told a tenant of the building that it was safe for the tenant to shut down the internal sprinkler system and go home. The fire destroyed the tenant's property and his insurance company sued for recovery of the losses.

LEGAL LESSONS LEARNED

The fire department should develop a "pre-plan" for structures, which include hazardous materials or metals such as titanium, when water should not be used.

CASE STUDY 4-3 Incident Command: Four Children Died—Allegations of IC Gross Negligence

Marie Dean v. Jeffrey Childs and Charter Township of Royal Oak, 705 N.W.2d 344 (Mich. Nov. 3, 2005), Michigan Supreme Court reverses Michigan Court of Appeals decision, 262 Mich. App. 48; 684 N.W.2d 894; Case No. 244627 (Mich. Ct. of App. May 13, 2004).

FACTS

On April 6, 2000, the plaintiff's home in Royal Oak Township was set on fire, apparently by an arsonist. Marie Dean, mother of four children, successfully escaped, while a firefighter attempted to save her four children who were trapped in the rear of the house. Jeffrey Childs was on duty as the shift supervisor and responded as incident commander. He was the most senior appointed member of the fire department (the fire chief's position was vacant). The lawsuit alleges he ignored the fire hydrant across the street from the house and instead hit the hydrant a block away. He was informed of the rescue attempt in the rear of the house, but he allegedly ordered firefighters to spray water on the fire in the front of the house. This forced the fire to the rear of the house, and all four children died.

The plaintiff filed suit, alleging that Childs was "grossly negligent" and his actions caused the death of the children. She also alleged that he deprived her of her constitutional right to life and liberty without due process, in violation of 42 U.S.C. 1983, and that the township had inadequately staffed and trained its fire department. The township and Childs filed motions for summary judgment, claiming governmental and qualified immunity. The trial judge dismissed the Sec. 1983 action against Childs, but not the township, and refused to dismiss the wrongful death claim against Childs. The township and Childs appeal.

The Court of Appeals (two to one) held that Childs and the township must stand trial. Childs appealed to the Michigan Supreme Court.

YOU BE THE JUDGE

Should the firefighter and the township be dismissed from the wrongful death lawsuit?

HOLDING

Yes. The Michigan Supreme Court reversed the Court of Appeals and held that the firefighter and the township enjoy governmental immunity, with a one-paragraph decision stating, "for the reasons stated by dissenting Judge Griffin, we REVERSE."

To understand the final decision, it will be helpful to understand the two-judge majority decision of the Michigan Court of Appeals and the dissenting opinion of Justice Griffin.

The two judges on the Court of Appeals wrote, "[G]overnmental action often relates to emergency service, which must, by necessity, be carried out quickly without deliberation. The standard [for governmental liability] is, therefore, much higher than **deliberate interference**. A plaintiff must prove the government action showed an affirmative intent to cause harm."

deliberate interference In lawsuits filed against state governments or political subdivisions alleging deprivation of constitutional rights, plaintiffs must prove willful misconduct, not merely interference to their duties.

Regarding the state law claim against the incident commander for wrongful death, under Michigan law a municipal employee enjoys "qualified immunity" if

> . . . the employee's conduct does not amount to gross negligence that is the proximate cause of the injury or damage.
>
> Gross negligence is defined in the statute as "conduct so reckless as to demonstrate a substantial lack of concern for whether injury results."
>
> Plaintiff presented evidence of Child's gross negligence through the affidavit of firefighter Joe Soave. Soave stated that Childs ignored a fire hydrant in the immediate area in favor of one a block away and ordered water shot at the front of the home, forcing fire and smoke into the rear of the house despite the knowledge that a firefighter was attempting to rescue the children from that area. This conduct is described as so reckless as to demonstrate a substantial lack of concern for whether any injury results.

Dissenting Judge Griffin wrote a dissent, which the Michigan Supreme Court held as the proper decision. Here is the entire decision by Judge Griffin:

GRIFFIN, J. (concurring in part and dissenting in part.)

I agree with the majority that plaintiff's third amended complaint fails to state a claim on which relief can be granted under 42 USC 1983. *Canton v Harris,* 489 US 378; 109 S Ct 1197; 103 L Ed 2d 412 (1989); *DeShaney v Winnebago Co Dep't of Social Services,* 489 US 189; 109 S Ct 998; 103 L Ed 2d 249 (1989); *Monell v New York City Dep't of Social Services,* 436 US 658; 98 S Ct 2018; 56 L Ed 2d 611 (1978). Accordingly, I concur in the result of reversing the partial denial of defendant Charter Township of Royal Oak's motion for summary disposition. MCR 2.116(C)(7) and MCR 2.116(C)(8).

However, in regard to plaintiff's state law claims, I respectfully dissent. In my view, township acting fire chief, defendant Jeffrey Childs, is immune from liability pursuant to MCL 691.1407(2) because the factual support for plaintiff's allegations are insufficient for a reasonable juror to conclude that the alleged gross negligence of defendant Childs was "*the* proximate cause" of the deaths of plaintiff's decedents. MCR 2.116(C)(7) and MCR 2.116(C)(10).

In *Robinson v City of Detroit,* 462 Mich 439; 613 NW2d 307 (2000), our Supreme Court held that the phrase "the proximate cause" contained in the governmental immunity act, MCL 691.1407(2), means *the* proximate cause, not *a* proximate cause. Further, the *Robinson* Court defined "the proximate cause" as "the most immediate, efficient, and direct cause preceding an injury, not "a proximate cause." *Id.* at 445-446.

The majority concludes, for purposes of withstanding the motion for summary disposition, plaintiff presented sufficient documentary evidence that defendant Childs' gross negligence in fighting the fire may have been "the proximate cause" of the children's deaths, rather than the fire itself. I disagree.

In opposing the motion for summary disposition, plaintiff relied heavily on the affidavit of John Soave. In his affidavit, Soave stated that defendant Childs drove the fire engine to a fire hydrant approximately one block away from the burning house, although there was another fire hydrant located directly across the street from the fire scene. Further, "had Mr. Child's [sic] hooked up to that fire hydrant directly in front of the house, time would have been saved and I would likely have had more time to save the plaintiff's decedents." In his affidavit, Soave was also critical of the firefighting strategy employed by Childs once the fire hose was finally connected at the scene. In hindsight, Soave opined that the fire suppression efforts directed at the front of the house "caused the fire in the front of the house to be pushed towards the rear of the house where plaintiff's decedents were located." Finally, he speculated that "the actions of

Jeffrey Childs increased the danger to the plaintiff's decedents and prevented me from rescuing the children."

After reviewing the facts in a light most favorable to plaintiff, I conclude "the most immediate, efficient, and direct cause," *Robinson, supra* at 458-459, of the tragic deaths of plaintiff's children was the fire itself, not defendant's alleged gross negligence in fighting it. Plaintiff's original, amended, and second amended complaints, in effect, concede that the fire was the proximate cause of the deaths:

"That on April 6, 2000, a fire occurred at the residence of plaintiff and decedents. *That the decedents died as a result of said fire.*" (Emphasis added.)

Although the alleged gross negligence of defendant Childs in fighting the fire may have been a "substantial factor," *Brisboy v Fibreboard Corp,* 429 Mich 540, 547-548; 418 NW2d 650 (1988), in the deaths, in my view, its causal connection is insufficient to meet the governmental immunity threshold standard of "the" proximate cause.

In this regard, the present case bears similarities to *Kruger v White Lake Twp,* 250 Mich App 622; 648 NW2d 660 (2002), where our Court held, as a matter of law, that the alleged gross negligence of the township police department was not the proximate cause of the plaintiff's decedent's death. In *Kruger,* the plaintiff called the township police department and requested that her daughter be taken into custody because her daughter was intoxicated and could pose a danger to herself and others. Thereafter, the plaintiff's daughter was transported to the township police department and placed alone in a holding cell. She later escaped, fled from the police, and ran into heavy traffic on Highway M-59, where she was tragically struck and killed by an automobile. In affirming the grant of summary disposition in favor of the defendant police department, we held:

In the instant case, there were several other more direct causes of Katherine's injuries than defendant officers' conduct, e.g., her escape and flight from the police station, her running onto M-59 and into traffic, and the unidentified driver hitting plaintiff's decedent. Any gross negligence on defendant officers' part is too remote to be "the" proximate cause of Katherine's injuries. As a result, the officers are immune from liability. *Id.* at 624.

See also, *Poppen v Tovey,* 256 Mich App 351; 357, n 2; 664 NW2d 269 (2003) and *Curtis v City of Flint,* 253 Mich App 555; 655 NW2d 791 (2003).

For these reasons, I would hold that defendant Childs is immune from tort liability for plaintiff's state law claims pursuant to the **governmental immunity statute**, MCL 691.1407(2). The circuit court erred in denying the motion for summary disposition on this basis.

I would remand for dismissal of all claims.

governmental immunity statute Some states, including Michigan, have enacted laws that protect fire and other governmental officials, and their political subdivisions, from liability unless they intentionally disregarded their responsibility.

LEGAL LESSONS LEARNED

There is a legal phrase, "bad facts make bad law." These facts, if true, are tragic. Fortunately, Michigan has a governmental immunity statute that protects the incident commander and the township from liability.

■■■

Chapter Review Questions

1. In Case Study 4-1, *City of Muncie v. Thomas Weidner and Lauren Weidner,* the court addressed the issue of power lines that are down during a storm. Discuss "best practices" for a fire department SOP concerning the handling of

"wires down" runs, including steps to be taken when faced with numerous runs after a storm for trees into wires, and similar incidents.

2. In Case Study 4-2, *State Automobile Mutual Insurance Company v. Titanium Metals Corporation and Oakwood Village Fire Department,* the Ohio Supreme Court ordered the case back to the trial judge for pretrial discovery (depositions) on the issue of liability of the fire department for putting water on a titanium fire. Discuss how fire department "pre-plans," prepared in cooperation with the owners of businesses in the fire district, will help eliminate this type of

liability. Share with the class an example of a well-written pre-plan.

3. In Case Study 4-3, *Marie Dean v. Jeffrey Childs and Charter Township of Royal Oak,* the court addressed the issue of incident command and personal liability of incident commanders. Contact the insurance agent for a fire department in your state, and discuss whether the insurance policy allows the insurance company to decline to pay a judgment entered against an incident command who is found by a jury to be personally liable for willful or wanton misconduct. Share with the class the actual language from the insurance policy.

Expand Your Learning

Read the following and complete the individual or group assignment, as directed by your instructor.

1. In June, 2001, NIOSH issued a report on incidents involving firefighters being struck and killed by motor vehicles, "Traffic Hazards to Fire Fighters While Working Along Roadways" (*www.cdc. gov/niosh/hid12.html*). The case studies illustrate the clear need to use fire engines to block highways to protect emergency responders. Occasionally, state police and local law enforcement disagree with the incident commander's desires to shut down a highway completely. In 1994, the Ohio Attorney General issued Opinion 94-076 (*www.ag.state.oh.us/sections/opinions/1994/94-076.html*), which held that when a motor vehicle injury occurs on a state highway, the senior fire department officer has legal authority over placement of emergency apparatus.

   Describe who in your state is "in command" of placement of emergency vehicles at a highway MVA (include the source of this authority). Discuss how fire, EMS, and law enforcement officials might develop a joint SOP that can help ensure that we all "play nice" at highway accident scenes.

2. The federal Occupational Safety and Health Administration (OSHA) has jurisdiction over safety and health conditions in most private industries. About 25 states have OSHA-approved "state plans" in which a state safety agency has authority, often including public employers and fire departments. OSHA has a Respiratory Protection Standard, 29 Code of Federal Regulations 1910.134 (go to *www.osha.gov* and search for "respiratory protection"). This standard applies to private fire brigades and incorporated fire departments. 29 CFR 1910.156. OSHA's guidance on this standard states that it applies to "private sector workers engaged in firefighting, including those working in industrial fire brigades and private incorporated fire companies, and to Federal employees." CPL 02-00-120—CPL 2-0.120.

   The OSHA Respiratory Protection Standard requires many safety actions important to the fire service, including an annual employee medical questionnaire that must be completed by each firefighter authorized to wear an SCBA (29 CFR 1910.134, Appendix C); fit testing (1910.134 (f)); and "two-in/two-out" for structural firefighting (1910.155). OSHA advises, "The two-in/two-out provision is not intended as a staffing requirement. It does not require fire departments to hire additional firefighters; it does not require four-person fire companies; it does not require four persons on a fire truck."

Discuss whether the fire service would be safer if federal OSHA and state OSHA agencies routinely conducted safety audits of public fire departments, issued citations for violations, and publicized these fines and correction action plans.

See Appendix A for additional Expand Your Learning activities related to this chapter.

Emergency Vehicle Operations

5 CHAPTER

Two fire trucks collide on the way to a structure fire on Christmas Day, 2004. The driver of Rescue 1 (pictured) turned westbound and collided into the side of Engine 3, traveling northbound on a parallel street. The three firefighters on Rescue 1 were not wearing seat belts and were seriously injured. Fortunately, all recovered and later returned to active duty. *Photo supplied by Chief Jack Stickradt, Newark (OH) FD.*

Key Terms

willful or wanton
  misconduct, p. 61

malicious purpose, p. 61

From the Headlines

Pittsburgh, PA—Fire Trucks Out of Maintenance

According to press reports, 12 of the city's 14 ladder trucks have not received annual maintenance certifications since at least 2003 and would likely fail nationally recommended tests. The 56-vehicle fleet has some engines that have been in service since 1984, with six engines with more than 100,000 miles. The fire chief is quoted as telling the city council, "Our equipment, when it arrives at the scene of a fire, has to work. I've got serious concerns about reliability."

Kansas City—Fire Engine Fatal Crash Related to Lack of Maintenance

A Kansas City fire engine crash on September 5, 2004, killed one firefighter and injured three others when a car turned left in front of the engine. The police report noted that Engine 33's brakes were out of adjustment and had not been serviced for 16 months.

Phoenix—Stopping at All Red Lights

Phoenix Fire Chief Alan Brunacini presented a seminar to the Fire Department Safety Officer Association, where he discussed the Phoenix Fire Department SOP, which requires all emergency apparatus to stop for red lights. There is also a sign in the cab of fire trucks, "Got a backer?" to remind drivers to have a firefighter serve as a backer.

Columbus (OH) aerial truck lost brakes while on a run, traveling down a hill, and crashed into a restaurant. Four firefighters were seriously injured, along with several patrons. The firefighters all eventually returned to work. *Photo supplied by Battalion Chief Mark R. Devine, Columbus Fire Department, who was at the scene.*

Deerfield Township (OH) Fire and Rescue responded to MVA in passing lane of I-71 and blocked the passing and center lanes with engines. Another motorist ran into the blocking engine. *Photo supplied by Chief Nathan Bromen, Deerfield Township Fire & Rescue.*

This chapter discusses emergency vehicle accidents and litigation. Many fire departments have adopted policies and SOPs that provide for "limited response" to certain emergency calls, such as alarm drops at residential structures. Other departments are requiring full stops at controlled intersections and imposing internal limits on the speed of emergency vehicles.

◆ **INTERSECTIONS—SLOW DOWN OR FULL STOP**

In most states, emergency vehicles are required to slow down at controlled intersections and to show "due regard" to the safety of others before entering the intersection against a red light or stop sign. These statutes, if followed, can protect firefighters and EMS from personal liability. For example, on April 20, 2006, the Court of Appeals of Georgia in *Wynn v. The City of Warner Robins,* 2006 Ga. App. LEXIS 443, held that the trial judge correctly refused to grant a new trial after the civil jury found in favor of the city. On August 27, 1999, at about 4:50 p.m., Perkins Nobles, a firefighter EMT driving an emergency van in response to a 9-1-1 call, slowed down when he entered the intersection controlled by a red light but did not come to a complete stop. He testified that it was a "rolling stop." His red lights and siren were operating, and he intended to turn left. He believed that all traffic had stopped to grant him the right of way. His partner, Firefighter Richard Senter, told him that the intersection was "clear." Unfortunately, the vehicle driven by Lorenzo Wynn did not stop and collided with the fire department vehicle, and Wynn was injured. The Court of Appeals agreed with the trial court's instruction to the jury,

> [T]he driver of an authorized emergency vehicle, when responding to an emergency call, may exercise the privilege set forth in the law. The driver of an authorized emergency vehicle may proceed past a red or stop signal or stop sign, but only after slowing down as may be necessary for safe operation. . . . Exceptions granted under Georgia law to an authorized emergency vehicle shall apply only when such vehicle is making use of a flashing or revolving red light visible under normal atmospheric conditions from a distance of 500 feet to the front of such vehicle.

The Court of Appeals also confirmed that the trial judge properly excluded from evidence the report of Fire Department Lt. Scott Renfroe, who stated that after the accident he cautioned both firefighters on the dangers of emergency driving. It was properly excluded because it constituted evidence of "subsequent remedial measures."

◆ **NFPA STANDARDS**

- *NFPA 1002: Standard for Fire Apparatus Driver/Operator Professional Qualifications* (2003 edition). This standard identifies the knowledge and skills required to function as a fire apparatus operator (FAO), including specific requirements for different types of fire apparatus.
- *NFPA 1071: Standard for Emergency Vehicle Technician Professional Qualifications* (2000 edition). This standard identifies the knowledge and skills required of all emergency vehicle technicians.

◆ *NFPA 1451: Standard for a Fire Service Vehicle Operations Training Program* (2002 edition). This standard focuses on the development and implementation of a comprehensive EVO program, including development of safe driving procedures, and the prohibition on driving emergency vehicles under the influence of alcohol or drugs.

◆ *NFPA 1901: Standard for Automotive Fire Apparatus* (2003 edition). This standard identifies apparatus design and specification features to allow operators to control their apparatus safely and give all firefighters riding on the apparatus a safer response environment.

◆ KEY STATUTES

COMMERCIAL DRIVERS LICENSES (CDLS)

The U.S. Department of Transportation regulations, 49 CFR 383 (1999), under the Federal Commercial Motor Vehicle Act, exempts emergency vehicle operators from the federal CDL requirements but authorizes states to implement a CDL requirement.

DUE REGARD STATUTE (OHIO REVISED CODE 4511.24)

Like many states, Ohio authorizes emergency vehicles to exceed the posted speed limits when responding to an emergency with red lights and sirens. "This section does not relieve the driver of an emergency vehicle or public safety vehicle from the duty to drive with due regard for the safety of all persons using the street or highway." Likewise, cities, townships, and other political subdivisions are not liable for injuries, deaths, or property damage arising out of emergency vehicle operations unless the operator drove in a "willful or wanton misconduct" manner. Ohio Rev. Code 2744.02 (B). (Go to *www.legislature.state.oh.us/,* and click on "Laws, Acts, and Legislation.")

◆ CASE STUDIES

CASE STUDY 5-1   Civilian Motorist Killed in Collision with Fire Engine

Perez v. Unified Government of Wyandotte County et al., 432 F.3d 1163 (10th Cir. 2005); 2005 U.S. App. LEXIS 28743.

FACTS

Anthony Mots, a firefighter with the Kansas City (KS) Fire Department, which is a division of the Unified Government of Wyandotte County/Kansas City (KS), was sued in U.S. District Court under 42 U.S.C. 1983 (violation of constitutional rights by government employee) for the death of a motorist, Aaron Becerra.

Mots was stationed at Pumper Station 9 when a call came in for a house fire. He responded to the fire and turned on his fire truck's emergency lights and siren as soon as he left the station. To reach the fire, Mots drove his fire truck westbound on Central Avenue, a major thoroughfare. According to an eyewitness, Mots was traveling down Central Avenue at 40 mph, and the posted speed limit was 30 mph. Where Central Avenue crosses 18th Street is a five-way intersection; another major thoroughfare, Park Drive, intersects to the southwest. The buildings on this corner can block emergency lights and the sound of a siren.

According to several witnesses, there was a red light when Mots approached the intersection. Mots testified at his deposition that he did not remember whether there was a red light when he reached the corner, but, if there had been, it would have been visible as he approached the intersection. Before he reached 18th Street, Mots moved into the eastbound lane of oncoming traffic on Central Avenue because westbound traffic would not move in response to his siren and emergency lights. He also blew his air horn to warn other cars about his approach.

Mots slowed down as he approached the intersection but did not stop. He then sped up as he went through it. The Kansas City emergency vehicle training documents recommend that emergency drivers should come to a complete stop before entering a negative right-of-way intersection, but the fire department's official policy does not require vehicles to come to a complete stop.

The civilian driver, Aaron Becerra, entered the intersection going eastbound on Central Avenue. Mots did not see Becerra's car until a split second before he slammed into it. Several eyewitnesses, including one in a car that Mots swerved around, said that Becerra's car was visible coming headlong into Mots' path. An accident reconstruction expert stated that the fire truck hit Becerra's car in a direct collision at 23 to 24 mph. Becerra died as a result of injuries sustained in the accident.

The firefighter filed a motion to dismiss, claiming qualified immunity. The trial judge denied the motion, and the firefighter took an immediate appeal to the 10th Circuit Court of Appeals.

YOU BE THE JUDGE

Should the firefighter be dismissed from this case?

HOLDING

Trial judge is reversed, and the case is dismissed. The three-judge panel wrote:

> Government officials who perform discretionary functions are entitled to qualified immunity if their conduct does not violate clearly established rights of which a reasonable government official would have known. Hulen v. Yates, 322 F.3d 1229, 1236 (10th Cir. 2003). When a defendant raises a qualified immunity defense, the plaintiff bears the burden of establishing that the defendant's conduct violated a constitutional or statutory right and that the right was clearly established at the time of conduct. *Id.* at 1237.

> That said, the Fourteenth Amendment is not a "font of tort law to be superimposed upon whatever systems may already be administered by the States." [U.S. Supreme Court decision in *County of Sacramento v. Lewis*, 523 U.S. 833 (1998), at 848 (quoting *Paul v. Davis*, 424 U.S. 693, 701 (1976)). Only government conduct that "shocks the conscience" can give rise to a substantive due process claim. *Id.*

> In Lewis, the Supreme Court clarified how courts should determine whether government action shocks the conscience. The Court held that a police officer who slammed his car into a motorcycle during a high-speed chase did not violate the Fourteenth Amendment because there was no allegation that the officer intended to harm the motorcycle driver. Lewis established a clear rule: When governmental officials face a situation "calling for fast action," only official conduct done with an intent to harm violates the Fourteenth Amendment. *Id.* at 853. "[W]hen unforeseen circumstances demand an officer's instant judgment, even precipitate recklessness fails to inch close enough to harmful purpose to spark the shock that implicates the large concerns of the governors and the governed." *Id.*

The Court acknowledged, however, that behavior that would not violate the Fourteenth Amendment if done in a time-sensitive, high-pressure situation may nevertheless shock the conscience if the official has time to deliberate before acting. Thus, the Court held that when a government official has enough time to engage in "actual deliberation," conduct that shows "deliberate indifference" to a person's life or security will shock the conscience and thereby violate the Fourteenth Amendment. *Id.* at 851. "[L]iability for deliberate indifference . . . rests upon the luxury . . . of having time to make unhurried judgments, upon the chance for repeated reflection, largely uncomplicated by the pulls of competing obligations." *Id.* at 853.

The Court cautioned that "actual deliberation" meant more than having a few seconds to think. It stated:

By "actual deliberation," we do not mean "deliberation" in the narrow, technical sense in which it has sometimes been used in traditional homicide law. See, e.g., *Caldwell v. State*, 203 Ala. 412, 84 So. 272, 276 (Ala. 1919) (noting that "deliberation here does not mean that the man slayer must ponder over the killing for a long time"; rather, "it may exist and may be entertained while the man slayer is pressing the trigger of the pistol that fired the fatal shot[,] even if it be only for a moment or instant of time").

Lewis, 523 U.S. at 851 n.11 [emphasis added]. The intent to harm standard is not limited to situations calling for split-second reactions. Rather, it applies whenever decisions must be made "in haste, under pressure, and frequently without the luxury of a second chance." *Id.* at 853. As the Eighth Circuit recently noted, "The intent-to-harm standard most clearly applies in rapidly evolving, fluid, and dangerous situations which preclude the luxury of calm and reflective deliberation." *Terrell v. Larson*, 396 F.3d 975, 978 (8th Cir. 2005) (en banc).

We have not had occasion to apply Lewis to a situation where a firefighter or police officer is involved in an automobile accident while responding to an emergency call. However, it is clear from Lewis that the intent to harm standard applies in this case. A firefighter responding to a house fire has no time to pause. He has no time to engage in calm, reflective deliberation in deciding how to respond to an emergency call. Doing so would risk lives. This case presents a paradigmatic example of a decision that must be made in haste and under pressure.

We may have remanded the case to the district court for application of the proper standard, but Becerra conceded at oral argument that there were no allegations and no facts in this case that supported a claim that Mots had an intent to harm. Because there is no such allegation in the complaint, we need not reach the question of what showing is necessary to evince an intent to harm. We simply hold that a bystander hit by an emergency response vehicle in the process of responding to an emergency call cannot sustain a claim under the substantive due process clause without alleging an intent to harm. As such, Mots should be granted qualified immunity.

LEGAL LESSONS LEARNED

Fire and EMS departments should at least annually conduct emergency vehicle refresher training. Experienced instructors with knowledge of braking distances can be very helpful. Bring in a certified emergency vehicle trainer and conduct training for all personnel.

CASE STUDY 5-2   Ambulance Exceeding Fire Department's 20 mph Speed Limit when Traveling against Traffic

Hunter v. City of Columbus, 139 Ohio App.3d 962, 746 N.E.2d 246 (10th District, 2000).

FACTS

Brenda Hunter was killed on February 7, 1998, when she was making a left turn and was struck by an ambulance. It was about noon, when an alarm was sounded for a woman having difficulty breathing. The Columbus Fire Department ambulance, with red lights and sirens on, approached Sullivan Avenue and Wilson Bridge Road. The road was dry and clear; the posted speed limit was 35 mph. The ambulance was traveling westbound, and both westbound lanes were blocked with vehicles. The ambulance driver pulled out into the eastbound lane to pass the blocked vehicles, about two car lengths prior to reaching the stopped vehicles. At the same time, Mrs. Hunter turned left, attempting to cross the eastbound lanes and enter a shopping center. She was struck and killed by the ambulance. Her husband filed a wrongful death lawsuit against the city and the two paramedics in the ambulance.

According to the plaintiff's expert witness, Dr. Weichel, Ph.D., who measured the ambulance's skid marks, the ambulance was traveling 61 mph (89.7 feet per second) when the driver applied the brakes, and 52 mph (76.5 feet per second) when he collided with Mrs. Hunter's vehicle. The fire department's SOP prohibited emergency vehicles from going over 20 mph when traveling against traffic. In the deposition of the ambulance driver, he testified he had no knowledge of the SOP. The other paramedic in the ambulance testified that he was aware of the SOP, but thought the driver was traveling safely, at about 45 mph.

Under Ohio law, the city or political subdivision is immune from liability for injuries during an emergency run, with red lights and siren activated, unless, so long as "the operation of the vehicle did not constitute **willful or wanton misconduct**." Ohio Revised Code 2744.02(A)(1).

Likewise, the emergency vehicle driver is immune from liability unless "the employee's acts or omissions were with **malicious purpose**, in bad faith, or in a wanton or reckless manner." Ohio Revised Code 2744.03(A)(6)(b).

The trial judge, relying on these statutes, granted the defendants' motion to dismiss the case. The plaintiff filed an appeal to the Ohio Court of Appeals for Franklin County.

YOU BE THE JUDGE

1. Should the paramedic driver be dismissed from the lawsuit, given his deposition testimony that he had no knowledge of the 20 mph SOP?
2. Should the paramedic passenger be dismissed from the lawsuit, given his testimony that he knew of the SOP, but he believed the driver was only going 45 mph and was driving safely for the conditions?
3. Should the city be dismissed?

HOLDING

The three-judge panel reversed the trial judge and held that all three defendants should stand trial.

The court wrote,

> The Columbus fire department rule is designed to protect fire personnel, other motorists, and the person to whom emergency aid is to be rendered, and it is a reasonable rule. Yet, in this instance, the driver of the vehicle was stated to have been proceeding at least sixty-one miles per hour, having veered out into the wrong lane only two or three hundred feet prior to the place where the accident occurred. The city argues that it should not be penalized because its rule of twenty miles per hour is not required by law and is more

willful or wanton misconduct Operating an emergency vehicle or performing other duties with gross negligence or failure to exercise any care towards the public.

malicious purpose In many states, the political subdivision enjoys "governmental immunity" unless the firefighter driving the emergency vehicle drove with willful or wanton misconduct.

strict than many other municipalities. That may be true and we agree that the rule is not *per se* determinative, but it is to be taken into consideration in determining what a reasonable speed is to protect the safety of all concerned.

[Note: The case never went to trial; the city settled it out of court.]

LEGAL LESSONS LEARNED

SOPs concerning emergency driving are of critical importance. Consider an annual training program reviewing all fire department safety-related SOPs. Consider also posting the emergency driving SOP in a prominent location in the report-writing room of each fire station.

CASE STUDY 5-3 Government Immunity—Police Chase in California/U.S. Supreme Court

County of Sacramento v. Lewis, 523 U.S. 833 (1998); U.S. Supreme Court.

FACTS

The parents of Philip Lewis, killed while a passenger on a motorcycle that crashed during a high-speed police chase, sued the County of Sacramento and the deputy sheriff, under 42 U.S.C. 1983, alleging deprivation of the constitutional rights of their son. The federal trial judge granted summary judgment for the county and the deputy sheriff. The three-judge U.S. Court of Appeals for the 9th Circuit reversed as to the deputy sheriff, and the deputy asked the U.S. Supreme Court to hear his appeal. The U.S. Supreme Court agreed (it takes four of nine justices to vote for a *writ of certiorari* and have the appellate file sent to the Supreme Court).

On May 22, 1990, at approximately 8:30 p.m., petitioner James Everett Smith, a Sacramento County sheriff's deputy, along with another officer, Murray Stapp, responded to a call to break up a fight. Upon returning to his patrol car, Deputy Stapp saw a motorcycle approaching at high speed. It was operated by 18-year-old Brian Willard and carried Philip Lewis, 16 years old, as a passenger. Neither boy had anything to do with the fight that prompted the call to the police. The deputy turned on his overhead rotating lights, yelled to the boys to stop, and pulled his patrol car closer to Smith's, attempting to hem the motorcycle in. Instead of pulling over in response to Stapp's warning lights and commands, Willard slowly maneuvered the motorcycle between the two police cars and sped off. Smith immediately switched on his own emergency lights and siren, made a quick turn, and began pursuit at high speed. For about 75 seconds over a course of 1.3 miles in a residential neighborhood, the motorcycle wove in and out of oncoming traffic, forcing two cars and a bicycle to swerve off the road. The motorcycle and patrol car reached speeds of up to 100 miles an hour, with Smith following at a distance as short as 100 feet. At that speed, his car would have required 650 feet to stop.

The Supreme Court described the tragic end of the chase. "The chase ended after the motorcycle tipped over as Willard tried a sharp left turn. By the time Smith slammed on his brakes, Willard was out of the way, but Lewis was not. The patrol car skidded into him at 40 miles an hour, propelling him some 70 feet down the road and inflicting massive injuries. Lewis was pronounced dead at the scene."

YOU BE THE JUDGE

1. Should the deputy sheriff be dismissed from this lawsuit?
2. Should the County be dismissed?

HOLDING

Both defendants should be dismissed from the suit.

Justice Souter wrote the opinion of the Court:

The issue in this case is whether a police officer violates the Fourteenth Amendment's guarantee of substantive due process by causing death through deliberate or reckless indifference to life in a high-speed automobile chase aimed at apprehending a suspected offender. We answer no, and hold that in such circumstances only a purpose to cause harm unrelated to the legitimate object of arrest will satisfy the element of arbitrary conduct shocking to the conscience, necessary for a due process violation.

Smith was faced with a course of lawless behavior for which the police were not to blame. They had done nothing to cause Willard's high-speed driving in the first place, nothing to excuse his flouting of the commonly understood law enforcement authority to control traffic, and nothing (beyond a refusal to call off the chase) to encourage him to race through traffic at breakneck speed forcing other drivers out of their travel lanes. Willard's outrageous behavior was practically instantaneous, and so was Smith's instinctive response. While prudence would have repressed the reaction, the officer's instinct was to do his job as a law enforcement officer, not to induce Willard's lawlessness, or to terrorize, cause harm, or kill. Prudence, that is, was subject to countervailing enforcement considerations, and while Smith exaggerated their demands, there is no reason to believe that they were tainted by an improper or malicious motive on his part.

Regardless whether Smith's behavior offended the reasonableness held up by tort law or the balance struck in law enforcement's own codes of sound practice, it does not shock the conscience, and petitioners are not called upon to answer for it under Sec. 1983. The judgment below is accordingly reversed.

LEGAL LESSONS LEARNED

The U.S. Supreme Court has adopted a "shock the conscience" standard of liability, which thankfully makes it very difficult for plaintiffs to establish liability for public safety officials, including firefighters and EMS, who make rapid decisions in an emergency.

■■■

Chapter Review Questions

1. In Case Study 5-1, *Perez v. Unified Government of Wyandotte County et al.,* the court addressed the issue of liability for a firefighter in the death of a motorist. In civil litigation, "pre-trial discovery" includes production of records and depositions. Discuss what records a fire department should maintain on a firefighter's training and actual experience driving emergency apparatus that could be very helpful in the defense of a civil suit.

2. In Case Study 5-2, *Hunter v. City of Columbus,* the court addressed the issue of a paramedic driving an ambulance beyond the fire department's speed limits in its internal policy. Many fire departments have speed limits in their policies, such as "Do not exceed the posted speed limit by 10 mph." Discuss the advantages and disadvantages of such internal speed limits.

3. In Case Study 5-3, *County of Sacramento v. Lewis,* the court addressed the issue of governmental liability for police chases. We have all seen the videos of high-speed chases that sometimes occur, and many police departments have

placed internal restrictions on the number of police vehicles that may join a chase, use of "stop sticks" to puncture the tires of the fleeing motorist, and other steps to reduce the risks to the public. In some fire departments, certain firefighters have a well-earned reputation for being a "heavy-footed" driver who drives way too fast in an effort to be one of the first to the scene. Describe what steps that a fire department can take to eliminate this behavior and discuss the advantages and disadvantages of documenting these efforts in the firefighter's personnel file (normally a "public record").

Expand Your Learning

Read the following cases and complete the individual or group assignment, as directed by your instructor.

1. The St. Louis Fire Department has experienced a drop in emergency apparatus accidents after issuing an SOP in February 1995 that requires "On the Quiet" responses (no red lights or sirens) for listed incidents, including automatic alarms, smoke detectors, sprinkler alarms, manual pull stations, natural gas leaks, and carbon monoxide detectors. Many fire departments limit "hot" responses to alarm drops to the first due engine and perhaps a battalion chief. Still other departments will send an engine company to accompany an EMS squad, but the engine company goes "On the Quiet."

   Describe what fire departments have done in your state to develop an SOP that limits emergency responses on certain runs. Share with the class a copy of the SOP.

2. NIOSH Fire Fighter Fatality Investigation Report F2004-43, September 23, 2005, (*www.cdc.gov/niosh/fire/*) concerns the death of a 34-year-old part-time firefighter, who died after the engine in which he was riding collided into another fire engine from another department. They collided at an intersection, both enroute to the same structure fire.

   Describe what steps can be taken by neighboring fire departments to avoid collisions at intersections where they are likely to meet on emergency responses. Identify a fire department in your state that has addressed this issue with a neighboring department, and share with the class their SOP.

See Appendix A for additional Expand Your Learning activities related to this chapter.

Module II: Employment and Personnel Issues

Employment Litigation
Age, Beards, Free Speech, and Promotions

6 CHAPTER

Warren County fire chiefs training on Rapid Intervention Team (RIT) skills prior to launching a countywide automatic RIT program. *Photo by author.*

Key Terms

bona fide occupational
  qualification, p. 67

disparate impact, p. 74
subterfuge, p. 75

retaliation, p. 76

From the Headlines

Nepotism—Firefighter Marries Captain's Daughter and Is Fired

A 25-year-old firefighter was fired by the City of Rock Hill (SC) after marrying the 21-year-old daughter of a FD captain. A federal judge granted a temporary injunction but later dissolved this, and a state judge refused to stop the city from firing him. The firefighter and his wife have moved to his hometown of Findlay (OH) and settled with the city for $5,000.

The Station Nightclub Fire in Rhode Island—Off-Duty Firefighter Denied "Injured-on-Duty" Coverage

Firefighter Steve Burgess, of the City of Cranston (RI) Fire Department, was off-duty and attending the concert at The Station nightclub in West Warwick on February 20, 2003, when the fire broke out that killed 100 patrons. He led his date out the back door and ran to the first responding engine from the West Warwick Fire Department. With permission of a lieutenant, Burgess kept water flowing on the front door as rescuers in turnout gear tried to pull patrons to safety. The fire singed his eyebrows off and burned his clothing. The trauma of this event led him to seek "injured-on duty" paid treatment for post-traumatic stress. The City of Cranston denied the request, the union filed a grievance, and arbitrator Lawrence Katz found for the city.

Residency Laws Eased in Ohio for Firefighters/Police Officers

The Ohio Legislature in 2006 has amended the Ohio Revised Code to strike down the requirement that full-time firefighters or police officers must reside in their city (Cleveland, Dayton, Toledo, Akron, and Youngstown require firefighters to live in the city; Cincinnati requires residency in Hamilton County). Senate Bill 82 provides that cities and other political subdivisions may only require that they reside "in the county where the political subdivision is located or in any adjacent county in this state" (see www.legislature.state.oh.us/). Several cities have filed lawsuits challenging the authority of the state legislature to overturn their "home rule" city ordinances.

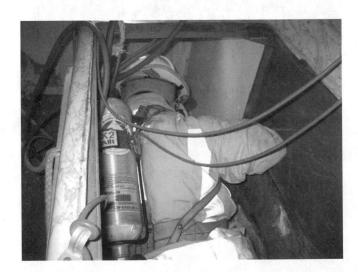

Warren County fire chief during RIT training, squeezing under wires. *Photo by author.*

◆   **INTRODUCTION**

This chapter reviews a variety of employment issues, including age, beards, free speech, and promotions in the fire service. There continues to be substantial litigation involving employment issues in the fire service. Case law is developing rapidly as a result of decisions of the U.S. Supreme Court and lower federal and state courts.

◆   **AGE DISCRIMINATION**

The Age Discrimination in Employment Act of 1967 (ADEA), 29 U.S.C. 623(a) and 631(a), protects individuals over age 40 from discrimination in employment. Under a 1996 amendment to the ADEA allowing state and local governments to have age restrictions on hiring and retiring of firefighters and police officers, and subsequent court decisions upholding this statutory exemption, plaintiffs have the difficult burden of proving that age restrictions are actually being used as a subterfuge to discriminate.

To understand the current application of the law to the fire service, a short history lesson is needed. As originally enacted in 1967, Congress did not apply its antidiscrimination provisions to state or local governments and therefore the statute did not apply to public fire departments. See Public Law 90-202, Sec. 11(b), 29 U.S.C. 630(b). In 1974, Congress amended the statute, however, to cover state and local governments. Public Law 93-259. Two years later, a U.S. Supreme Court decision brought into question whether Congress had the constitutional authority to extend this law to state and local governments, when the Supreme Court held that the 10th Amendment prohibited application of wage and hour provisions of the Fair Labor Standards Act (FLSA) to state and local governments. *National League of Cities v. Usery,* 426 U.S. 833 (1976). The Supreme Court put this issue to rest in 1983 when it held that the ADEA could be applied to state law enforcement officers. *EEOC v. Wyoming,* 460 U.S. 226 (1983). Following the *Wyoming* decision, state and local government employers (including public fire departments) were required to show that age was a **bona fide occupational qualification** (BFOQ), similar to the requirement on private employers.

In 1986, Congress amended the ADEA to provide state and local governments an exception on age restrictions for firefighters and law enforcement. Public Law 99-592, 29 U.S.C. 623 (i). This important amendment authorized state and local governments that had age restrictions in place for firefighters and police as of March 3, 1993, (the day after the Supreme Court handed down the decision in *EEOC v. Wyoming*) to keep those restrictions in place until December 31, 1993. Congress thereby provided governments time to review their age restrictions and ultimately to keep only those in which they could meet the BFOQ requirements.

In 1996, Congress again amended the ADEA and reinstated the firefighter/law enforcement exception, with no future expiration date. Public Law 104-208; 29 U.S.C. 623 (j). In effect, this law authorizes state and local governments (including public fire departments) to impose age restrictions regarding the hiring and retiring of firefighters and law enforcement officers. The exception states:

> (j) Employment as firefighter or law enforcement officer
> It shall not be unlawful for an employer which is a State [or] a political subdivision of a State . . . to fail or refuse to hire or to discharge any individual because of such individual's age if such action is taken—

bona fide occupational qualification   The Age Discrimination in Employment Act (ADEA) prohibits employment discrimination against individuals over age 40 unless the employer can show that age is a legitimate occupational qualification given the duties and risks of the job.

(1) with respect to the employment of an individual as a firefighter or as a law enforcement officer . . . and the individual has attained . . .

(A) the age of hiring or retirement, respectively, in effect under applicable State or local law on March 3, 1983; or (B) (I) If the individual was not hired, the age of hiring in effect on the date of such failure or refusal to hire under applicable State or local law enacted after September 30, 1996.

See, for example, *Alan Jay Feldman v. Nassau County,* 434 F.3d 177 (2nd Cir. January 9, 2006), in which the county refused to allow Mr. Feldman to take the civil service examination for police officer because he was over the age of 35, under New York Civil Service Law 58 (1)(a). He contended that the county failed to prove that age 35 was a bona fide occupational qualification. The U.S. District Court dismissed his lawsuit, and the three-judge Court of Appeals agreed, holding that the 1996 amendment to the ADEA "relieves law enforcement agencies of having to establish that age is a BFOQ in order to discriminate on the basis of age." The court explained, "The only required showing of bona fides under section 4(j) relates to the [governmental entity's] hiring plan itself and not whether age amounts to a proper qualification for the job." The court also quoted from a State of New York report that shared reasons why communities have established age restrictions on new hires, "Recent reports indicate that older officers leave the force at a higher rate than young officers and utilize more sick leave. With per officer training costs calculated at an estimated $100,000, officer long term retention becomes an economic issue for the State." New York State Assembly, Sponsor's Memorandum (1999).

The U.S. Supreme Court has issued two recent age decisions, *General Dynamics Land Sys. v. Cline,* 540 U.S. 581 (2004) and *Smith v. Jackson,* 125 S. Ct. 1536 (2005). (See www.supremecourtus.gov. and Case Study 6-1.)

In the *Cline* decision, the court held that the ADEA does not prohibit employers granting greater retirement health benefits in a collective bargaining agreement to employees over 50, and fewer benefits to those currently 40 to 49 years of age. The Court in *Cline* said that the EEOC was incorrect in recognizing so-called reverse discrimination claims by younger employees who are age 40 or older. In the *Jackson* decision, the Supreme Court held that a city's effort to recruit younger police officers by giving larger pay raises at the entry level could be a basis for a "disparate impact" claim by older workers, but the city could defend itself by showing that there was legitimate governmental need to justify higher pay for entry-level employees. These two decisions are expected to reduce age discrimination lawsuits, and the Equal Employment Opportunity Commission (EEOC) is expected to amend their regulations to reflect the court's ADEA decisions, including 29 CFR 1625.2–.5.

Mandatory retirement in the fire service has been the subject of litigation (see Case Study 6-2 concerning the Chicago Fire Department).

◆ **BEARDS**

There has been considerable litigation by African-American and Islamic firefighters seeking to challenge SCBA "no beard" safety rules. For example, on February 23, 2006, in *Curtis DeVeau v. City of Philadelphia,* 2006 Phila. Ct. Com. Pl. LEXIS 121, the court held that a Muslim firefighter's claim of religious discrimination was not proven. He had been suspended from the fire department without pay for refusing to shave his beard. The trial judge declined to "take judicial notice or to otherwise permit Plaintiff to introduce evidence of the District of Columbia's practice and experience permitting its firefighters to serve with beards."

Compare this decision with the District of Columbia firefighter case, *Brian R. Kennedy v. District of Columbia,* 654 A.2d 847 (D.C. 1995), and the prior decision of September 12, 1994, 654 A.2d 847, 1994 D.C. App. LEXIS 156, in which the District of Columbia Court of Appeals upheld the trial court's decision that Kennedy was improperly terminated for growing a handlebar mustache and a beard. The fire department's regulation prohibited mustaches more than ¼ inch beyond the corners of the mouth and also prohibited beards, except "members suffering from folliculitis barbae . . . may wear a beard ¼ inch in length and neatly trimmed." The Court of Appeals stated, "The evidence further demonstrated, however, that of the approximately twenty firefighters who were afflicted with pseudofolliculitis barbae ("PFB") and were allowed to wear short beards, there were no reported incidents resulting from improperly secured face masks in the past seven years."

In another District of Columbia case, *Calvert R. Porter et al. v. District of Columbia,* August 11, 2005, 382 F.Supp.2d 35 (D.D.C.), three Muslim firefighters who wore beards for religious reasons claimed that the fire department's grooming standards violated the Religious Freedom Restoration Act, 42 U.S.C. 2000bb et seq. Federal district judge James Robertson issued a preliminary injunction in their favor on June 22, 2001, ordering the fire department not to discipline them while the case was pending. The court commented that for the past four years, the case "seemed to put itself to sleep" because the fire department adopted a new policy allowing beards for religious reasons. One of the plaintiffs allowed his beard to grow and was transferred out of the HazMat unit in 2002, after he failed the computerized fit test. On June 7, 2005, the fire department issued Special Order 20, which prohibits all firefighters from having facial hair that "comes between the sealing surface of the facepiece and the face or interferes with the valve function." Firefighters who for religious reasons insist on wearing a beard will be removed from active firefighting and placed in administrative duties. The court found that this tougher new rule is consistent with the OSHA standard, 29 CFR 1910.134 (g)(1)(i), which prohibits employers from allowing employees to wear tight-fitting facepieces if the employees have "(A) facial hair that comes between the sealing surface of the facepiece and the face or that interferes with valve functions." Judge Robertson held the fire department may not terminate the Muslim firefighters with beards but may remove them from active firefighting positions and reassign them to administrative duties.

Many courts have upheld the "no beard rule" based on the business necessity of complying with safety requirements. The "no beard" rule was upheld in the Atlanta Fire Department, see *Walter Fitzpatrick v. City of Atlanta,* September 27, 1993, U.S. Court of Appeals for the 11th Circuit, 2 F.3d 1112 (11th Cir. 1993); and City of Jacksonville (FL) in the federal district court decision in *Robert Yarborough v. City of Jacksonville,* 363 F. Supp. 1176 (D. Fl. 1973).

California has also upheld the "no beard" rule. For example, in *Harry K. Vernon v. State of California,* February 25, 2004, 116 Cal. App. 4th 114, 2004 Cal. App. LEXIS 224, the Court of Appeals of California upheld the dismissal of a lawsuit filed by an African-American firefighter with the hereditary skin disorder *pseudofolliculitis barbae,* in which facial hairs curl back into the facial skin. He had been a firefighter with the City of Berkeley since 1978, and the city allowed him to have facial hair between 1984 and 1999. The city implemented a new facial hair policy in 1999, in compliance with the State of California Cal-OSHA requirements for firefighters, which followed the 1977 federal OSHA standards. The Court of Appeals upheld the dismissal of his lawsuit against the state, because it was the city that took employment action against the plaintiff. "And while the State demanded compliance with the Cal-OSHA regulations, as it does with any legislation, implementation was directly undertaken by the City, which ultimately determined the means and manner of discipline and the nature of the enforcement mechanisms within the Fire Department."

On May 30, 2006, the Court of Appeals of California also enforced the "no beard" rule against another African-American firefighter in *Harry K. Vernon v. City of Berkeley,* 2006 Cal. App. Unpub. LEXIS 4620. The Court stated:

> Appellant has proceeded on a theory of disparate impact rather than disparate treatment. Disparate impact discrimination has been recognized as actionable under both Title VII of the Civil Rights Act of 1964 (Title VII) (42 U.S.C. § 2000e et seq.), and the California Fair Employment and Housing Act (FEHA) (Gov. Code, § 12900 et seq.). (*Carter v. CB Richard Ellis, Inc.* (2004) 122 Cal.App.4th 1313, 1321.) To prevail on a theory of disparate *impact,* the employee must show that regardless of motive, a facially neutral employer practice or policy, bearing no manifest relationship to job requirements, in fact had a disproportionate adverse effect on certain employees because of their membership in a protected group. [D]isparate impact is not proved merely because all members of a disadvantaged subgroup are also members of a protected group. Proof must be offered, "usually through statistical disparities, that facially neutral employment practices adopted without a deliberately discriminatory motive nevertheless have such significant adverse effects on protected groups that they are in operation . . . functionally equivalent to intentional discrimination.
>
> As the parties have done, we assume appellant proved a prima facie case of disparate impact from the City's enforcement of regulations that prohibited him from passing the SCBA respirator fit test, and proceed to resolve this appeal by examining the issue of respondent's proof of a legitimate, nondiscriminatory reason for its employment decision.
>
> A business necessity is recognized as a legitimate, nondiscriminatory reason that operates as a defense in disparate impact cases under the FEHA (Cal. Code Regs., Tit. 2, § 7286.7(b)). Under the business necessity defense, the business purpose must be sufficiently compelling to override any racial impact; the challenged practice must effectively carry out the business purpose it is alleged to serve; and there must be available no acceptable alternative policies or practices which would better accomplish the business purpose advanced, or accomplish it equally well with a lesser differential racial impact.
>
> The City claims that its actions were "required by California law," and therefore were "permissible as a matter of law." While we do not agree with the City that compliance with Cal-OSHA regulations establishes per se a defense to any disparate impact case under the FEHA as a matter of law without further inquiry (*Fitzpatrick v. City of Atlanta* (11th Cir. 1993) 2 F.3d 1112, 1121(*Fitzpatrick*)), we find upon examination of the record that a legitimate, nondiscriminatory reason for the City's practice has been established in the case before us. We know from the evidence presented that respondent did not take any adverse action against appellant until after DOSH adopted regulations in 1997 that prohibited the use or testing of SCBA respirators upon employees with visible hair. Thus, the City did not evince any inclination to subject appellant to any disparate impact before the State regulations were enacted. Once the SCBA respirator regulations became effective, they were binding upon the City as governing standards for maintaining and enforcing employee safety. (Lab. Code, § 6304.5; *Elsner v. Uveges* (2004) 34 Cal.4th 915, 927–928.) The City had neither control over the formulation or enactment of the safety regulations, nor the authority to cancel or disregard them. (See *Harris v. Civil Service Com., supra,* 65 Cal.App.4th 1356, 1363.)

The City's subsequent implementation of the "Respiratory Protection Policy" to conform to State requirements demonstrates a legitimate, nondiscriminatory reason for the prohibition against any person "who had visible facial hair" from taking a SCBA mask fit test. Not only did the City act to comply with State standards, but also to safeguard the safety of its employees as specified in the Cal-OSHA regulations. Employment tests or requirements "that are discriminatory in effect" are nonetheless valid if "the employer meets the burden of showing that any given requirement [has] . . . a manifest relationship to the employment in question." (*Harris v. Civil Service Com., supra,* 65 Cal.App.4th 1356, 1366.) Imposition of standards of employment by the employer to promote employee safety is recognized as a legitimate basis for job discrimination.

◆ FREE SPEECH—U.S. SUPREME COURT LIMITS PROTECTION OF PUBLIC EMPLOYEES

On May 30, 2006, the U.S. Supreme Court decided an important decision concerning First Amendment rights of public employees. This case may have a direct effect on public employees, including firefighters, and their public comments about their employers. In *Gil Garcetti v. Richard Cebalos,* No. 04-473 (S.Ct. 3/27/06) (*www.supremecourtus.gov*) the Supreme Court held that the free speech rights in the First Amendment to the Constitution do not fully protect a government employee from discipline based on speech made pursuant to the employee's official duties. This is a "must read" court decision. The following is a brief summary of the facts.

In 1989, Richard Ceballos was hired as a deputy district attorney for the Los Angeles County District Attorney's Office. He was a supervisor in the Pomona (CA) branch in February 2000, when a defense attorney told him that a Los Angeles County deputy sheriff had filed an affidavit in support of a search warrant, and this affidavit contained significant inaccuracies. Ceballos closely examined the affidavit, even visited the residence described in the affidavit, and determined the affidavit contained "serious misrepresentations." For example, the affidavit stated the residence had a long driveway, and Ceballos thought it should have stated that the residence had a separate driveway. The court also wrote, "Ceballos also questioned the affidavit's statement that the tire tracks led from a stripped-down truck to the premises covered by the warrant. His doubts arose from his conclusion that the roadway's composition in some places made it difficult or impossible to leave visible tire tracks."

Ceballos spoke to the deputy sheriff who had filed the affidavit, but he was not satisfied with the deputy's explanations. Ceballos then spoke to his two supervisors and prepared a memorandum recommending the criminal case be dismissed. He later followed this up with a second memo to his immediate supervisor, describing a second troubling phone conversation he had with the deputy sheriff. His two supervisors set up a meeting with Ceballos, the deputy sheriff, and officials of the L.A. County Sheriff's Department. The meeting became heated, with an L.A. County Sheriff's Department lieutenant sharply criticizing Ceballos. After the meeting, the lead prosecutor decided the criminal case would not be dismissed. Criminal defense counsel had filed a motion to suppress the evidence seized based on the inaccuracies in the affidavit, and the defense attorney called Ceballos to testify in a hearing on his motion. Ceballos told the court of his observations at the premises. The trial judge denied the motion to dismiss.

Ceballos claimed that he thereafter suffered retaliation, including being removed as a supervising calendar deputy and sent back to being just a trial deputy. He was transferred to another courthouse and was denied a promotion. He filed an internal grievance, which was denied. He then filed a federal lawsuit against his employer and his two supervisors, claiming his First and Fourteenth Amendment constitutional rights were denied in violation of 42 U.S.C. 1983.

The federal district judge granted the L.A. County district attorney's motion for summary judgment, on the basis that Ceballos' two memos were written pursuant to his employment duties and were therefore not protected by the First Amendment. The court also held that the two supervisors enjoyed "qualified immunity."

The U.S. Court of Appeals for the 9th Circuit (based in San Francisco) reversed the dismissal, holding that Ceballos' memos constituted protected speech under the First Amendment, because they addressed a matter of "public concern" and the operations of the prosecutor's office.

The U.S. Supreme Court reversed the 9th Circuit, holding that when public employees make statements pursuant to their official duties, they are not speaking as citizens for First Amendment purposes, and the U.S. Constitution does not protect them from discipline. Justice Kennedy wrote the majority opinion (five to four decision), "When a citizen enters government service, the citizen must accept certain limitations on his or her freedom. . . . Government employers, like private employers, need a significant degree of control over their employee's words and actions; without it, there would be little chance for efficient provision of public services."

This is a decision that should be carefully read, including the dissenting opinions, because it has enormous potential application to the fire service and other public employers.

◆ FREE SPEECH—RETALIATION FOR UNION ACTIVITY

Another area of employment litigation concerns retaliation claims by firefighters, claiming they were denied promotions because they had favored forming a union. These cases often involve factual disputes that are critical to the court's decision. For example, on December 14, 2005, the U.S. Court of Appeals for the 6th Circuit, in *James R. Edgar v. City of Collierville, TN,* 160 Fed. App. 440, 2005 U.S. App. LEXIS 27776, upheld the district court's dismissal of this firefighter's lawsuit. Edgar joined the fire department in 1995, and in 1998 he helped submit paperwork that led to recognition of Local 3864, in which he served as interim secretary. He was out of work for a time, recovering from an automobile accident, and he claimed that the fire department had adopted a rule penalizing firefighters seeking promotion that had used substantial sick leave. Unfortunately for his claim, the assistant chief in charge of the promotion process said he was not even aware of Edgar's earlier union activities. The trial court held, and the Court of Appeals agreed, "This lack of knowledge was fatal to the claimant's assertions, because defendants must know about the protected activity in order to have motivated the adverse action."

◆ PROMOTIONS

There has been substantial litigation in the fire service concerning promotions. They often arise in the context of retaliation for free speech, or having filed a lawsuit involving the fire department (see Case Study 6-3).

Many of the promotions lawsuits concern allegations of retaliation by African-American firefighters for having filed an EEOC charge or by Caucasian firefighters alleging "reverse" discrimination (see case studies in Chapter 8; "Race Discrimination").

◆ **NFPA STANDARDS**

- *NFPA 1001: Standard for Fire Fighter Professional Qualifications* (2002 edition). This standard identifies the knowledge and skills required of all firefighters. It is an effective standard for developing an evaluation about a firefighter's ability to safely and efficiently meet job requirements.
- *NFPA 1021: Standard for Fire Officer Professional Qualifications* (2003 edition). This standard identifies the knowledge and skills required for fire officers at all levels to perform their job effectively. It is a performance-based standard that can be used to measure the effectiveness of officers.
- *NFPA 1582: Comprehensive Occupational Medical Program for Fire Departments* (2003 edition). This standard is intended to reduce the risk to firefighters by identifying medical conditions that, under stressful conditions, may present significant risks to the health and safety of the firefighter or anyone operating with the firefighter. Each condition is related to a specific firefighting skill or function, which the firefighter may be unable to perform.

◆ **KEY STATUTES**

AGE DISCRIMINATION IN EMPLOYMENT ACT OF 1967 (ADEA) (29 U.S.C. 631)

Congress enacted the ADEA in 1967, and this statute has been revised several times. The ADEA originally did not apply to the federal government, to the states or their political subdivisions, or to employers with fewer than 25 employees. In 1974, Congress extended coverage to federal, state, and local governments and to employers with at least 20 workers. §§ 630(b), 633a. Also, while the Act initially covered employees only up to age 65, in 1978 Congress raised the maximum age to 70 for state, local, and private employees and eliminated the cap entirely for federal workers.

ADEA AMENDMENTS OF 1978 [29 U.S.C. 631(b)]

The 1978 Amendments eliminated substantially all federal age limits on employment, but they left untouched several mandatory retirement provisions of the federal civil service statute applicable to specific federal occupations, including firefighters, air traffic controllers, and law enforcement officers, as well as mandatory retirement provisions applicable to the Foreign Service and the Central Intelligence Agency. Among the provisions that were left unaffected by the 1978 Amendments is 5 U.S.C. § 8335(b), which requires certain federal law enforcement officers and firefighters to retire at age 55 if they have sufficient years of service to qualify for a pension and their agency does not find that it is in the public interest to continue their employment.

UNIFORMED SERVICES EMPLOYMENT AND REEMPLOYMENT RIGHTS ACT OF 1994 (USERRA) (38 U.S.C. 4301)

Following the Gulf War (Desert Storm One), returning reservists and National Guard members reported numerous problems in their attempts to return to their civilian jobs. Congress passed this law to provide greater reemployment rights to these returning veterans. On December 16, 2005, the U.S. Department of Labor issued their first regulations under the law (*www.dol.gov/vets*).

◆ CASE STUDIES

CASE STUDY 6-1   Age Discrimination

Smith v. City of Jackson, Mississippi, 544 U.S. 228 (U.S. March 30, 2005); U.S. Supreme Court.

FACTS

Thirty police officers and dispatchers over age 40 sued the City of Jackson, claiming age discrimination. The city was attempting to make starting pay and the pay of junior officers with less than five years more competitive with other police departments in the region and gave a proportionally larger raise to those with less than five years of service.

The city's new pay plan divided employees into five basic positions: police officer, master police officer, police sergeant, police lieutenant, and deputy police chief. The new wage scale had a series of steps and half-steps. Employees were assigned to a step in which they would receive a two percent pay raise. As would be expected, most of the police employees were in the three lowest ranks: officer, master officer, and sergeant. The employees in lieutenant and deputy police chief positions were all over 40. There were some employees over 40 in each rank, but all officers in lieutenant and deputy chief positions were over 40.

The plaintiffs sued, claiming that the pay raises had a "disparate impact" on employees over 40. Although these older employees received more money in their raises, they claimed they received a smaller percentage of their salary. Their statistical data showed that 66.2 percent of the officers under age 40 received raises of more than 10 percent, while only 44.5 percent of those over 40 received that much.

YOU BE THE JUDGE

Does this data show a "disparate impact" on the plaintiffs?

HOLDING

disparate impact   In 2005, the U.S. Supreme Court held that age discrimination could be proved by showing an adverse impact on older workers, without need to prove the employer acted with bad intent.

The Supreme Court ruled against the plaintiffs, upholding earlier decisions by the federal U.S. District Court judge, and the three-judge panel of the U.S. Court of Appeals.

The case is important, however, because the Supreme Court established a new legal approach to analyzing age cases, in which plaintiffs can sue employers not only if there is a *deliberate* attempt to favor younger workers, but also if there is a **disparate impact** (an adverse impact, even without bad intent) on older workers.

The Supreme Court said that in this case the City of Jackson had a legitimate governmental reason to increase the pay of new employees, writing:

> Thus, the disparate impact is attributable to the City's decision to give raises based on seniority and position. Reliance on seniority and rank is unquestionably reasonable given the City's goal of raising employees' salary to match those in surrounding communities. In sum, we hold that the City's decision to grant a larger raise to lower echelon employees for the purpose of bringing salaries in line with the surrounding police forces was a decision based on a reasonable factor other than age that responded to the City's legitimate goal of retaining police officers.

LEGAL LESSONS LEARNED

Age discrimination cases often include the presentation of statistical evidence in order to prove that there has been an adverse impact on older workers.

CASE STUDY 6-2   Firefighter Mandatory Retirement Age

Minch v. City of Chicago, 363 F.3d 615 (7th Cir. April 9, 2004).

FACTS

On May 17, 2000, the City of Chicago adopted in its Municipal Code a Mandatory Retirement Ordinance that provided,

> Effective December 31, 2000, the age 63 shall be the maximum age for employment of any member of the uniformed service of the fire department, the duties of whose position are primarily to perform work directly connected with the control and extinguishment of fires or the maintenance and use of firefighting apparatus and equipment, including any employee engaged in the activity who is transferred or appointed to a supervisory or administrative position. . . .

Four plaintiffs, including firefighters and police officers, filed a lawsuit in federal district court challenging the mandatory retirement ordinance as a violation of the ADEA. The plaintiffs argued that the Chicago's age 63 ordinance amounts to a **subterfuge** because one city council member said it was designed to get rid of the "old-timers" and "deadbeats." Also, plaintiffs allege it was designed to make room for younger, more racially and ethnically diverse individuals, and was therefore not a bona fide plan. The plaintiffs also noted that the City did not pass the age 63 Mandatory Retirement Ordinance until a firefighter friend of the mayor retired at age 68.

The federal district judge denied the city's motion to dismiss the case, based on the plaintiff's allegations, and the city appealed.

subterfuge Congress has authorized mandatory retirement for firefighters age 55 or older, as long as there is a state or local bona fide retirement plan, and it is not used in a discriminatory manner to force only certain employees to retire.

YOU BE THE JUDGE

Can the city impose a mandatory age 63 retirement age on "front-line" firefighters?

HOLDING

Yes. Plaintiffs' case is ordered dismissed.

The three-judge panel of the U.S. Court of Appeals wrote:

> Historically, Chicago, like many other state and local governments, has placed age limits on the employment of its police and firefighting personnel. As early as 1939, for example, Chicago's municipal code required city firefighters to retire at age 63.

As it was originally enacted in 1967, the ADEA by its term did not apply to the employees of state and local governments. Congress amended the statute to include those employees in 1974.

Responding to the concerns expressed by state and local governments, Congress in 1986 amended the ADEA to exempt the mandatory retirement of state and local police and fire fighting personnel from the statute's coverage.... In 1988, Chicago took advantage of the exemption and reinstated a mandatory retirement requirement age of 63 for its firefighters and police officers.

The congressional statute of 1986 had a "sunset provision," meaning that it expired automatically at the end of 1993.

In 1996, however, Congress reinstated the exemption, this time without any sunset provision, and retroactivity to the date the prior exemption had expired in 1993.... [T]he exemption, codified at 29 U.S.C. 623 (j), permits a public employer to discharge a police officer or firefighter based on his age.... [pursuant to] to a bona fide retirement plan that is not a subterfuge to evade the provisions of the statute.

[The plaintiffs'] sole contention is that in exercising the City's prerogative to reinstate mandatory retirement, certain City legislators and other officials were motivated by a desire to remove from the police and firefighting forces older workers whom they felt were not up to the job and/or to create openings for younger workers. Yet, the statutory exemption expressly permits the City to reinstate its mandatory retirement program, and the inevitable result of any such program will be to force older workers from the workplace and create openings for younger workers. That some City officials affirmatively wished for that result, perhaps because of unfortunate stereotypes about the ability of older workers, is immaterial as section 623(j)(2) is concerned.

LEGAL LESSONS LEARNED

Congress has authorized mandatory retirement for firefighters over age 55, "pursuant to a bona fide hiring or retirement plan that is not a subterfuge to evade the purposes of this chapter."

[Note: The U.S. Supreme Court addressed mandatory age retirements in Johnson v. Mayor *and City Council of Baltimore, 472 U.S. 353 (1985), in which the court noted that neither the language in the ADEA, nor the legislative history, indicated that Congress determined that age 55 was a bona fide occupational qualification (BFOQ) for firefighters. Instead, the age 55 language in the ADEA represents nothing more than a congressional decision that federal fire fighters must retire, as a general matter, at age 55. Read the U.S. Department of Justice brief asking the Supreme Court to take this appeal and declare the City of Baltimore's municipal ordinance of mandatory retirement of firefighters at age 55 to be a violation of ADEA,* EEOC v. Mayor & City Council of Baltimore *(go to* www.usdoj.gov *and search for "firefighter").*

CASE STUDY 6-3 Alleged Retaliation after Winning Lawsuit for Failure to Promote

retaliation Under Title VII of the federal Civil Rights Act of 1964, an adverse employment action for retaliation is an act or harassment that results in an adverse effect on employment term, conditions, or benefits of employment.

Gallipo v. City of Rutland, 2005 VT 83 (July 29, 2005); 882 A.2d 1177; 2005 Vt. LEXIS 173; Vermont Supreme Court.

FACTS

Raymond Gallipo is a former firefighter who filed a lawsuit for **retaliation**. He claimed the fire department began closely monitoring him, including videotaping the training he was giving on September 13, 1996, after he had won a jury verdict in an earlier discrimination suit for the department's failure to promote him based on his religious practices and a learning disability.

Gallipo had a heated exchange with the person who was instructed to videotape the training session, left the training, and never returned to work. Instead, he filed a workers' compensation claim for anxiety and depression as a result of his treatment by the fire department. He articulated four stressors in his workers' compensation claim that caused him mental injury: "ongoing ridicule from fellow firefighters; use of profane language by other employees; lack of a computer password to perform his role as a computer specialist; and the confrontation over the videotaping."

The Vermont Commission of Labor and Industry awarded him interim benefits while his claim was pending. Ultimately, on July 12, 2000, the commission denied his claim, finding that his four stressors were no greater than for any other firefighter.

Gallipo filed his lawsuit claiming retaliation, and the fire department filed a counter-claim seeking reimbursement of the workers' compensation interim payments he had received, on the legal doctrine of "unjust enrichment."

The fire department's attorney took the plaintiff's deposition. The fire department also served the plaintiff with a Request to Admit (pleading where the other part must "Admit" or "Deny" questions), to which he never replied. The fire department then filed a motion to dismiss his suit. The trial judge granted the motion, holding that the finding of the Commission of Labor and Industry was preclusive on the issue of mental stress, and "his perceived mistreatment was subjective." The trial judge denied the fire department's "unjust enrichment" claim for reimbursement of the interim workers' compensation payments, because they never filed any medical evidence with the State Commission in opposition to his claim. Both sides appeal.

YOU BE THE JUDGE

1. Should Gallipo be permitted to get a jury trial on his retaliation claim?
2. Should the fire department be reimbursed for the interim workers' compensation payments?

HOLDING

Gallipo's lawsuit was properly dismissed. The fire department's counter-claim was also properly dismissed.

The Vermont Supreme Court wrote, "Under Title VII (of the federal Civil Rights Act of 1964), an adverse employment action for retaliation is generally an 'act or harassment [that] results in an adverse effect on 'term, conditions, or benefits' of employment.'" *Von Guten v. Maryland,* 3 F.3d 858, 866 (94th Cir. 2001).

The Supreme Court then cited that the plaintiff never replied to the Request to Admit, including the following: "[Aside from the videotaping incident] you are unable to offer any specific incident of retaliatory discrimination attributable to any specific individual, that occurred subsequent to the jury trial." Under Rule of Civil Procedure 36, a failure to respond is treated as an admission.

Regarding the profane language in the firehouse, the court wrote:

> Similarly, the plaintiff fails to offer any conclusive proof that the profane language was directed specifically at him, was a conscious effort to harass him, or was an objectively real stress.

> Because plaintiff is unable to establish a prima facie case under the facts by which he is bound, the court properly granted defendant summary judgment on the merits of plaintiff's case.

Regarding the fire department's claim of "unjust enrichment," the court wrote, "We initially point out that, as the defendant notes, the workers' compensation statute does not address the question of interim-payment reimbursement. . . . In examining our workers' compensation statute, we cannot find a legislative intent to grant employers such a reimbursement right."

LEGAL LESSONS LEARNED

The courts apply an objective standard to determine retaliation, not an employee's subjective feelings. If an employee complains about workplace retaliation or harassment, investigate it thoroughly, fix any problems, and thoroughly document your conclusions.

■■■

Chapter Review Questions

1. In Case Study 6-1, *Smith v. City of Jackson, Mississippi,* the court addressed the issue of the city's increase in pay for starting positions to attract more applicants. There are many positions in the fire service that call for "young bodies" in great physical shape. In many departments having a collective bargaining agreement, firefighters with longer seniority can elect to bid for a position at a "slower" fire station. Discuss whether management should reserve the right to deny such a transfer if it appears that the station is over-staffed with older firefighters.

2. In Case Study 6-2, *Minch v. City of Chicago,* the court addressed the issue of mandatory age retirements for firefighters. Discuss whether older firefighters should be required to pass physical examinations more often than younger firefighters. Also discuss whether physical examinations of all firefighters should be required on an annual basis.

3. In Case Study 6-3, *Gallipo v. City of Rutland,* the court addressed the issue of retaliation. In the fire service, there are virtually "no secrets" concerning employee complaints, particularly if a firefighter has filed a charge of discrimination with the EEOC or state agency. Discuss what steps a station officer should take to avoid co-workers harassing or otherwise retaliating against a firefighter who has filed a charge that must be investigated.

■■■

Expand Your Learning

1. Phoenix Fire Chief Alan Brunacini, who retired in June 2006, is famous for his seminars on customer service and serving "Mrs. Smith." He wrote in his book *Fire Command* that this focus on service delivery has led the Phoenix Fire Department to "think of, and treat, our own department members as internal customers."

   Discuss what actions that fire departments can take to treat their firefighters as "internal customers."

2. The U.S. Equal Employment Opportunity Commission lists settlements in lawsuits filed by the EEOC on their website (*www.eeoc.gov,* "Litigation Settlement Monthly Reports"). For example, see *EEOC v. Oberto Sausage Co.,* No. CV 05-25C (W.D. Wash, July 1, 2005) regarding treatment of Muslim workers.

   Describe a recent settlement that has "legal lessons learned" for the fire service. Explain the lessons.

See Appendix A for additional Expand Your Learning activities related to this chapter.

Sexual Harassment
Hostile Work Atmosphere, Pregnancy Discrimination, and Gender Discrimination

7 CHAPTER

Many women in the fire service have become leaders known nationwide. This includes Fire Chief Trish Brooks, City of Forest Park, near Cincinnati (OH). Her fire department has been an outstanding example of recruiting of female and minority firefighters. *Photo by author.*

Key Terms

From the Headlines

Chicago—First Female Battalion Chief

On December 6, 2005, Chicago Fire Department promoted 36 battalion chiefs, including the city's first female chief.

FDNY/First Female Rescue Lieutenant

On December 20, 2005, FDNY promoted 34 officers, including a female lieutenant who made city history when she became the first female to join a FDNY rescue company. There are only 32 women on the 11,000 FDNY.

Nationwide/Slow Progress in Hiring Female Firefighters

As of 2005, only 6,500 of the 296,000 career firefighters in the nation are women, less than 2.5 percent.

Some women in rural fire service have risen to the top.
Fire Chief Lisa Reeves, Mt. Orab (OH). *Photo by author.*

Other women in the fire service have become role models for their community, including firefighter and Girl Scout leader Pia Washington. She is shown here at the "National Day of Prayer" ceremony honoring fallen firefighters and police officers at the Hamilton County Courthouse, along with Cincinnati Police Officer Jeff Ruberg and Rev. Ted Kalsbeek. *Photo by author.*

pregnancy discrimination
Title VII was amended by the Pregnancy Discrimination Act of 1978, which prohibits discrimination on the basis of pregnancy, childbirth, and related medical conditions.

Title VII   Refers to Title VII of the Civil Rights Act of 1964. Title VII protects individuals against employment discrimination on the basis of sex as well as race, color, national origin, and religion. Applies to employers with 15 or more employees, including state and local governments. It also applies to employment agencies and to labor organizations, as well as to the federal government.

◆ INTRODUCTION

This chapter concerns men and women in the fire service who want to do their jobs, free of sexual harassment and other forms of gender discrimination. Gender discrimination is prohibited by Title VII of the Civil Rights Act of 1964. This is a very broadly written statute that prohibits many forms of discrimination, including discrimination based on race, color, religion, sex, or national origin. The statute also prohibits retaliation against individuals who file claims of discrimination or otherwise oppose unlawful practices.

The fire service has been the subject of much litigation involving sexual harassment. Although most will agree that progress has been made, there are still many lawsuits being filed. This chapter will examine some of these past issues and the legal lessons learned from the court decisions in those cases. Two areas of continuing controversy concern **pregnancy discrimination** and physical fitness testing.

◆ PREGNANCY DISCRIMINATION

Title VII of the Civil Rights Act of 1964 prohibits discrimination based on sex, as well as race, color, religion, or national origin. The Pregnancy Discrimination Act, 42 U.S.C. 2000e(k) amended Title VII on October 31, 1978, to specify that sex discrimination under Title VII includes discrimination based on pregnancy and pregnancy-related conditions. Federal EEOC regulations on pregnancy discrimination, 29 CFR Part 1604.10(a), and U.S. Supreme Court decisions such as *International Union et al.*

v. Johnson Controls, Inc., 499 U.S. 187 (1991), have applied Title VII of the Civil Rights Act of 1964 to pregnancy.

◆ PHYSICAL FITNESS TESTS

Generally, physical performance tests, which may have an adverse impact on female applicants, must be validated through a study in accordance with EEOC regulations. See the article by Deputy Chief Mike Cardwell in Appendix A: Expand Your Learning Activities (Chapter 7), entitled "Equality of Performance and NFPA."

◆ NFPA STANDARDS

- *NFPA 1001: Standard for Fire Fighter Professional Qualifications* (2002). This standard identifies knowledge and skills required of all firefighters. It is a performance-based standard, which can apply equally to both men and women.
- *NFPA 1021: Standard for Fire Officer Professional Qualifications* (2003). This standard identifies the knowledge and skills of officers at all levels. It can serve as a basis of promotions based solely on an individual's skills and ability to meet the standards.
- *NFPA 1582: Standard on Comprehensive Occupational Medical Program for Fire Departments* (2003 edition). This standard is intended to reduce the risk to firefighters due to medical conditions. Pregnancy is addressed as a medical condition within the standard.
- *NFPA Fire Protection Handbook* (2004 edition). This handbook includes the latest fire service leadership thoughts on critical fire service issues, including legal issues involving the fire service.

◆ KEY STATUTES

TITLE VII OF THE CIVIL RIGHTS ACT OF 1964 (42 U.S.C. SECTION 2000e)

This important federal statute prohibits unlawful employment practices, including sexual harassment. It was signed into law by President Lyndon B. Johnson and is codified in the U.S. Code at 42 U.S.C. Sec. 2000e-2(a)(1). The law is enforced by the federal Equal Employment Opportunity Commission and by private lawsuits filed in federal court by individuals seeking damages and attorney's fees.

Title VII states in part:

It shall be an unlawful employment practice for an employer—
(1) to fail or refuse to hire or to discharge any individuals, or otherwise to discriminate against any individual with respect to [her] compensation, terms, conditions, or privileges or employment because of such individual's sex. Section 2000e-2(a)(1), Title 42, U.S. Code.

PREGNANCY DISCRIMINATION ACT OF 1978

This statute amended Title VII of the Civil Rights Act of 1964, 42 U.S.C. 2000e-5, to specifically protect pregnant employees and applicants from discrimination. It is enforced by the Equal Employment Opportunity Commission and also the U.S. Department of Justice. Private lawsuits may also be filed, after "exhausting administrative remedies" by first filing a charge with the EEOC within 180 days of the alleged unlawful employment practice and receiving a "right to sue" letter. Fire departments and other employers do have a bona fide occupational qualification (BFOQ) defense under 42 U.S.C. 2000e-2(e) and can adjust work assignments of pregnant firefighters based on the safety risks to the pregnant firefighter and her unborn child. *Caution:* The EEOC has expressly stated that this BFOQ defense is to be narrowly construed. See also U.S. Supreme Court's decision in *United Auto, Aerospace and Agr. Implement Workers of America v. Johnson Controls, Inc.,* 499 U.S. 187, 201 (1991).

EQUAL PAY ACT OF 1963

This statute requires that men and women be given equal pay for equal work in the same establishment. The jobs need not be identical, but they must be substantially identical.

STATE LAWS

Each of the 50 states has laws that also prohibit discrimination. Most states have their own version of the EEOC and also provide for private enforcement through lawsuits filed in state court.

◆ **CASE STUDIES**

CASE STUDY 7-1   Hostile Work Atmosphere—PA State Police Barracks/U.S. Supreme Court

Pennsylvania State Police v. Nancy Drew Suders, 542 U.S. 129 (2004); U.S. Supreme Court (June 14, 2004).

FACTS

Nancy Drew Suders alleged sexually harassing conduct by her supervisors, officers of the Pennsylvania State Police (PSP), of such severity that she was forced to resign.

Suders was hired in March 1998 as a police communications operator for the McConnellsburg PSP barracks. Her supervisors included Sergeant Eric D. Easton, station commander at the barracks, Patrol Corporal William D. Baker, and Corporal Eric B. Prendergast. Suders alleges that these three supervisors subjected her to a continuous barrage of sexual harassment that ceased only when she resigned from the force.

According to the complaint, Sergeant Easton would bring up the subject of "people having sex with animals" each time she entered his office. He told Corporal Prendergast, in front of Suders, that young girls should be given instruction in "how to gratify men with oral sex." Sergeant Easton also would sit down near Suders, wearing Spandex shorts, and spread his legs apart, apparently imitating a move "popularized by television wrestling."

According to the complaint, Corporal Baker repeatedly made an obscene gesture in Suders' presence by grabbing his genitals and "shouting out a vulgar comment inviting oral sex." Baker made this gesture as many as "five-to-ten times per night" throughout Suders' employment at the barracks. Suders once told Corporal

Baker that she did not think he should be doing this. Baker responded by jumping on a chair and again performing the gesture, with the accompanying vulgarity. Corporal Baker would also "rub his rear end" in front of her and remark, "I have a nice ass, don't I?"

Corporal Prendergast allegedly told Suders that "the village idiot could do her job." He would also wear black gloves and "pound on furniture" to intimidate her.

In June 1998, Prendergast accused Suders of taking a missing accident file home with her. After that incident, Suders approached the PSP's Equal Employment Opportunity Officer, Virginia Smith-Elliott, and told her she "might need some help." Smith-Elliott gave Suders her telephone number, but neither person followed up on the conversation. On August 18, 1998, Suders contacted Smith-Elliott again, this time stating that she was being harassed and was afraid. Smith-Elliott told Suders to file an internal complaint but did not tell her how to obtain the necessary form. According to Suders, Smith-Elliott's response and the manner in which it was conveyed appeared to be insensitive and unhelpful.

Two days later, Suders' supervisors arrested her for theft, and Suders resigned from the force. The theft arrest occurred after Suders had several times taken a computer-skills exam to satisfy a PSP job requirement. Each time, Suders' supervisors told her that she had failed. Suders one day found her exams in a desk drawer in the women's locker room. She concluded that her supervisors had never forwarded the tests for grading and that their reports of her failures were false.

Regarding the tests as her property, Suders removed them from the locker room. When her supervisors found the exams had been removed, they purportedly devised a plan to arrest her for theft. The officers dusted the drawer in which the exams had been stored with a theft-detection powder that turns hands blue when touched. Suders attempted to return the tests to the drawer, and her hands turned telltale blue. The supervisors then apprehended and handcuffed her, photographed her blue hands, and commenced to question her. Suders had previously prepared a written resignation, which she tendered soon after the supervisors detained her.

The supervisors initially refused to release her. Instead, they brought her to an interrogation room, gave her Miranda warnings, and continued to question her. Suders again told them that she wanted to resign, and Sergeant Easton then let her leave. The PSP never brought theft charges against her in a court of law.

In September 2000, Suders sued the PSP in federal District Court, alleging that she had been subjected to sexual harassment and constructively discharged, in violation of Title VII of the Civil Rights Act of 1964. After depositions and other discovery, the federal trial judge granted the PSP's motion for summary judgment. The judge concluded that while Suders' testimony was sufficient for a jury to conclude that her supervisors had created a hostile work atmosphere, the PSP was not vicariously liable for the supervisors' conduct.

Plaintiff appealed to the U.S. Court of Appeals, and after getting no relief there, asked the U.S. Supreme Court to take the case. The Supreme Court agreed to hear the appeal. (In order for a case to be heard, it needs the vote of at least four of the nine justices.)

YOU BE THE JUDGE

1. Is the plaintiff entitled to a jury trial?
2. Should the Pennsylvania State Police be potentially liable for the misconduct of their supervisors when Suders never filed an internal complaint?

HOLDING

Suders may be entitled to a jury trial to prove constructive discharge, but her employer may assert an affirmative defense that she failed to make an internal complaint.

Justice Ruth Bader Ginsburg wrote the majority opinion. The Supreme Court cited prior case law, which held that to establish a **hostile work environment,** plaintiffs must show harassing behavior "sufficiently severe or pervasive to alter the conditions of [their] employment."

The court set new law, stating,

> We hold, to establish **"constructive discharge,"** the plaintiff must make a further showing: She must show that the abusive working environment became so intolerable that her resignation qualified as a fitting response. An employer may defend against such a claim by showing both (1) that it had installed a readily accessible and effective policy for reporting and resolving complaints of sexual harassment, and (2) that the plaintiff unreasonably failed to avail herself of that employer-provided preventive or remedial apparatus. This affirmative defense will not be available to the employer, however, if the plaintiff quits in reasonable response to an employer-sanctioned adverse action officially changing her employment status or situation, for example, a humiliating demotion, extreme cut in pay, or transfer to a position in which she would face unbearable working conditions.

[Note: Suders' case was therefore sent back to the U.S. District Court judge to hold a hearing on whether she met these standards for constructive discharge.]

LEGAL LESSONS LEARNED

Fire departments should instruct all employees on how to file an internal complaint concerning sexual harassment. Every complaint, including informal verbal complaints, should be documented by an officer, sent to the fire chief, and promptly and thoroughly investigated.

CASE STUDY 7-2   Employer "Mixed Motive" When Terminating Female—Circumstantial Evidence Admissible to Prove Anti-Female Workplace/U.S. Supreme Court

Desert Palace, Inc. dba Caesars Palace Hotel and Casino v. Costa, 539 U.S.90 (2003); U.S. Supreme Court (June 9, 2003).

FACTS

Desert Palace, Inc., doing business as Caesars Palace Hotel and Casino of Las Vegas (NV), employed Catharina Costa as a warehouse worker and heavy equipment operator. She was the only woman in this job and in her local Teamsters bargaining unit.

She experienced a number of problems with management and her coworkers that led to an escalating series of disciplinary sanctions, including informal rebukes, a denial of privileges, and suspension. The casino eventually terminated her after she was involved in a physical altercation in a warehouse elevator with fellow Teamsters Member Herbert Gerber. The casino disciplined both employees because the facts surrounding the incident were in dispute, but Gerber, who had a clean disciplinary record, received only a five-day suspension, whereas plaintiff's thick disciplinary file led to her firing.

hostile work environment
Prohibited by Title VII of the Civil Rights Act of 1964, it includes practices ranging from direct requests for sexual favors to workplace conditions that create a hostile environment for persons of either gender, including same sex harassment.

constructive discharge
Describes a work atmosphere that is sufficiently severe or pervasive to alter the conditions of employment. A plaintiff must show that the abusive working environment became so intolerable that resignation qualified as a fitting response.

sex discrimination The EEOC definition is as follows: "It is unlawful to discriminate against any employee or applicant for employment because of his/her sex in regard to hiring, termination, promotion, compensation, job training, or any other term, condition, or privilege of employment. Title VII also prohibits employment decisions based on stereotypes and assumptions about abilities, traits, or the performance of individuals on the basis of sex."

mixed motive Employers who discipline a female employee more harshly than male employees can be sued for sexual harassment and will be liable if the jury concludes the employer's conduct was motivated by the plaintiff's sex, even if the employer's conduct was also motivated by a lawful reason.

She filed this lawsuit in federal court, asserting claims of **sex discrimination** and sexual harassment under Title VII. The District Court dismissed the sexual harassment claim but allowed the claim for sex discrimination to go to the jury. At trial, her attorney presented evidence that (1) she was singled out for "intense stalking" by one of her supervisors, (2) she received harsher discipline than men for the same conduct, (3) she was treated less favorably than men in the assignment of overtime, and (4) supervisors repeatedly "stack[ed]" her disciplinary record and "frequently used or tolerated" sex-based slurs against her.

The District Court denied the casino's motion for judgment as a matter of law and submitted the case to the jury with instructions, two of which are relevant here. First, without objection from the casino, the District Court instructed the jury that "[t]he plaintiff has the burden of proving . . . by a preponderance of the evidence" that she "suffered adverse work conditions" and that her sex "was a motivating factor in any such work conditions imposed upon her."

Second, the District Court gave the jury the following **mixed-motive** instruction:

> You have heard evidence that the defendant's treatment of the plaintiff was motivated by the plaintiff's sex and also by other lawful reasons. If you find that the plaintiff's sex was a motivating factor in the defendant's treatment of the plaintiff, the plaintiff is entitled to your verdict, even if you find that the defendant's conduct was also motivated by a lawful reason. However, if you find that the defendant's treatment of the plaintiff was motivated by both gender and lawful reasons, you must decide whether the plaintiff is entitled to damages. The plaintiff is entitled to damages unless the defendant proves by a preponderance of the evidence that the defendant would have treated plaintiff similarly even if the plaintiff's gender had played no role in the employment decision.

The casino unsuccessfully objected to this instruction, claiming that the plaintiff had failed to adduce "direct evidence" that sex was a motivating factor in her dismissal or in any of the other adverse employment actions taken against her.

The jury awarded her $64,377.74 for back wages, $200,000 in emotional pain, and $100,000 punitive damages, plus attorney's fees and court costs.

The employer appealed to the U.S. Court of Appeals, and not obtaining relief, asked the U.S. Supreme Court to take the appeal.

YOU BE THE JUDGE

1. Should the jury verdict in favor of the plaintiff be set aside?
2. Is it appropriate for a plaintiff's attorney to introduce "circumstantial evidence" to show a jury that senior management's true motive was to drive out the only female in the warehouse?

HOLDING

The jury verdict remains; circumstantial evidence is admissible.

Justice Clarence Thomas wrote the majority decision, writing, "In order to obtain an instruction under §2000e-2(m), a plaintiff need only present sufficient evidence for a reasonable jury to conclude, by a preponderance of the evidence, that 'race, color, religion, sex, or national origin was a motivating factor for any employment practice.'"

Because direct evidence of discrimination is not required in mixed-motive cases, the Court of Appeals correctly concluded that the District Court did not abuse its discretion in giving a mixed-motive instruction to the jury.

LEGAL LESSONS LEARNED

Fire departments should keep records on discipline imposed, so that personnel will know that their discipline is similar to others. It is a good practice normally to follow a progressive discipline policy, with the first offense being a verbal warning (documented); second offense, a written warning; and third offense, a three-day suspension, and so on.

CASE STUDY 7-3 City Failed to Train Employees and Supervisors/U.S. Supreme Court

Faragher v. City of Boca Raton, 524 U.S. 775 (1998); U.S. Supreme Court (June 26, 1998).

FACTS

Between 1985 and 1990, during the summers and also part-time while attending college, Beth Ann Faragher worked as a lifeguard for the Marine Safety Section of the Parks and Recreation Department of the City of Boca Raton (FL). Her supervisors included Captain David Silverman, chief of the Marine Safety Division, and Training Captain Robert Gordon. In June 1990, she resigned without having made any complaint of sexual harassment.

In 1992, she filed a lawsuit in federal court against the city and her two supervisors, Bill Terry and David Silverman, asserting claims under Title VII, constitutional violations under 42 U.S.C. § 1983 and also under Florida law. The complaint alleged that Terry and Silverman created a "sexually hostile atmosphere" at the beach by repeatedly subjecting Faragher and other female lifeguards to "uninvited and offensive touching," by making lewd remarks, and by speaking of women in offensive terms.

The complaint contained specific allegations that Terry said that he would never promote a woman to the rank of lieutenant and that Silverman had said to Faragher, "Date me or clean the toilets for a year." Asserting that Terry and Silverman were agents of the city and that their conduct amounted to discrimination in the "terms, conditions, and privileges" of her employment, 42 U.S.C. § 2000e-2(a)(1), Faragher sought a judgment against the city for "nominal damages, costs, and attorney's fees."

Following a bench trial (before a judge, but not a jury), the U.S. District Court for the Southern District of Florida found that throughout Faragher's employment with the city, Terry served as chief of the Marine Safety Division with authority to hire new lifeguards (subject to the approval of higher management), to supervise all aspects of the lifeguards' work assignments, to engage in counseling, to deliver oral reprimands, and to make a record of any such discipline. Silverman was a Marine Safety lieutenant from 1985 until June 1989, when he became a captain. Gordon began the employment period as a lieutenant and at some point was promoted to the position of training captain. In these positions, Silverman and Gordon were responsible for making the lifeguards' daily assignments and for supervising their work and fitness training.

The lifeguards and supervisors were stationed at the city beach and worked out of the Marine Safety headquarters, a small one-story building containing an office, a meeting room, and a single, unisex locker room with a shower. Their work routine was structured in a "paramilitary configuration," with a clear chain of command. Lifeguards reported to lieutenants and captains, who reported to Terry. He was supervised by the Recreation superintendent, who in turn reported to a director of Parks and Recreation, answerable to the city manager. The lifeguards had no significant contact with higher city officials, such as the Recreation superintendent.

In February 1986, the city adopted a sexual harassment policy, which it stated in a memorandum from the city manager addressed to all employees. In May 1990, the city revised the policy and reissued a statement of it. Although the city may actually have circulated the memos and statements to some employees, it completely failed to disseminate its policy among employees of the Marine Safety section, with the result that Terry, Silverman, Gordon, and many lifeguards were unaware of it.

During Faragher's tenure at the Marine Safety section, approximately 4 to 6 of the 40 to 50 lifeguards were women. During that five-year period, Terry "repeatedly touched the bodies of female employees without invitation, would put his arm around Faragher, with his hand on her buttocks, and once made contact with another female lifeguard in a motion of sexual simulation." He made crudely demeaning references to women generally and once commented disparagingly on Faragher's shape. During a job interview with a woman that he hired as a lifeguard, Terry said that the female lifeguards had sex with their male counterparts and asked whether she would do the same.

Captain Silverman reportedly behaved in similar ways. He once tackled Faragher and remarked that, but for a physical characteristic he found unattractive, he would readily have had sexual relations with her. Another time, he pantomimed an act of oral sex. Within earshot of the female lifeguards, Silverman made frequent, vulgar references to women and sexual matters, commented on the bodies of female lifeguards and beachgoers, and at least twice told female lifeguards that he would like to engage in sex with them.

Faragher did not complain to higher management about Terry or Silverman. Although she spoke of their behavior to Training Captain Gordon, she did not regard these discussions as formal complaints to a supervisor but as conversations with a person she held in high esteem. Other female lifeguards had similarly informal talks with Gordon, but because Gordon did not feel that it was his place to do so, he did not report these complaints to Terry, his own supervisor, or to any other city official. Gordon responded to the complaints of one lifeguard by saying that "the city just [doesn't] care."

In April 1990, however, two months before Faragher's resignation, Nancy Ewanchew, a former lifeguard, wrote to Richard Bender, the city's personnel director, complaining that Terry and Silverman had harassed her and other female lifeguards. Following investigation of this complaint, the city found that Terry and Silverman had behaved improperly, reprimanded them, and required them to choose between a suspension without pay or the forfeiture of annual leave.

The trial judge found the city liable, as well as the two supervisors. They appealed to the U.S. Court of Appeals, and a three-judge panel reversed, finding the city not liable. The plaintiffs asked for an "en banc" ruling (all of the Court of Appeals judges), and they also reversed, finding the city not liable. The plaintiffs then asked the U.S. Supreme Court to hear their appeal.

YOU BE THE JUDGE

1. Should the city be liable for the conduct of their supervisors, if no internal complaint had ever been filed by the plaintiff?
2. Does it matter that the city never told the plaintiff or others where to file a complaint?

HOLDING

The city is "vicariously liable" for the misconduct of their supervisors.

If the city had instructed all employees about how to file an internal complaint, then the employer would have a defense to the lawsuit if an employee failed to make an internal complaint.

Justice David H. Souter wrote the majority decision, "We hold that an employer is vicariously liable for actionable discrimination caused by a supervisor, but subject to an affirmative defense looking to the reasonableness of the employer's conduct as well as that of a plaintiff victim."

The Supreme Court described two categories of hostile work environment claims: (1) harassment that "culminates in a tangible employment action" (such as a demotion, or suspension) for which an employer is strictly liable, and (2) harassment that takes place in the absence of a tangible employment action, to which employers may assert an affirmative defense.

The Supreme Court distinguished between supervisor harassment unaccompanied by an adverse official act and supervisor harassment attended by "a tangible employment action." An employer is strictly liable for supervisor harassment that "culminates in a tangible employment action, such as discharge, demotion, or undesirable reassignment." But when no tangible employment action is taken, the employer may raise an affirmative defense to liability, subject to proof by a preponderance of the evidence:

> The defense comprises two necessary elements: (a) that the employer exercised reasonable care to prevent and correct promptly any sexually harassing behavior, and (b) that the plaintiff employee unreasonably failed to take advantage of any preventive or corrective opportunities provided by the employer or to avoid harm otherwise.

LEGAL LESSONS LEARNED

Fire departments should provide every new hire, sworn or civilian, with a written policy on sexual harassment, including clear steps on how to file a complaint. Every officer should be directed to stop any acts of harassment immediately, and to report up the chain of command any complaints immediately, whether the complaint is written or verbal.

■■

Chapter Review Questions

1. In Case Study 7-1, *Pennsylvania State Police v. Nancy Drew Suders,* the court addressed the issue of hostile work atmosphere. Describe the steps that a fire station officer should take to avoid any charges of hostile work atmosphere by female or male firefighters.

2. In Case Study 7-2, *Desert Palace, Inc. dba Caesars Palace Hotel and Casino v. Costa,* the court addressed the issue of use of "circumstantial evidence" to prove to a jury that the employer's true motive was to force out the only female in the casino warehouse. If a female firefighter has complained to the fire chief about ongoing harassment, and the fire chief has promptly investigated and verbally cautioned her male crew members to stop the harassment, describe what steps that the chief should take when he receives complaints from crew members about her lack of cooperation in performing station cleanup duties.

3. In Case Study 7-3, *Beth Ann Faragher v. City of Boca Raton,* the court addressed the issue of an employer's obligation to inform employees about how to make a complaint of sexual harassment. Describe what steps that a fire department should take with all new hires concerning sexual harassment complaints.

■■

Expand Your Learning

Read and complete the individual student or group assignment, as directed by your instructor.

1. On September 5, 2005, the U.S. Department of Justice announced a Settlement Agreement of a lawsuit against the District of Columbia, to prevent pre-employment pregnancy testing of female EMTs (go to *www.usdoj.gov* and search under "EMTs"). The three original plaintiffs are EMTs with the FD; they filed suit using the pseudonyms Jane Doe 1, Jane Doe 2, and Jane Doe 3. The U.S. Department of Justice intervened in the lawsuit. Acting Assistant Attorney General for the Civil Rights Commission Bradly J. Scholzman is quoted, "Using pregnancy as a barrier to a woman's entry into the workplace and continued job security is illegal and inexcusable." The District of Columbia agreed to compensate each of the three Jane Does in the amount of $101,000 plus attorney's fees. The settlement includes a requirement that the fire department must within one year complete training on the rights of pregnant applicants and employees to "all Fire and Emergency Medical Services personnel who are involved in the hiring, evaluation and disciplining of employees in the Fire Department."

   Discuss what key points should be covered in this training.

2. Discuss how a fire department should respond when a firefighter advises that he is a transsexual and seeks to stop workplace harassment. Before answering, read the following case.

   In the case of *Jimmie L. Smith v. City of Salem, Ohio,* 378 F.3d 566 (6th Cir. 2004), the U.S. Court of Appeals for the 6th Circuit held that a transsexual firefighter was entitled to a jury trial to prove hostile work atmosphere. The court wrote:

   > Smith is—and has been, at all times relevant to this action—employed by the city of Salem, Ohio, as a lieutenant in the Salem Fire Department (the "Fire Department"). Prior to the events surrounding this action, Smith worked for the Fire Department for seven years without any negative incidents. Smith—biologically and by birth a male—is a transsexual and has been diagnosed with Gender Identity Disorder ("GID"), which the American Psychiatric Association characterizes as a disjunction between an individual's sexual organs and sexual identity. *American Psychiatric Association, Diagnostic and Statistical Manual of Mental Disorders 576-582* (4th ed., 2000). After being diagnosed with GID, Smith began "expressing a more feminine appearance on a full-time basis"—including at work—in accordance with international medical protocols for treating GID. Soon thereafter, Smith's co-workers began questioning him about his appearance and commenting that his appearance and mannerisms were not "masculine enough." As a result, Smith notified his immediate supervisor, Defendant Thomas Eastek, about his GID diagnosis and treatment. He also informed Eastek of the likelihood that his treatment would eventually include complete physical transformation from male to female. Smith had approached Eastek in order to answer any questions Eastek might have concerning his appearance and manner and so that Eastek could address Smith's co-workers' comments and inquiries.

   > Smith specifically asked Eastek, and Eastek promised, not to divulge the substance of their conversation to any of his superiors, particularly to Defendant Walter Greenamyer, Chief of the Fire Department. In short order, however, Eastek told Greenamyer about Smith's behavior and his GID. Greenamyer then met with Defendant C. Brooke Zellers, the Law Director for the City of Salem, with the intention of using Smith's transsexualism and its manifestations as a basis for terminating his employment.

   > The city ordered Smith to undergo three separate psychological evaluations with physicians of the city's choosing. They hoped that Smith would either resign or refuse to comply. If he refused to comply, the defendants reasoned, they could terminate Smith's employment on the ground of insubordination.

   > [O]n April 26, 2001, [Fire Chief] Greenamyer suspended Smith for one twenty-four hour shift,

based on his alleged infraction of a City and/or Fire Department policy. At a subsequent hearing before the Salem Civil Service Commission (the "Commission") regarding his suspension, Smith contended that the suspension was a result of selective enforcement in retaliation for his having obtained legal representation in response to Defendants' plan to terminate his employment because of his transsexualism and its manifestations. The Commission ultimately upheld Smith's suspension. Smith appealed to the Columbiana County Court of Common Pleas, which reversed the suspension [because the policy he allegedly violated was not in effect].

Smith then filed suit in the federal district court, but the federal district judge dismissed the lawsuit. The 6th Circuit reversed and held that he is entitled to a jury trial to prove his allegations. The Court of Appeals wrote:

In his complaint, Smith asserts Title VII claims of retaliation and employment discrimination "because of . . . sex." The district court dismissed Smith's Title VII claims on the ground that he failed to state a claim for sex stereotyping pursuant to *Price Waterhouse v. Hopkins, 490 U.S. 228, 109 S.Ct. 1775, 104 L.Ed.2d 268 (1989).* The district court implied that Smith's claim was disingenuous, stating that he merely "invokes the term-of-art created by *Price Waterhouse,* that is, 'sex-stereotyping,' as an end run around his 'real' claim, which, the district court stated, was 'based upon his transsexuality.'" In *Price Waterhouse,* the plaintiff, a female senior manager in an accounting firm, was denied partnership in the firm, in part, because she was considered "macho."

The Supreme Court made clear that in the context of Title VII, discrimination because of "sex" includes gender discrimination: "In the context of sex stereotyping, an employer who acts on the basis of a belief that a woman cannot be aggressive, or that she must not be, has acted on the basis of gender." The Court emphasized that "we are beyond the day when an employer could evaluate employees by assuming or insisting that they matched the stereotype associated with their group."

Having alleged that his failure to conform to sex stereotypes concerning how a man should look and behave was the driving force behind Defendants' actions, Smith has sufficiently pleaded claims of sex stereotyping and gender discrimination.

[Note: On November 8, 2005, the *Cincinnati Enquirer* wrote a story about a transsexual Cincinnati police officer. "City may pay cop $1M-plus in lawsuit. Transsexual officer fought discrimination." The story was about a lawsuit filed in 1999 by a Cincinnati police officer who was demoted from sergeant. He dressed like a man while on duty but like a woman while off duty and changed his first name from "Phillip" to "Phelecia." The U.S. Supreme Court refused to hear the city's appeal of the federal jury verdict and award of $300,000 to the officer and $550,000 in attorney fees.]

See Appendix A for additional Expand Your Learning activities related to this chapter.

CHAPTER 8

Race Discrimination

Opportunities to Succeed. Captain Mike Washington, Cincinnati Fire Department, is working on his Bachelor of Science degree at the University of Cincinnati, Fire Science Department. The author was at the Cincinnati Reds Opening Day Parade, with his oldest grandson, Jasper Dean Bennett, when a Cincinnati FD ladder company responded with red lights and siren for a person who had fainted. Captain Washington was the officer in charge (OIC). After treating the patient, Mike stopped to say hello. *Photo by author.*

Key Terms

reverse discrimination, p. 94
**affirmative action
    programs, p. 94**

front pay, p. 98
prejudgment interest, p. 101

prima facie case, p. 106

From the Headlines

Nooses on Turn-out Gear of Two African-American Firefighters

The fire chief of Jacksonville (FL) Fire-Rescue has asked the sheriffs' office to launch an investigation into the incident and is quoted as stating that the department is "extremely embarrassed" by the incident and will terminate whoever is responsible.

Austin, Texas Seeks More Fire Cadet Diversity

Austin's goal is to make the cadets class more diverse, to reflect the community, which is 31 percent Hispanic, 10 percent African-American, and 5 percent Asian. Firefighters won collective bargaining rights in 2004, and the three-year contract recently negotiated gives the city the tools for diversity hiring, including giving less weight to the civil service written test and more weight to oral interviews. There will also be more community outreach, including recruiting in Houston, Detroit, and South Texas.

Race Discrimination Jury Verdicts—NJ Fire Department/Michigan Police Department

A federal jury awarded two African-American officers $1.04 million; they both had over 20 years on the fire department. They alleged that they had been subjected to repeated racial slurs and taunting, and their promotions were delayed. After they publicly criticized city officials and the fire department policies, they asserted that they suffered further retaliation, including delayed promotions. A federal jury in Michigan awarded a Mexican-American police officer $4.14 million, including $2 million in punitive damages to be paid by the city, and $1 million to be paid by the police chief, after being subjected to discipline for incidents in which white officers were not disciplined, and for being arrested for an off-duty altercation with his wife, while Caucasian officers under similar circumstances were not arrested.

Captain Mike Washington married firefighter Pia Young on September 11, 2004, as a tribute to 9/11. Mike is affectionately known as "Bike Nine," from his days as a child when he would joyfully ride his bike to Fire Station No. 9. *Photo by author.*

◆ INTRODUCTION

This chapter concerns race discrimination, including consent decrees by federal judges to correct prior discrimination and litigation by minorities complaining of discrimination. It also discusses Caucasians complaining of **reverse discrimination**.

◆ TITLE VII AND EEOC

reverse discrimination A city or other political subdivision may breach the law by providing affirmative action for minorities, unless there has been a history of prior race discrimination and the steps taken were reasonable to overcome that history.

Title VII of the Civil Rights Act of 1964, and implementing regulations of the Equal Employment Opportunity Commission, prohibit discrimination in employment based on race, sex, color, religion, or national origin. Employers are generally defined to include all governmental agencies, including publicly owned fire departments. Private employers with 15 or more employees are also covered by the federal EEOC regulations, as are all state and local public employers without regard to the number of employees. Many states have their own equal employment enforcement agencies and regulations.

◆ AFFIRMATIVE ACTION

affirmative action programs Federal court orders that require fire departments and other public employers to adopt programs to hire, retain, and promote minority employees.

Affirmative action is a controversial subject, but it has long been used by the federal government regarding contractors seeking public contracts. In 1965, President Lyndon B. Johnson issued Executive Order 11246, which prohibited racial and other discrimination by federal contractors and subcontractors and also required these contractors to adopt affirmative action policies to seek out, hire, and promote minority employees. The executive order was amended in 1967 to require similar affirmative action for female employees. Affirmative action has also been extended to veterans. In 1973, Congress required covered contractors and subcontractors to have **affirmative action programs** to hire and promote disabled veterans and Vietnam veterans. Section 503 of the Rehabilitation Act of 1973, 29 U.S.C. 793. This was later expanded to include veterans of the Gulf War I and, in 1998, to any veteran of any battle in which a campaign badge has been authorized. 38 U.S.C. 4212, and Pub. L. 105-339. All of these programs are enforced by the U.S. Department of Labor, Office of Federal Contract Compliance Programs. These laws are enforced only by governmental action, not by private lawsuits or complaints to the EEOC.

◆ CONSENT DECREES

Consent decrees have been issued for many years to correct prior racial discrimination in the fire service. For example, the Supreme Court of California on May 26, 2006, held that the San Francisco Fire Department could implement a new promotion program, without first having to bargain to impasse and then go to binding arbitration. The court stated,

> The San Francisco Fire Department hired no African-American firefighters before 1955. . . . In 1970, only four of 1,800 uniformed fire personnel were African American. The department allowed no women to apply before 1976 and hired no women until 1987. Between 1970 and 1973, a federal district court ruled that three successive versions of firefighter entry-level examinations had adverse impact on minority applicants and had not been professional validated as an accurate measure of the knowledge, skills and ability needed for the job. The court ordered affirmative action, requiring the City to hire one minority for each non-minority hired from the entry-level eligibility list until all minority applicants on the list had been hired. *San Francisco Firefighters Local 798 v. City and County of San Francisco,* 2006 Ca. LEXIS 5932, May 18, 2006.

BOSTON CONSENT DECREE SET ASIDE

Caucasian firefighters and applicants have filed lawsuits to have consent decrees set aside when there is no longer a need for affirmative action. In *Joseph E. Quinn et al. v. City of Boston and NAACP,* 2003 1st Cir. 116 (2003), five Caucasian applicants to join the Boston Fire Department sued, alleging reverse discrimination. The federal district judge dismissed their suit, citing the almost 30-year-old consent decree. The applicants appealed, and the U.S. Court of Appeals for the First Circuit decided in favor of the Caucasian applicants and set aside the 1974 affirmative action consent degree. The three-judge court wrote,

> For over a quarter of a century, the hiring of firefighters in the City of Boston (the City) has taken place in the [wake] of a federal court consent decree designed to remedy the effects of past discrimination against African-Americans and Hispanics. On April 11, 2001, five candidates for employment (the Candidates) brought suit in the federal district court alleging that the City had discriminated against them on the basis of race when hiring new firefighters in the fall of 2000. The City defended its hiring practices as compliant with, and compelled by, the terms of the consent decree. The district court granted summary judgment in the defendants' favor. The Candidates appeal. For the reasons that follow, we reverse the judgment and remand to the district court for further proceedings consistent with this opinion.
>
> In the early 1970s, two suits were brought against a number of municipalities subject to the Massachusetts Civil Service law (now codified at Mass. Gen. Laws ch. 31, §§ 1-77). The suits alleged that the municipalities engaged in discriminatory recruitment and hiring practices whilst staffing their respective fire departments. These actions culminated in the entry of an omnibus consent decree that influenced the manner in which the affected municipalities could recruit and hire firefighters. The decree was affirmed on appeal. It has been in effect for nearly thirty years.
>
> Unlike some forty-five other fire departments that heretofore have met the goals of the decree and gained release from its constraints, the Boston Fire Department (BFD) has operated under the auspices of the Beecher decree since 1974. A decade ago, a group of non-minority men, aspiring to appointments as firefighters within the BFD, endeavored to bring the City out from under the umbrella of the decree. See Mackin, 969 F.2d at 1275. Although we affirmed the district court's rejection of that attempt, we noted that the decree was not meant to operate in perpetuity. To the contrary, it would remain in force as to any particular municipality only until its stated

goal had been achieved. *Id.* at 1278. A decade has passed, but—despite increased diversity within the BFD—the City still hires firefighters in accordance with the Beecher decree.

We conclude, therefore, that the City's continued resort to race-based preferences from and after the time when parity was achieved fails the second prong of the strict scrutiny analysis. See Bakke, 438 U.S. at 309. Thus, the City's adherence to the Beecher decree during the 2000 hiring cycle was unconstitutional. Consistent with the foregoing, we reverse the district court's entry of summary judgment in favor of the City and direct the court to enter judgment in the Candidates' favor.

The goal of the Beecher decree was to eliminate discriminatory practices in the recruitment and hiring of firefighters in communities subject to the Massachusetts Civil Service laws, and, relatedly, to remedy the effects of past discrimination in recruitment and hiring. Remediation has taken more than a quarter-century. At long last, however, that objective has been achieved with respect to the BFD; parity has been reached between the percentage of minority firefighters in the BFD and the percentage of minorities in the City as a whole.

Although this is a significant landmark along the road to equality, we add a word of caution. We are not Pollyannas, and we recognize that achieving parity at the firefighter level is not tantamount to saying that all is well in regard to racial and ethnic issues within the BFD as a whole. To the extent that inequalities remain, however, they are not within the compass of either the Beecher decree or this litigation. Nor will we reach out for them—issues of constitutional magnitude should not be the subject of speculation, but, rather, should be litigated fully by parties with standing to represent various pertinent points of view. For today, we fulfill our responsibility by holding that the City's appointment of firefighters ought no longer be subject to the strictures of the Beecher decree. We need go no further.

LOUISIANA CONSENT DECREE NOT SET ASIDE

On January 25, 2006, the U.S. Court of Appeals for the Fifth Circuit held in *Jeffrey Tood Dean et al. v. City of Shreveport, Louisiana,* 438 F.3d 448 (5th Cir. 2006), reversed a federal district judge who had dismissed a challenge to a 20-year-old consent decree, and ordered that applicant Dean be given an opportunity to present evidence that the "race-conscious hiring process" was no longer necessary when he applied to the fire department between 2000 and 2002. The Court of Appeals ordered the city to "provide reliable statistical data showing the percentages of blacks in its work force and in its qualified labor pool between 2000 and 2002. Only when the district court has this information can it properly decide whether a sufficient disparity still existed." The Court of Appeals referred to the reasons why the original consent decree was filed in 1980, including the fact that, prior to 1974, the fire department had never hired a black firefighter. In that year, the department was sued by several black applicants. The city settled and hired three minority firefighters. In 1997, the U.S. Department of Justice filed a lawsuit because only 10 of the City's 270 firefighters were black, yet the city was 40 percent minority. The consent decree required the city to adopt a hiring goal of firefighters of at least 50 percent minority and 15 percent females. The Court of Appeals was told by the city that it voluntarily changed its hiring process in 2004 to make it race-neutral, but "Much to this Court's dissatisfaction, the City has kept secret the details of its new hiring process." For these reasons, the 5th Circuit remanded the case for a full hearing before the federal trial judge.

◆ NEW EEOC GUIDANCE ON RACE DISCRIMINATION

On April 19, 2006, the EEOC issued new guidelines on race discrimination to broaden the protections to include ancestry, physical characteristics, raced-linked illness, and cultural characteristics (including grooming practices and manner of speech) (*www. eeoc.gov*). Charges of race discrimination filed with the EEOC are therefore expected to grow. In FY 2005, over 35 percent of the 75,400 charges filed with the EEOC were race-based.

◆ PROMPT INVESTIGATIONS

Fortunately, elected officials in many communities promptly investigate internal complaints of discrimination, whether made by African-American or Caucasian employees. Prompt investigations, including consultation with legal counsel, can help avoid both discrimination against minorities and reverse discrimination against Caucasians. For example, in *Kevin Golden v. Town of Collierville,* __ F.3d __ (6th Cir., January 23, 2006), 2006 U.S. App. LEXIS 1712, a Caucasian firefighter with this Tennessee fire department filed a lawsuit alleging the fire chief passed him over for promotion to lieutenant in favor of an African-American firefighter. After pre-trial discovery, the U.S. District Court granted summary judgment for the fire department, and the Court of Appeals agreed. In December 2000, the town opened a new firehouse and approved three new lieutenant positions. Ten applied, including two African-Americans. After the testing and interview process, Golden was ranked fourth. The fire chief told Golden that he was considering promoting him, instead of the third-ranked candidate (African-American), because Golden had greater experience "riding out of rank." Word of this intention spread throughout the fire department, and the African-American candidate filed a grievance with the town administrator, because "riding out of rank" was never a criterion for minimum qualifications for the promotion. The promotions were put on hold. The administrator consulted legal counsel and advised the fire chief to promote the top three candidates. The federal district court dismissed Golden's lawsuit. The Court of Appeals agreed,

> Golden has not provided any evidence of background circumstances to support his assertion that defendants discriminate against white employees. The entire upper management of the Collierville fire department is white. Of the fifteen lieutenants, only two (including Boone [who was one of the three lieutenants promoted over Golden] are black. Of the fifteen drivers, only two are black. Further, it is undisputed that Collierville does not have any type of affirmative action plan. Additionally, the record [including the pre-trial deposition of Golden] is rich with Golden's admissions that he had no reason to believe that [the fire chief] did anything in the promotional process to discriminate against white employees.

The Court of Appeals (three to zero) concluded, "We agree with the district court's conclusion that defendants' desire to avoid discrimination against minorities is insufficient to support an inference that they discriminate against the majority."

- *NFPA 1001: Standard for Fire Fighter Professional Qualifications* (2002 edition). This standard identifies knowledge and skills required of all firefighters. It is a performance-based standard that can be applied to all firefighters, without discrimination concerning their race or gender.
- *NFPA 1021: Standard for Fire Officer Professional Qualifications* (2003 edition). This standard identifies the knowledge and skills of all officers. It serves as a basis for promotion without regard to race or gender.

◆ KEY STATUTES

CIVIL RIGHTS ACT OF 1964 (42 U.S.C. 2000e ET SEQ.)

This statute, including Title VII, is the most significant federal statute protecting individuals from discrimination, including race, national origin, or gender discrimination. President John F. Kennedy in the early 1960s used the U.S. Department of Justice, under Attorney General Robert Kennedy, to enforce the voting rights laws in the South. Both were assassinated before there was enough political support to get this statute enacted. It took the political might of a Southerner, President Lyndon B. Johnson from Texas, to get the necessary votes to pass the law, which gave individuals the right to file complaints of discrimination with the EEOC and to file private lawsuits in federal court.

Significant to the fire service, federal judges were also authorized under Section 706(g) of the statute to correct prior race discrimination in hiring minority firefighters. Affirmative action lawsuits were filed in many cities, resulting in federal court consent decrees (many with dual hiring lists and dual promotion lists) to make up for past discrimination. Section 706(g) authorized federal judges to enter such orders:

> If the court finds that the respondent has intentionally engaged in or is intentionally engaging in an unlawful employment practice charged in the complaint, the court may enjoin the respondent from engaging in such unlawful employment practice, and order such affirmative action as may be appropriate, which may include, but is not limited to, reinstatement or hiring of employees, with or without back pay (payable by the employer, employment agency, or labor organization, as the case may be, responsible for the unlawful employment practice), or any other equitable relief as the court deems appropriate.

CIVIL RIGHTS ACT OF 1991 [42 U.S.C. 1981a (b)(3)]

In response to some jury verdicts awarding very large amounts to individuals suing for discrimination, Congress limited compensatory damages in federal lawsuits (back pay awards) to $300,000 in 1991. This cap on liability has been the subject of much litigation, including lower court decisions holding that the cap does not apply to **front pay** (future wages).

front pay In race and sex discrimination lawsuits, and in so-called reverse discrimination lawsuits, damages can be awarded not only for lost back pay for failure to promote, but also for equitable relief in the form of expected lost future earnings.

42 U.S.C. 1983

This important Civil War-era statute, referred to as "1983 lawsuits," has been used frequently in current litigation in federal and state courts, in which individuals seek to sue state and local governments, alleging that the rights under the U.S. Constitution have been deprived by improper conduct of governmental officials. These 1983 lawsuits are frequently filed in police officer shootings, alleging that the police officer was inadequately trained or supervised by the city in the use of firearms. In the fire service, "section 1983" lawsuits have been filed by the surviving families of firefighters killed in the line of duty, alleging improper conduct of incident commanders and inadequate training. These 1983 suits have also been filed by firefighters alleging they were denied promotion or disciplined for exercising their First Amendment rights to publicly criticize the fire chief or to support the fire union.

◆ **CASE STUDIES**

**CASE STUDY 8-1   Fire Apprentice Program to Recruit Minorities—
Added Points on Entrance Test**

IAFF Local 136 v. City of Dayton, 107 Ohio St. 3d 10, 2005-Ohio-5826 (November 16, 2005).

FACTS

The City of Dayton is governed by a charter. Under this charter, it has a Civil Service Commission, which administers competitive examinations for positions within the competitive class of the classified service. Under Rule 6 of the Commission, the fire department had a Fire Apprentice Program, in which successful graduates would receive preference points on the firefighter entrance examination. EMT-Bs would get five points; paramedics would receive an additional five points.

The fire department designed the Fire Apprentice Program, conducted at Sinclair Community College, to increase female and minority firefighters. The Diversity Plan stated that 6 percent of current members of the department were female, and only 5 percent were minority (the city is 42 percent minority).

On September 13, 2002, a competitive examination was held for recruit firefighters. Some of the candidates included members of the Fire Apprentice Program. Local 136 of the IAFF and the Fraternal Order of Police filed a lawsuit in Ohio Court of Common Pleas, seeking a court order declaring the added points program to be a violation of the city charter and an injunction prohibiting the Civil Service Commission from awarding the added points.

The trial judge issued the injunction, declaring the program violated the law. The city appealed, and the Ohio Court of Appeals reversed. The IAFF and FOP then appealed to the seven-member Ohio Supreme Court.

YOU BE THE JUDGE

Does the added points program violate the rights of the other candidates for appointment to the Dayton Fire Department?

HOLDING

No.

The Ohio Supreme Court (six to one), in an opinion by Chief Justice Moyer, wrote:

The Fire Apprentice Program educates and trains participants for the position of firefighter recruit and provides unique exposure to the Dayton Fire Department by pairing each apprentice with a firefighter for a mentoring relationship. Persons who successfully complete one or two years in the program are likely to have increased their merit, fitness, efficiency, character, and industry. And the nature of that experience may be immeasurable by written, oral, or performance examination. In that respect, it is not unreasonable for city officials to conclude that a person who has successfully completed the Fire Apprentice Program should receive some credit for that training when competing for a position with the Dayton Fire Department.

The civil service board is not authorized to award preference points based solely on the sex or race of the applicant. The charter does not permit implementing the goal of diversity by those means.

The Fire Apprentice Program is a recruitment mechanism through which the Dayton Fire Department may attract a diverse group of potential firefighters. Its purpose is to reach a cross-section of the population of the city so that men, women, and members of all racial groups will be aware of and have an opportunity to pursue employment with the fire department. The record does not indicate that the Fire Apprentice Program operates exclusively for, or even that it prefers, minorities or women. Accordingly, Rule 6, Section 11 does not violate the charter of the city of Dayton on that basis.

LEGAL LESSONS LEARNED

Recruiting programs such as the Dayton Fire Department program may be a very effective and lawful way to increase females and minorities on the department.

CASE STUDY 8-2 Caucasian Lieutenants Alleging "Reverse Discrimination"

Biondo v. City of Chicago, 382 F.3d 680 (7th Cir. 2004).

FACTS

Nineteen Caucasian candidates for promotion to lieutenant on the Chicago Fire Department filed suit in U.S. District Court for reverse discrimination.

The fire department has five ranks: firefighter, engineer, lieutenant, captain, and battalion chief. If developed a promotion exam in 1986 for lieutenant with care to ensure that it was non-discriminatory and a valid test of skills. The results were 29 percent of the firefighters and engineers who took the exam were African-American or Hispanic, but only 12 percent of this group scored in the top 300 candidates.

The fire department therefore decided to depart from "rank-order" promotions from one list and created two lists instead. Lieutenants were promoted by taking the highest scorer on the Caucasian list, then the highest scorer on the African-American/Hispanic list. It used these two lists from 1986 until 1991, promoting a total of 209 lieutenants.

Nineteen Caucasian candidates who were passed over for promotion by minority candidates with lower scores filed suit under 42 U.S.C. 1983 and Title VII of the Civil Rights Act of 1964. Five of 19 got delayed promotions to lieutenant on the 1986 list, 10 were promoted on later exams to lieutenant, and four had never been promoted. Of the 15 who made Lieutenant, nine took later exams for captain but were never promoted to captain.

The city did not argue that its two-list approach was lawful because it had a "compelling interest" in eliminating past discrimination and in its quest for diversity of its officers. Instead, the city asserted it was required by federal EEOC regulations that "frown on using tests to make promotions in strict sequence" of scores. The city's statistical expert said the 1986 promotion test had a "standard error" of 3.5, meaning that a person who scored 80, if they took a similar test, could score as high as 83.5 or as low as 76.5. The fire department concluded that it could not justify the disparate impact of the test on minorities and had to establish two promotion lists. The city relied on the EEOC's "Uniform Guidelines on Employee Selection Procedures," 29 CFR 1607.4.

The District Court judge conducted a trial before an "advisory jury." This jury concluded that the 1986 promotion exam was fair, and therefore the city violated the law by establishing a "two hiring list" promotion process. The district judge told the jury he agreed with their conclusion.

The district judge then held two separate jury trials for 19 of the Caucasian plaintiffs to establish their damages. The juries were asked to decide, for each plaintiff, the probability they would have been promoted to lieutenant, and if so, the probability they would later be promoted to captain and then battalion chief. Based on these probabilities, the jury awarded back pay and front pay (expected lost future income because of failure to promote), plus damages for emotional distress, and **prejudgment interest** on the back pay and attorney's fees.

prejudgment interest   A plaintiff who proves discrimination can be awarded interest on the money that he or she should have been paid if he or she were promoted timely.

The jury returned the following verdict for lead plaintiff Peter Biondo:

- Probability of promotion to captain: 100 percent
- Actual date of promotion to captain, based on later test: August 16, 1993
- Probability of promotion to battalion chief: 100 percent
- Actual date of promotion to battalion chief, based on later test: April 16, 2000
- Back pay award: $112,000
- Damages for emotional distress: $125,000
- Years of front pay award: 12 years
- Prejudgment interest on back pay: $35,125
- The promotion to lieutenant was made retroactive to November 1, 1990, for additional seniority and pension funds; he paid as a captain as of August 16, 1993; and also he paid as a battalion chief as of March 1, 2002, until March 1, 2014, unless he retires earlier.

Plaintiff Michael Timothy was a battalion chief by the time his case went to trial. The jury understandably awarded him no front pay. It did award him back pay of $24,000 (presumably for the delay in promotion to lieutenant) and damages for emotional distress of $10,000.

Some of the jury verdicts were questionable. For example, Michael Gacki, originally passed over in the 1986 exam, was later promoted to lieutenant and then captain. The jury awarded him no front pay based on their estimate that he had no probability of being promoted to captain, but awarded him $5,992 in back pay and $94,100 for emotional distress.

The city appealed to the 7th Circuit Court of Appeals.

YOU BE THE JUDGE

1. Should the jury verdicts be confirmed or set aside?
2. Assuming there has been a history of past racial discrimination, can a city employ a "two-list" promotion process?

HOLDING

Jury verdicts are set aside.

A two-list promotion process may be lawful. Prior court decisions confirm that governments have a "compelling interest" in having a diverse population.

The three-judge panel set aside the jury verdicts:

> Unfortunately, the juries' estimates do not reflect a plausible appreciation of the lost chances: the juries concluded that every one of the plaintiffs who had not yet received a captaincy were *certain* to have done so—even though four plaintiffs had been unsuccessful in post-1986 attempts to become lieutenants, and 9 of the 15 plaintiffs who reached lieutenant had tried and failed to achieve higher ranks.
>
> One must take into account the possibility of sitting for multiple exams: 132 lieutenants who were not from the 1992 test for captain applied again in 1998; of these, 29 (or 22%) have been promoted. This suggests that of the lieutenants who take the captain's exam at all, about 33% are promoted either from the first exam or a second try. . . . Still this is a long way from the 100% success rate that the juries calculated for our 19 firefighters.

The court was also critical of the plaintiff having offered no evidence to compare themselves to minority or Hispanic candidates who were promoted ahead of the plaintiffs.

> [T]he plaintiffs bypassed their opportunity to introduce comparative evidence (comparing themselves to candidates who won promotion); and what evidence the record does contain—that these plaintiffs did not excel on the 1968 exam; that they have not done well on tests for promotion since then; that only 33% of those who were actually promoted to lieutenant by 1992—undermines the verdict. Even giving all of the evidence and inferences a reading favorable to the plaintiffs, no reasonable juror could find that all were destined to be captains. The verdicts therefore cannot stand.

The court also criticized the city's "two-list" promotion program:

> Public employers [when addressing disparate impact] have other options. One would suffice here. Instead of making rank-order promotions, Chicago could have created bands reflecting the standard error of management. For example, the Department could have treated all scores in the range 96–100 as functionally identical and made promotions by lot from that band; when all test-takers with those scores had been promoted, that Department could have pooled scores in the range 91–95 and promoted randomly from that group, and so on.

The court also criticized the federal district judge for not imposing the $300,000 cap in the back pay awards, under the Civil Rights Act of 1991.

Circuit Judge Williams wrote a concurring decision. He agreed the jury verdicts should be set aside, but noted that the courts have "recognized on numerous occasions that a governmental agency has a compelling interest in remedying its past unlawful discrimination." Judge Williams cited prior precedent, *McNamara v. City of Chicago,* 138 F.3d 1219 (7th Cir. 1998) (finding that non-rank order promotions in the Chicago Fire Department were justified by the city's past racial discrimination in employment of firefighters; *Chicago Firefighters Union Local 2 v. Washington,* 1999 U.S. Dist. LEXIS 20310 (N.D. Ill. Dec. 30, 1999) (defending non-rank order promotions as part of policy to remedy effects of past racial discrimination). In another case,

> This court found a compelling interest in having a diverse population at the rank of sergeant. *Petit v. City of Chicago,* 352 F.3d 1111, 115 (7th Cir. 2003); cf. *Grutter [v. Bollinger,* 539 U.S. 306 (2003)], 539 U.S. at 329 (finding law school had a compelling interest in a diverse student body). We similarly recognized an operational need for persons of different races in the corrections department. *Wittmer v. Peters,* 87 F.3d 916, 919 (7th Cir. 1969).

LEGAL LESSONS LEARNED

Diversity programs have been the subject of substantial litigation. Targeted recruiting of minorities, with focused ads, are both lawful and an excellent manner of increasing interest for minority candidates. See the Cincinnati Fire Department's diversity goals in Appendix D.

CASE STUDY 8-3   African-American Battalion Chiefs Allege Race Discrimination

Lee Belton et al. v. City of Charlotte, __ F.3d ____ (4th Cir. May 23, 2006); 2006 U.S. App. LEXIS 12747.

FACTS

Lee Belton, Jerome Frederick, and Larry Mackey are African-Americans and battalion chiefs with the City of Charlotte (NC). They have each worked for the fire department for over 30 years and have outstanding employment records. In 2004, each filed a separate lawsuit, alleging employment discrimination, including racially disparate treatment, racially hostile work environment, and retaliation for protected activity under Title VII of the Civil Rights Act of 1964.

The federal district judge granted summary judgment for the city against each of the three plaintiffs, and they filed an appeal to the 4th Circuit Court of Appeals. The Charlotte Fire Department has 1,015 employees, including a fire chief, four deputy chiefs, and 25 battalion chiefs. If a deputy chief retires, only battalion chiefs are eligible for promotion to deputy chief by personal selection of the fire chief and concurrence of the city manager.

The current fire chief is Caucasian and has served as chief since 1987. In his first 17 years as chief (1987–2004), he never selected an African-American to be deputy chief. In each promotion, he selected a Caucasian male. Of the current 25 battalion chiefs, there are five African-Americans and two Caucasian females. The annual pay differential between battalion chief and deputy chief is about $17,000.

The fire chief used various selection processes for filling the deputy chief positions, first by simply hand-picking the battalion chief to be promoted; starting in 2001, he used a more formal selection process, including written eligibility criteria and a written test. The plaintiffs presented some evidence that the fire chief "grew up on the other side of the tracks" and does not relate well to minorities. He is reportedly accused of having been upset when his son was up for a fire department promotion in the late 1990s, and the three members of the assessment panel were African-American. By his reasoning, no panel should have members of only one race, because "every point of view" will not be represented.

Plaintiff Lee Belton joined the fire department in 1977 and rose through the ranks to battalion chief. Five times he applied for promotion to deputy chief between 1996 and 2001. The fire chief even selected battalion chiefs with less experience, and the chief would never tell Belton why he was not selected.

Plaintiff Jerome Frederick also joined the fire department in 1977 and also sought promotion to deputy chief five times. He asked the fire chief what he could do to improve his chances, and the chief reportedly said, "If you don't know, I don't know."

Plaintiff Larry Mackey also joined the department in 1977 and competed for the rank of battalion chief approximately nine times. In 1996, he filed charges with the

EEOC alleging the department's use of in-house assessors for the battalion chief's promotion was discriminatory. The department agreed to use outside assessors, in return for Mackey dropping his EEOC charge. In 1998, he was promoted to battalion chief. He was never promoted to deputy chief. He was not eligible for several years because he did not yet have the minimum of three years experience, but he nonetheless contends that he should have had the "opportunity to put [his] cards on the table and display [his] abilities just like everyone else."

At some point in 2001, prior to the final promotion in question, the five African-American battalion chiefs (including plaintiffs) met with [the fire chief] to discuss his promotional process and the need for diversity within the department. [The fire chief] addressed their concerns [*15] by "allow[ing] an outside contractor to come in, do a diversity training session for the Battalion Chiefs, and that was it." M.J.A. 260.

The court described the final promotion process in question,

> For the 2001 opening [the fire chief] reinstituted the formal selection process because he had received a complaint about his informal process from Battalion Chief Willie Summers (also African American). The formal process consisted of a written test, possibly an interview, and a review of the candidates' records. At some point before the written test, [the fire chief's] assistant mailed out an official announcement of the eligibility criteria: (1) at least four years' experience as Battalion Chief; (2) at least an "Expected" rating in the prior two performance appraisals; (3) a two-year degree (after high school); and (4) no disciplinary record in the prior two years.
>
> On the morning in question the five interested and eligible candidates (including Belton and Frederick) arrived at the General Office. The candidates had three hours to provide written answers to twelve questions. Because there were only four available laptops but five candidates, [the fire chief] directed Battalion Chief Dulin to take the exam on a desktop computer in a nearby office that Dulin had been using as his temporary office. The remaining candidates took the test together in a conference room on laptops. These laptops were old and finicky machines that frequently malfunctioned. Computer technicians at times interrupted the testing process to make repairs. Nonetheless, Belton, Frederick, and the other two candidates were able to complete the exam in the allotted time.
>
> [The fire chief] evaluated each of the candidates in his "Deputy Chiefs Process Score Sheet—2001. He scored the three white candidates higher than the two African-American candidates, Belton and Frederick. He graded Belton's leadership as "average," his written communication skills as "lacking," his oral communication skills as "average," and his written exercise as "below average." He further noted that Belton had received a "G" rating ("good performance which meets and periodically exceeds requirements") in the prior two appraisal evaluations. Similarly, [the fire chief] graded Frederick's leadership as "average," his communication skills as "[n]ot very good," and his written exercise as "less than average." B.J.A. 750. [The fire chief] noted that Frederick had also received a "G" rating in the prior two performance appraisals. By contrast, [the fire chief] graded Dulin as "excellent" in all categories (leadership skills, communication skills, and written exercise), and noted that Dulin had received an "E" rating ("consistently far exceeds requirements") in the prior two appraisals. B.J.A. 714. Dulin had fewer years experience on the force than either Belton, Frederick, or Mackey.
>
> [The fire chief] later testified that, when evaluating the eligible candidates, he favored "competency and skills" over "seniority and experience," particularly because Deputy Chiefs interact with the City Manager, make presentations to the budget director and city council, and help craft city policy. Further, he testified that he considered an applicant's

complete disciplinary record even though only recent disciplinary action disqualified a candidate. The fire chief also considered each candidate's list of special projects, such as involvement with emergency response teams and charitable campaigns. The fire chief however, "handpicks who he wants to do the special projects" and thereby "groom[s]" candidates for promotion. Indeed, in 2000 the fire chief ignored Belton's request to lead a special recruitment project, and on another occasion similarly "blew . . . off" Mackey's request to lead a special project involving emergency medical services. (Belton and Frederick have performed a number of other special projects over the years, though.)

The Fire Department formally announced Dulin's promotion on December 26, 2001, roughly two months after the competition day. Belton testified that, upon learning of Dulin's promotion, he immediately felt that Dulin had been promoted for "racial reasons" and that "the precedent had been set that the Chief was not going to promote a black man." Sometime in January 2002 Belton asked the fire chief why "was a younger, less experienced chief being leapfrogged over [Belton] again into the position when [Belton] had basically been in this position [roughly] fifteen years." [The fire chief] gave a vague answer and then elaborated with seeming contradictions. [The fire chief] said that Belton wrote less on the test than Dulin but there was "absolutely nothing" that Belton could have written that would have changed [the fire chief's] mind. Frederick also questioned [the fire chief] about not being promoted. [The fire chief] responded that Frederick "wasn't the type of candidate [that the Fire Department was] looking for that position because he had the inability to write and deliver a message."

In September 2002 Belton, Frederick, Mackey, and Summers met with David Sanders in the City's human resources department. They complained that [the fire chief] excluded them from the promotion process, that the process kept changing, and that they had no idea what the qualifications were. In a subsequent meeting Sanders indicated that he would investigate their claims. On November 4, 2002, Belton and Frederick filed a charge of discrimination with the EEOC, alleging in part a "failure to promote, discriminatory remarks, and disparate treatment." Mackey filed a similar charge with the EEOC on November 7.

In December 2003 the Fire Department instituted a requirement that candidates for Deputy Chief have a four-year degree instead of a two-year degree, effective January 2005. As of 2004 Belton was pursuing a four-year degree and expected to earn it by the end of that year (though he did not meet that goal); Frederick was not enrolled in a program but indicated his intention to do so; and Mackey had already earned his four-year degree. On March 18, 2004, Belton and Frederick each filed a second charge of discrimination with the EEOC, alleging retaliation through the adoption of the four-year-degree requirement.

The City's investigation lasted ten months, from September 2002 through June 2003. The City never once advised Belton, Frederick, or Mackey on the progress or results of the investigation. Ultimately, Sanders found no evidence that [the fire chief] promoted based on race but concluded that [the fire chief's] selection process could be perceived as racially discriminatory. In his deposition testimony Sanders conceded that [the fire chief] controlled the selection process and could, if he wished, change the criteria to "make sure" that an individual whom he favored (presumably a white individual) was promoted over other interested candidates (including African Americans). At the close of the investigation, Sanders recommended to the City's human resources director that [the fire chief] receive counseling on the potential perception of racial bias, though Sanders does not know if [the fire chief] ever received such counseling. The human resources director thereafter met with [the fire chief], presumably to relay the results of the investigation.

YOU BE THE JUDGE

Have the African-American plaintiffs alleged sufficient facts to be entitled to a jury trial?

HOLDING

No. The three-judge panel split, with two judges affirming the dismissal of the lawsuit and one dissenting.

Failure to Promote—EEOC Charge Filed Late

The two-judge majority of the court wrote:

> The district court determined that plaintiffs' five alleged failures to promote were all discrete acts, the last of which occurred either on December 26, 2001 (when the Department announced Dulin's promotion) or on January 2, 2002 (when Dulin assumed the position). Plaintiffs filed their charges with the EEOC roughly nine months later, well over the six-month (180-day) period: Belton and Frederick filed theirs on November 4, 2002, and Mackey filed his on November 7. The court concluded that "there is no legally cognizable argument that [plaintiffs] filed the [c]harge within the required time period." We agree with the district court's conclusion.
>
> The Supreme Court has specifically identified the failure to promote as a "discrete act" that "occur[s] on the day that it happen[s]." *National R.R. Passenger Corp. v. Morgan,* 536 U.S. 101, 110-14, 122 S. Ct. 2061, 153 L. Ed. 2d 106 (2002). Each discrete act of discrimination "starts a new clock for filing charges alleging that act," meaning that only incidents that occur within the timely filing period are actionable; prior acts can only be used as background evidence in support of an otherwise timely claim. *Id.* at 113-14; see also *Williams v. Giant Food Inc.,* 370 F.3d 423, 428 (4th Cir. 2004). Here, because all of the challenged acts occurred outside of the period for timely filing, the entire claim is time-barred.

No Disparate Treatment

prima facie case A plaintiff must present sufficient evidence of race discrimination, and then the burden shifts to the employer to prove race was not a factor in the employment decision.

Belton and Mackey also allege disparate treatment based on a general failure to provide equal opportunities. To establish a **prima facie case**, plaintiffs must show that they are members of a protected class, they applied for the positions in question, they were qualified for the positions, and they were "rejected for the position[s] under circumstances giving rise to an inference of unlawful discrimination." *McNairn v. Sullivan,* 929 F.2d 974, 977 (4th Cir. 1991). The district court determined that neither Belton nor Mackey specifically identified the opportunities allegedly denied them. After combing the record, the district court identified certain opportunities that could be the subject of Belton's and Mackey's complaints.

Belton, for his part, presumably challenges: (i) the Department's failure to assign him unspecified special projects; (ii) the Department's failure to select him to attend a training session in Washington, D. C., in 2001; and (iii) [the fire chief's] failure to grant his request to lead recruiting efforts in 2000. The district court determined that Belton fails to establish a prima facie case of discrimination on the first allegation because he does not provide sufficient evidence that he was qualified for unspecified special projects and that other white employees in similar circumstances were given such opportunities. Indeed, Belton has participated in a number of special projects throughout his career, including work on the citywide Workforce Committee and on fund-raising campaigns for the United Way and the Arts and Sciences Council. Even if [the fire chief] selectively assigns special projects to favored employees and thereby grooms them for promotion, favoritism alone does not prove racial discrimination. See *Blue v. United States Dept. of the Army,* 914 F.2d 525, 541 (4th Cir. 1990) ("If one employee was unfairly preselected for the job, the preselection would work to the detriment of all applicants for the job, black and white alike."); see also F.J.A. 94 ([The fire chief] selected a close friend's son to lead safety committee). Regarding the remaining two allegations (the 2001 training session and the 2000 recruitment project), the district court deemed them time barred because they occurred well over 180 days prior to Belton's EEOC filing on November 4, 2002 (meaning that they occurred before May 7, 2002).

No Hostile Work Environment

Plaintiffs each allege that they have suffered a racially hostile work environment in violation of Title VII. 42 U.S.C. §2000e-2(a)(1). Title VII prohibits racial discrimination that "has created a hostile or abusive work environment." *Meritor Sav. Bank, FSB v. Vinson,* 477 U.S. 57, 66, 106 S. Ct. 2399, 91 L. Ed. 2d 49 (1984). To establish a prima facie case, a plaintiff must show that (1) the harassment was unwelcome, (2) based on race, (3) sufficiently severe or pervasive to alter the conditions of employment and create an abusive environment, and (4) there exists some basis for imposing liability on the employer. *Spriggs,* 242 F.3d at 183–84. Whether an environment is hostile or abusive depends on factors such as: "the frequency of the discriminatory conduct; its severity; whether it is physically threatening or humiliating, or a mere offensive utterance; and whether it unreasonably interferes with an employee's work performance." *Harris v. Forklift Sys. Inc.,* 510 U.S. 17, 23, 114 S. Ct. 367, 126 L. Ed. 2d 295 (1993). The environment must be both subjectively and objectively offensive to be actionable under Title VII. *Id.* at 22.

In granting summary judgment to the City on the hostile work environment claims, the district court concluded that each plaintiff fails to proffer evidence of discriminatory conduct that is sufficiently severe or pervasive to sustain such a claim. Reviewing de novo, we affirm on this ground.

Belton's Allegations

Belton's hostile environment claim hinges on six main allegations that span his nearly thirty-year career: (i) a fellow firefighter used the word "N . . ." sometime between 1977 and 1981; (ii) Belton felt isolated as the only African-American Captain in meetings; (iii) Battalion Chief Flowe did not communicate with him when he was Captain (between 1981 and 1990); (iv) [the fire chief] has become angry during meetings and used profanity to address Belton and a white firefighter; (v) other firefighters have reported to Belton since he became Battalion Chief about hearing discriminatory remarks within their own battalions; and (vi) Belton has not been promoted to Deputy Chief. As the district court indicated, these incidents were all temporally remote from each other. Some cannot be characterized as based on race (notably (iv)), and some of the racially hostile remarks were not said in Belton's presence ((v) is hearsay). The most egregious and unambiguously race-based incident is (i), when Belton's fellow firefighter said "N . . ." over twenty years ago in his presence. The firefighter's use of the racial slur "N . . ." was reprehensible and inexcusable. But it was isolated and is remote in time. The single incident did not permeate Belton's work environment with discriminatory insult and abuse and for this reason did not amount to "discriminatory changes in the terms and conditions of employment." *Faragher v. City of Boca Raton,* 524 U.S. 775, 788, 118 S. Ct. 2275, 141 L. Ed. 2d 662 (1998) (internal quotation marks and citation omitted). It is therefore insufficient to satisfy the "severe or pervasive" element of Belton's claim. *See id.* at 788 (1998) (noting that isolated incidents of abusive language will generally not meet requisite threshold of severity or pervasiveness) (citing *Oncale v. Sundowner Offshore Servs.,* 523 U.S. 75, 80-82,

118 S. Ct. 998, 140 L. Ed. 2d 201 (1998)). Indeed, Belton testified that he loves his job, respects his colleagues, and has even steered his son toward a career in the Department.

Frederick's Allegation

Frederick's hostile environment claim hinges on five main allegations, including the chief's alleged use of the term "N . . ." when describing a neighborhood to Frederick and told Frederick to "go and cook [me] some grits," F.J.A. 97; (ii) at some point since 1988, Deputy Chief Weaver drew an imaginary line across his desk and warned Frederick that "he was going to get [Frederick] if Frederick 'cross[ed] that line,'" F.J.A. 101; (iii) in 1998 he was told secondhand that a lower-ranking firefighter in a different battalion made a racist remark, though Frederick did not know if this report was "truth or hearsay," F.J.A. 87; (iv) [the fire chief] has raised his voice when speaking with Frederick on at least two occasions; and (v) Frederick has not been promoted to Deputy Chief. Frederick also alleges that in 2003, subsequent to his filing charges, [the fire chief] once ignored Frederick's greeting and told Frederick that he [the fire chief] made Frederick's hair turn gray. While all of these incidents were unwelcome, several cannot be characterized as based on race. The glaring exception is (i). [The fire chief's] use of the racial slur "N . . ." to describe a Charlotte neighborhood was repugnant and inexcusable. "Perhaps no single act can more quickly alter the conditions of employment and create an abusive working environment than the use of an unambiguously racial epithet such as "N . . ." by a supervisor in the presence of his subordinates." *Spriggs,* 242 F.3d at 185. The slur is made even more offensive by [the fire chief's] subsequent comment, "[G]o and cook [me] some grits" F.J.A. 97. As odious as these comments were, however, they were isolated and are now remote in time. They occurred during one exchange with [the fire chief] in 1989, fourteen years before Frederick filed his lawsuit. As such, they cannot sustain Frederick's hostile environment claim. See *Diggs v. Town of Manchester,* 303 F. Supp. 2d 163, 180-82 & n.11 (D. Conn. 2004) holding that racially derogatory remarks made in the early 1980s, more than fifteen years prior to plaintiff's filing suit, were "too remote to support an actionable hostile work environment claim" and that remaining incidents were "too isolated and sporadic to have created a subjectively and objectively hostile work environment."

Mackey's Allegation

Mackey's hostile environment claim consists of five main allegations occurring since the late 1970s: (i) when he was first promoted to Firefighter-Engineer, his boss did not allow him to drive; (ii) also while he was Firefighter-Engineer, another boss wrote him up as "AWOL" for missing work because of a thumb injury; (iii) he was twice reprimanded for allegedly inappropriate conduct; (iv) a superior questioned the quality of Mackey's instruction while Mackey taught at the Fire Academy; and (v) Mackey has not been promoted to Deputy Chief. Mackey has never, however, heard a racially discriminatory remark "to his face" during his career at the Department. M.J.A. 213-24. Again, these events are temporally remote from each other and some cannot be fairly characterized as based on race. While unwelcome and in some cases offensive, the challenged conduct is insufficiently severe or pervasive to create a racially abusive work environment as a matter of law. See *Faragher,* 524 U.S. at 788.

Dissent

The one dissenting judge wrote,

In view of these precedents, I am persuaded that the district court cut its inquiry short. The record provides evidence from which a jury could conclude that the City took adverse action against the Plaintiffs after they complained of discrimination in September 2002. According to the Plaintiffs, during the two months following their complaints, the City failed to respond to their allegations. Although the African-American Battalion Chiefs told the City that [the fire chief] had withheld opportunities from them because of their race, see, e. g., B.J.A. 165, the City disregarded the Plaintiffs' phone calls, *see id.* at 828-29, and failed to initiate an investigation, *see id.* at 171. Given the City's decade-long practice of ignoring accusations of race discrimination in the Fire Department, *see id.* at 974, 999, the Plaintiffs reasonably believed that their complaints had once again fallen on deaf ears. This belief prompted them to file charges with the EEOC. Only then, did they hear from the City. See *B.J.A.* 1028. Accordingly, on this record, a jury could reasonably find the City's inaction in the face of multiple race discrimination complaints sufficient to support a claim of retaliation.

LEGAL LESSONS LEARNED

Promotion processes should be formal, with graded evaluations and suggested areas of improvement. Internal complaints of racial discrimination should be thoroughly investigated. Officers must be very conscious of their speech practices and should eliminate completely any racially sensitive words.

■■

Chapter Review Questions

1. In Case Study 8-1, *IAFF Local 136 v. City of Dayton,* the court addressed the issue of a Fire Apprentice Program. Although the city has recently cancelled the program because of the budget, several fire departments have used similar programs to attract minority candidates. Describe other efforts that fire departments might use to recruit minority firefighters.

2. In Case Study 8-2, *Biondo v. City of Chicago,* the court addressed the issue of reverse discrimination, in which Caucasian firefighters seeking promotions alleged they were mistreated by the city. Describe the steps that a fire department might take to construct a promotion test of the skills needed for the job that will be fair to all candidates.

3. In Case Study 8-3, *Belton v. City of Charlotte, N.C.,* describe what changes the fire department could have added to the promotion process for deputy fire chief to provide greater feedback to the candidates who were not selected, including areas they might improve to become more competitive for the next promotion.

■■

Expand Your Learning

Read and complete the individual student or group assignment, as directed by your instructor.

1. Discuss affirmative action programs that federal courts have ordered fire departments to take in your state to make up for past discrimination. See the following case.

   In *San Francisco Fire Fighters Local 798 v. City and County of San Fran-* *cisco,* Case No. A104822 (Court of Appeals of California; January 20, 2005), the court described a history of affirmative action federal court orders.

   The San Francisco Fire Department hired no African-American firefighters before 1955. In 1970, only four of 1,800 uniformed fire personnel were African-American. The department also allowed no women to apply before 1976 and hired their first female in 1987. Between 1970 and 1973, a federal district court ruled that three successive firefighter entrance examinations had an adverse impact on minority applicants and had not been professionally validated as an accurate measure of the knowledge, skills, and ability needed for the job. The federal court ordered the city to hire one minority for each non-minority hired from the entry-level eligibility list. As of 1987, more than 55 percent of the minority firefighters were hired pursuant to this program.

A 1977 federal consent decree set a goal of 40 percent representation of minorities on the list of candidates but did not require strict ratios or quota hiring. This consent decree expired in 1982. In 1986, a federal district court found that entry-level and promotion exams used between 1982 and 1984 had an adverse impact on minorities. In 1988, a consent decree set long-term hiring goals of 40 percent minority and 10 percent female representation in the fire department. In 1991, the federal district court approved the use of "banding" for hiring and promotions. Candidates within the designated band or group of scores were considered equally qualified. Promotions were made from those in the band using secondary criteria, including race. In 1998, the district court terminated the consent decree, and the city agreed to start a Cadet Program for entry-level positions. As of 2003, the uniformed force was 57.7 percent Caucasian, 9.6 percent African-American, 13.9 percent Hispanic, 18.4 percent Asian/Pacific Islander/Filipino, and 12.8 percent female.

2. The federal Equal Employment Opportunity Commission (EEOC) lists recent litigation settlements on their website (*www.eeoc.gov,* "Litigation Settlement Monthly Reports"). Identify a recent settlement involving race discrimination and describe the "legal lessons learned" for the fire service.

See Appendix A for additional Expand Your Learning activities related to this chapter.

Americans with Disabilities Act (ADA)

9 CHAPTER

Assistant Chief Preston Moore, City of Whitehall (OH) Fire Division, helps "catch a hydrant" as firefighters from three departments arrive for a fire at a shopping center. Chief Moore was undergoing treatment for a problem in one eye but was able to drive his command vehicle safely and perform his duties. *Photo by author.*

Key Terms

disability, p. 112

failure-to-accommodate, p. 117

major life activity, p. 118

class of jobs, p. 118

From the Headlines

Assistant Fire Chief with Diabetes Wins $1.97 Million Verdict/ Settles for $500,000

A federal jury awarded $1.97 million to a 29-year veteran of an Ohio fire department, who is insulin dependent and was forced to retire after two on-duty vehicle accidents. The jury specifically found he could perform his duties with reasonable accommodation, and the city failed to prove that any accommodation would constitute an undue hardship (the city appealed, and the parties settled for $500,000).

Major Overhaul of ADA

The U.S. Department of Justice has announced a major overhaul in Americans with Disabilities Act (ADA) rules for the first time in 15 years, including improved access to buildings for people in wheelchairs.

EEOC Enforcement

In 2004, the EEOC received 15,376 charges of disability discrimination and recovered $47.7 million in monetary benefits for charging parties and other aggrieved individuals.

◆ INTRODUCTION

disability Under the ADA, a physical or mental impairment that substantially limits one or more major life activities.

This chapter addresses the Americans with Disabilities Act (ADA) and the duty of fire departments to reasonably accommodate firefighters who have a **disability** if the accommodation will permit the firefighters to perform the essential functions of their job. Fire departments should have job descriptions. Essential job functions are determined through a job analysis that meets the requirements of the U.S. Department of Justice in 29 CFR 1630.2(n)(3).

Aerial truck responds from the Federal Defense Supply Agency of Columbus (OH). *Photo by author.*

Columbus Fire Fighters climb aerial. *Photo by author.*

In the fire service, there has been considerable litigation concerning firefighters and applicants to fire departments with medical conditions. In order to establish a violation under the ADA, the plaintiff must prove (1) he or she has a disability; (2) that he or she is otherwise qualified for the job; (3) that the employer either refused to make reasonable accommodation for his or her disability or made an adverse employment decision regarding him or her solely because of the disability.

Whitehall (OH) FD engine responding to a fire. *Photo by author.*

◆ U.S. SUPREME COURT NARROWS DEFINITION OF DISABILITY

The U.S. Supreme Court has narrowly defined the term *disability,* and many firefighters will now not fall within the protection of ADA. For example, in *Toyota Motor Mfg. v. Williams,* 534 U.S. 184 (2202), the court held that employees may not be able to perform their job but still not be considered disabled unless they are also restricted from doing activities of central importance in their daily life. In addition, even if firefighters can prove they are disabled if their job duties include fire suppression and they have a physical limitation such as a bad back, there may be very little a fire department can do to reasonably accommodate such firefighters because of the very physical functions of the job.

FIREFIGHTERS WITH DISABILITIES

Much of the litigation in the fire service concerns firefighters seeking to prove that they are in fact "disabled." The ADA defines *disability* as "a physical or mental impairment that substantially limits one or more major life activities of such individual," or "being regarded as having such an impairment." In *Toyota Motor Mfg. v. Williams,* the court held that Ella Williams was not disabled even though her carpal tunnel prevented her from assembling or inspecting automobiles. The court held that Toyota did not need to accommodate her because she was not "disabled" in her normal life functions. The new test is "whether the claimant is unable to perform the variety of tasks central to most people's daily lives, not whether the claimant is unable to perform the tasks associated with her specific job." 534 U.S. at 200-201.

FIREFIGHTERS WITH ATTENTION DEFICIT HYPERACTIVITY DISORDER

The Supreme Court's decision has had an impact on firefighter litigation, including firefighters with Attention Deficit Hyperactivity Disorder (ADHD). In *Gary L. Knapp et al. v. City of Columbus,* __ F.3d __ (6th Cir; July 6, 2006, unpublished opinion), 2006 U.S. App. LEXIS 17081, three Columbus firefighters with ADHD had requested various accommodations from the city's Civil Service Commission when taking promotion examinations for captain and lieutenant exams. They sought an alternative, quieter test site, more time to complete the test, and even a verbal explanation of the test directions. Each of the firefighters took Ritalin® for their ADHD condition. The city refused to accommodate them. The three-judge court of appeals held that the city did not need to accommodate them, and the federal district judge had properly dismissed their lawsuit, because each firefighter admitted in their depositions that the Ritalin allowed them to perform their family and work obligations. "Plaintiffs have not shown that Ritalin is anything but an effective treatment for this disorder. . . . We therefore conclude that Plaintiffs have failed to establish that ADHA substantially limits their ability to learn. Accordingly, the judgment of the district court is **AFFIRMED.**"

◆ NFPA STANDARDS

- *NFPA 1001: Standard for Fire Fighter Professional Qualifications* (2002 edition). This standard identifies the knowledge and skills required of all firefighters. It is a performance-based standard applicable to all firefighters, including those with physical limitations.

- *NFPA 1451: Standard on Fire Service Vehicle Operations Training Program* (2002 edition). This standard identifies qualifications for emergency vehicle operators, including limitations on any condition, that might impair or inhibit an individual's ability to meet the driving performance standards.
- *NFPA 1500: Standard on Fire Department Occupational Safety and Health* (2002 edition). This standard identifies steps that departments should take in an emergency and also in non-emergency environments to minimize hazards and prevent firefighter injuries and death.
- *NFPA 1582: Standard on Comprehensive Occupational Medical Program for Fire Departments* (2003 edition). This standard sets criteria for pre-existing medical conditions, which may preclude individuals from performing as firefighters. Reference materials are included in the Appendix to ADA regulations.

◆ **KEY STATUTES**

AMERICANS WITH DISABILITIES ACT OF 1990 (42 U.S.C. 12101 ET SEQ.)

Title I prohibits private employers, state and local governments, and labor unions from discriminating against qualified individuals in job application procedures, hiring, firing, advancement, compensation, job training, and other terms and conditions of employment. The ADA covers employers with 15 or more employees, including state and local governments.

REHABILITATION ACT OF 1973 (29 U.S.C. 791 ET SEQ.)

Section 504 states, "[N]o otherwise qualified individual with a disability in the United States shall, solely by reason of her or his disability, be excluded from the participation in, be denied the benefits of, or be subjected to discrimination under any program or activity receiving Federal financial assistance."

◆ **CASE STUDIES**

CASE STUDY 9-1   ADA and Access to Courthouses/Implications for Fire Stations/U.S. Supreme Court

Tennessee v. Lane, 541 U.S.509 (May 17, 2004).

FACTS

George Lane, and his wife, Beverly Jones, both paraplegics, filed a lawsuit in federal court against the State of Tennessee and a number of Tennessee counties, alleging violations of Title II of the ADA. Both need wheelchairs and claim they were denied access to the state court system by reason of their disabilities. George Lane crawled up two flights of stairs at the Polk County (TN) courthouse to face reckless driving charges. He has been unable to walk without crutches since a motor vehicle accident eight years earlier. The state filed a motion to dismiss, arguing that under the Eleventh Amendment to the U.S. Constitution, the federal government had no authority over state courthouses. The U.S. District judge denied their motion, and they appealed. The

U.S. Court of Appeals for the 6th Circuit affirmed. The state then appealed to the U.S. Supreme Court, and the court agreed to hear the case.

YOU BE THE JUDGE

Do state courthouses have to build elevators or otherwise provide access to courtrooms for citizens in wheelchairs?

HOLDING

Yes.

Justice Stevens wrote the opinion, stating:

The ADA was passed by large majorities in both Houses of Congress after decades of deliberation and investigation into the need for comprehensive legislation to address discrimination against persons with disabilities.

The Eleventh Amendment renders the States immune from "any suit in law or equity, commenced or prosecuted . . . by Citizens of another State, or by Citizens or Subjects of any Foreign State." Our cases have held that Congress may abrogate the States' Eleventh Amendment immunity.

Congress did this in the ADA, where it provided that "A State shall not be immune under the Eleventh Amendment to the Constitution of the United States from an action in Federal or State court of competent jurisdiction for a violation of this chapter. 42 U.S.C. 12202.

With respect to the particular services at issue in this case, Congress learned that many individuals, in many States across the country, were being excluded from the courthouse and court proceedings by reason of their disability. A report before Congress showed that some 76% of public services and programs housed in state-owned buildings were inaccessible to or unusable by persons with disabilities. . . .

LEGAL LESSONS LEARNED

The court did not address whether fire houses must be equipped to accommodate disabled citizens. If the upper floors of the fire station are used by local governmental officials to conduct public meetings, they should read this decision carefully.

CASE STUDY 9-2 Firefighter with Back Injury Offered Light Duty in Police Department

Pflanz v. City of Cincinnati, 149 Ohio App.3d 743, 778 N.E.2d 1073 (Ohio 1st District Court of Appeals (2002)).

FACTS

Paul Pflanz was a firefighter/EMT with the Cincinnati Fire Division from 1973 until July 7, 1995. He injured his back in September 1989, while on an EMS run, when he attempted to catch a woman being transported on a stretcher when suddenly the stretcher latch failed. He was off work under workers compensation and eventually was able to return to work but continued to have back problems. In early 1994, the fire department had him examined by their physician and determined he was no longer qualified to perform the duties of a firefighter. He was therefore eligible for medical separation and retirement under the Ohio Police and Fire Pension fund.

Pflanz requested reasonable accommodation for his back problems, and in May 1994, the fire department put him in a temporary light-duty position. In August 1994, the city's public safety director offered him a light-duty position in the Police Division as a property room technician, but the job paid $10,000 less in salary. He was given until June 18, 1995, to either accept the position or take a medical separation.

On June 19, 1995, Pflanz filed a discrimination charge with the EEOC. The city also extended the time for his decision, at the union's request, until July 7, 1995. He was medically separated on that day.

The EEOC dismissed his charge. In March 2000, he filed a lawsuit in the Ohio Court of Common Pleas in Cincinnati, alleging that the city had failed to accommodate his disability. The court granted the city's motion to dismiss, and he appealed to the Ohio Court of Appeals.

YOU BE THE JUDGE

Was the firefighter "disabled" under ADA?

HOLDING

No.

The three-judge Court of Appeals wrote:

In order to succeed on his **failure-to-accommodate** claim, Pflanz had to show the following: (1) that he was disabled; (2) that his employer was aware of the disability; and (3) that he was an otherwise qualified individual with a disability in that he satisfied the prerequisites for the position and could perform the essential functions of the job with or without accommodation.

In addition, the regulations identify factors that a court should consider when determining whether an individual is substantially limited in a major life activity. Those include "(1) the nature and severity of the impairment; (2) the duration or expected duration of the impairment; and (3) the permanent or long-term impact or the expected permanent or long-term impact of or resulting from the impairment." Section 1630.(2)(j)(2), Title 29 CFR.

Pflanz first contends that his chronic and ongoing back condition substantially limited the major life activity of lifting. Pflanz relies on a note from his treating physician, Thomas R. Sullivan, D.C., dated February 22, 1994, in which Dr. Sullivan stated that Pflanz could perform "no lifting over 10 lbs, no sitting longer than 60 minutes at a time. Same for walking/standing for 4 weeks 3-22-94." The record also contains an undated memorandum entitled "Physician's Limited Duty Work Restrictions," initialed by the city's physician, Dr. Berenson, stating that Pflanz "c[ould] perform limited light lifting and bending," but "for heavy objects he sh[ould] have assistance from on duty personnel."

Pflanz next argues that he was disabled because he suffered from constant pain. Pflanz relies on an unsworn medical report by Steven Skurow, D.C., which was dated April 2, 1996, to support his argument. In his report, Dr. Skurow stated that Pflanz had indicated the following on an "Oswestry Pain Questionnaire: 1. His pain comes and goes and is very severe. 2. Washing and dressing increase the pain and he finds it necessary to change his way of doing [them]. 3. He can lift only very light weights at the most. 4. He cannot walk at all without increasing pain. 5. Pain prevents him from sitting more than 1/2 hour. 6. He cannot stand for longer then [sic] 1/2 hour without increasing pain. 7. Because of pain, his normal night['s] sleep is reduced by less than 3/4. He has hardly any social life because of the pain. * * * 9. Pain restricts all forms of travel. 10. His pain is gradually worsening (when questioned in more detail, he states that the

failure-to-accommodate An employer must reasonably accommodate an employee with a disability if this will assist the employee in performing the essential functions of the job.

pain has been fairly stable over the last several years). He rates his pain at approximately a 6/10 level."

We begin our analysis by noting that "living with pain" is not listed as a major life activity under Ohio Rev. Code 4112.01(A)(13). Further, we have been unable to find any decisions by the Ohio Supreme Court or intermediate Ohio appellate courts that have held that "living with pain" constitutes a major life activity under Ohio Rev. Code 4112.01 (A)(13).

Pflanz also argues that his back condition constituted a disability because this condition substantially limited the **major life activity** of working. With reference again to federal law, ADA regulations provide that when an individual claims that a physical impairment substantially limits the major life activity of working, "[t]he term substantially limits means significantly restricted in the ability to perform either a class of jobs or a broad range of jobs in various classes as compared to the average person having comparable training, skills and abilities. The inability to perform a single, particular job does not constitute a substantial limitation in the major life activity of working."

ADA regulations further provide that courts may consider the following additional factors:

(A) The geographical area to which the individual has reasonable access; (B) The job from which the individual has been disqualified because of an impairment, and the number and types of jobs utilizing similar training, knowledge, skills or abilities, within that geographical area, from which the individual is also disqualified because of the impairment (class of jobs); and/or (C) The job from which the individual has been disqualified because of an impairment, and the number and types of other jobs not utilizing similar training, knowledge, skills or abilities, within that geographical area, from which the individual is also disqualified because of the impairment (broad range of jobs in various classes). Section 1630.2(j)(3)(ii), Title 29, CFR.

Although ADA regulations do not define the terms **"class of jobs"** and "broad range of jobs," the Appendix to Part 1630 provides notable examples of relevant limitations:

"An individual who has a back condition that prevents [him] from performing any heavy labor job would be substantially limited in the major life activity of working because [his] impairment eliminates his * * * ability to perform a class of jobs. This would be so even if the individual were able to perform jobs in another class, e.g., the class of semi-skilled jobs." *Id.* at Appendix to Part 1630.

Because Manges's report stated that Pflanz's back condition limited his ability to work only in the fire-suppression industry and because such positions constituted only a narrow class of jobs, a reasonable factfinder could not have concluded that Pflanz was limited in the major life activity of working. Consequently Pflanz failed to present a genuine issue of material fact as to whether he was substantially limited in the major life activity of working. Thus, Pflanz has failed to demonstrate that he is disabled under Ohio's handicap-discrimination law.

Even if we were to assume that Pflanz was disabled, summary judgment was still appropriate on his failure-to-accommodate claim because Pflanz did not meet the third element of the accommodation test—that is, he did not show that despite his disability he could perform the essential function of his former position with reasonable accommodations.

Ohio law requires an employer to make reasonable accommodations to a disabled employee or applicant, unless the employer can demonstrate that such an accommodation would impose an undue hardship on the conduct of the employer's business. A reasonable accommodation is "a reasonable adjustment made to a job and/or the work environment that enables a qualified disabled person to safely and substantially perform the duties of that position." Reasonable accommodations may include job restructuring, the acquisition or modification of equipment or devices, a realignment of duties, the re-

major life activity ADA requires persons claiming to be disabled to show that their condition prevents them from performing major life activities.

class of jobs ADA and the U.S. Supreme Court require employees to prove that their disability prevents them from performing not only their current job but also a broad range of work.

vision of job descriptions, or a modified part-time work schedule, transfer, reassignment, or hire into a vacant position.

LEGAL LESSONS LEARNED

It is difficult for a line firefighter with a permanent back injury to prove that he or she is "disabled" as defined by ADA and court decisions, and it is very hard to prove that with reasonable accommodation he or she can perform the essential functions of the job. Fire departments should address these issues as a medical question, relying on the department's physician to determine the firefighter's ability to do the job.

CASE STUDY 9-3 Applicant for Recruit Firefighter with 20/100 Vision

Columbus Civil Service Commission v. McGlone, 82 Ohio St.3d 569, 697 N.E.2g 204 (1998); Ohio Supreme Court.

FACTS

James McGlone took the test to become a Columbus firefighter on April 2, 1990. At that time, the application process consisted of a written examination and a physical capability test. Candidates who passed those tests were ranked on an eligible list based on their combined scores. The next phases of the process included an aerial ladder climb, a background review, and a medical examination, which included a vision test. Candidates who failed any portion of the application process could not be considered for appointment to the firefighter training academy.

McGlone was ranked number 156 on the eligible list after the written examination and physical capability test. He then successfully completed the ladder climb and background review portions of the process. However, McGlone failed the vision test portion of his medical examination. The city's visual acuity standard requires a firefighter applicant to have not less than 20/40 vision in both eyes without correction, acuity of not less than 20/20 in both eyes with correction, and normal color vision. McGlone's vision was 20/100 in both eyes without correction. A person with 20/100 vision can see an object from twenty feet only as well as a person with 20/20 vision can see an object at one hundred feet. Because he failed the vision portion of the medical examination, McGlone was removed from the eligible list on June 9, 1992.

On November 17, 1992, McGlone filed a charge with the Ohio Civil Rights Commission (OCRC) alleging that the city had discriminated against him on the basis of a handicap—his visual impairment. The OCRC investigated the charge, issued a complaint, and held a hearing. The OCRC hearing examiner found that the city had discriminated against McGlone on the basis of a perceived handicap and recommended that he be reinstated to the eligible list. The hearing examiner did not recommend any back pay. The OCRC adopted the hearing examiner's finding of discrimination but also awarded back pay and ordered the city to offer McGlone employment as a firefighter.

The city appealed that decision to the Franklin County Common Pleas Court. The common pleas court upheld the discrimination finding but reversed the remedy. The OCRC appealed the remedial portion of the court's decision; the city cross-appealed on the discrimination finding. The appellate court affirmed the finding of discrimination, holding that the city perceived McGlone to be handicapped and removed him from the eligible list because of that handicap, despite the fact that he could safely and substantially perform the essential functions of a firefighter with the reasonable accommodation of being allowed to wear contact lenses while on duty. With respect to the remedy, the appellate court reversed the trial court, holding that the remedy ordered by the OCRC was supported by reliable, probative, and substantial evidence.

The city filed an appeal to the Ohio Supreme Court.

YOU BE THE JUDGE

1. Is this firefighter "disabled"?
2. Does the fire department have an obligation to reasonably accommodate him?

HOLDING

Not disabled, and therefore no obligation to reasonably accommodate.

The seven-member Ohio Supreme Court wrote,

> We hold that a person denied employment because of a physical impairment is not necessarily "handicapped" pursuant to former R.C. 4112.01(A)(13).
>
> To establish a prima facie case of handicap discrimination, the person seeking relief must demonstrate (1) that he or she was handicapped, (2) that an adverse employment action was taken by an employer, at least in part, because the individual was handicapped, and (3) that the person, though handicapped, can safely and substantially perform the essential functions of the job in question.
>
> This case revolves around the first element, *i.e.*, whether McGlone was handicapped. At the time this case arose, the predecessor to the current R.C. 4112.01(A)(B) was in effect, and it defined "handicap" as follows:
>
> "Handicap" means a medically diagnosable, abnormal condition which is expected to continue for a considerable length of time, whether correctable or uncorrectable by good medical practice, which can reasonably be expected to limit the person's functional ability, including, but not limited to, seeing, hearing, thinking, ambulating, climbing, descending, lifting, grasping, sitting, rising, any related function, or any limitation due to weakness and significantly decreased endurance, so that he cannot perform his everyday routine living and working without significantly increased hardship and vulnerability to what are considered the everyday obstacles and hazards encountered by the non-handicapped. 143 Ohio Laws, Part III, 4156.
>
> In the current version of R.C. 4112.01 (A)(13), even if a person is not handicapped, he can gain the protection of handicap discrimination laws if he is "regarded [by an employer] as having a physical or mental impairment." While the "regarded as handicapped" language was not part of the statute when this case arose, the pertinent Administrative Code section in effect at the time, Ohio Adm. Code 4112-5-02(H), included in its definition of a "handicapped person" "any person who is regarded as handicapped by a respondent." We therefore find that it was appropriate for the OCRC and the reviewing courts to consider whether the city perceived McGlone as handicapped.
>
> The question before this court then is whether a person can be foreclosed from a particular job based upon a physical impairment without at the same time being handicapped, or perceived as handicapped, under former R.C. 4112.01(A)(13), and therefore due the protections of the Ohio Civil Rights Act. We find that McGlone was neither handicapped nor perceived as handicapped by the city.
>
> To find that McGlone was handicapped, we would have to conclude that his nearsightedness was a "medically diagnosable, abnormal condition which is expected to continue for a considerable length of time * * * which can reasonably be expected to limit [his] functional ability * * * so that he cannot perform his everyday routine living and working without significantly increased hardship and vulnerability to what are considered the everyday obstacles and hazards encountered by the non-handicapped.

There is no dispute that McGlone's 20/100 vision is a medically diagnosable condition that is expected to continue. Whether that condition limits his functional ability so that he cannot perform his everyday routine living and working without significantly increased hardship is another matter. The record shows that McGlone leads a normal life. The fact that he wears eyeglasses or contact lenses is not a significant hardship. It is a common burden shared by millions, including a majority of this court.

McGlone's nearsightedness has led to one major hardship in his life, his inability to become a firefighter. But the statute speaks in terms of "everyday routine living and working." It is a broad reference to a general quality of life. The handicap discrimination statute was designed to protect those who live with a handicap that significantly affects the way they live their lives on a day-to-day basis.

The federal Americans with Disabilities Act ("ADA") is similar to the Ohio handicap discrimination law. It defines a disability as a "physical or mental impairment that substantially limits one or more of the major life activities of [an] individual." Section 12102(2)(A), Title 42, U.S. Code. We can look to regulations and cases interpreting the federal Act for guidance in our interpretation of Ohio law.

In its interpretation of the ADA, Section 1630.2(j)(3), Title 29, C.F.R. discusses what factors should be considered in determining whether an individual is substantially limited in a major life activity:

With respect to the major life activity of *working*—

(i) The term *substantially limits* means significantly restricted in the ability to perform either a class of jobs or a broad range of jobs in various classes as compared to the average person having comparable training, skills, and abilities. *The inability to perform a single, particular job does not constitute a substantial limitation in the major life activity of working.* (Emphasis added.)

There is no evidence that McGlone's vision disqualified him from a class of jobs or a wide range of jobs. The city merely precluded him from one position, firefighter. In *Bridges v. Bossier* (C.A.5, 1996), 92 F.3d 329, the court held that an applicant who was disqualified from performing firefighting jobs for the city based on a mild form of hemophilia was not disabled under the ADA, since the field of firefighting jobs was too narrow a field to constitute a "class of jobs." We agree with the *Bridges* court's interpretation that the position of firefighter does not constitute a class of jobs, but is merely one job. We further conclude that the inability to perform a single job does not present significantly increased hardship to a person's everyday routine living and working.

Other federal courts have refused to find that nearsightedness constitutes a disability. In *Sutton v. United Air Lines, Inc.,* (C.A. 10, 1997), plaintiffs, twin sisters, were denied employment by United Air Lines for failure to have uncorrected vision of 20/100 or better in each eye. The *Sutton* court found that the impairment did not substantially limit a major life activity, and that the sisters were not disabled.

We conclude that McGlone's 20/100 vision is not a handicap under the statute. His vision problem did not create significantly increased hardship in McGlone's functional ability to perform his everyday living and working.

We further conclude that the city did not perceive McGlone as handicapped. We stated above that the inability to perform a single job because of an abnormal condition does not transform that condition into a handicap. The city in this case considered McGlone nearsighted, not handicapped, merely lacking a single physical requirement for a single job. For McGlone to succeed on a theory of perceived handicap, the city would have had to consider McGlone's nearsightedness as foreclosing him from a class of jobs. There is no evidence that the city had such a perception.

LEGAL LESSONS LEARNED

Poor eyesight is not a disability. Fire departments can require reasonable eyesight for all recruits.

■■■

Chapter Review Questions

1. In Case Study 9-1, *Tennessee v. Lane,* the court addressed the issue of access to public buildings. Many fire departments have meeting rooms on the second floor of the fire station where residents can arrange to hold meetings of public interest. Discuss whether the department has an obligation to make the room accessible to residents in wheelchairs or can simply agree to move the meeting to a ground-floor location.

2. In Case Study 9-2, *Pflanz v. City of Cincinnati,* the court addressed the issue of offering a firefighter a light-duty assignment. Many large fire departments have a light-duty policy for firefighters who are temporarily unable to perform the essential functions of their job; smaller departments do not have light-duty opportunities. Discuss the obligations of employers and employees to enter into discussions about reasonable accommodation opportunities and whether fire departments who do offer light-duty temporary assignments should impose a time limit when the firefighter must either return to full duty, take disability retirement, or otherwise leave the department.

3. In Case Study 9-3, *Columbus Civil Service Commission v. McGlone,* the court addressed the issue of poor eyesight. Discuss what "reasonable accommodations" a fire department can take to assist a firefighter who needs to wear glasses for daily activities, including while wearing SCBAs or driving emergency apparatus.

■■■

Expand Your Learning

Read and complete the individual student or group assignment, as directed by your instructor.

1. A firefighter/paramedic learns he has Type II diabetes, and you are his station captain. You become increasingly concerned about his health and his ability to perform his duties, because he is gaining a lot of weight, not working out with the other on-duty personnel, and you have heard rumors about performance issues on ALS runs. You have requested a meeting with your battalion chief and the fire chief. Prepare a one-page memo on whether the firefighter should be considered disabled under the ADA, what steps the FD should take to confirm that he is healthy enough to run, and what steps the FD might have to take to "reasonably accommodate" him.

2. Read the EEOC's list of recent ADA settlements (go to *www.eeoc.gov,* "Litigation Settlement Monthly Reports"). Identify a recent settlement and the "legal lessons learned" for the fire service.

See Appendix A for additional Expand Your Learning activities related to this chapter.

Family Medical Leave Act (FMLA)

10 CHAPTER

Battalion Chief Vickie Koch, Mason (OH) Fire Department, returns to work after a period of needed leave. She was in New Orleans attending an EMS Conference when Hurricane Katrina hit and was in the Super Dome for three days trying to aid victims with very limited supplies. *Photo by author.*

Key Terms

intermittent leave, p. 125

restored to prior position, p. 127

equivalent position, p. 127

estoppel, p. 128

From the Headlines

Texas Battalion Chief Wins $1.01 Million/Florida Firefighter Wins $1 Million Verdict

A 22-year veteran battalion chief in Arlington (TX) was fired for staying home with his sick wife during the Y2K weekend after all leaves had been cancelled. A federal judge upheld the jury verdict, finding a violation of the Family Medical

Leave Act (FMLA), including $395,000 in back pay and benefits, $300,000 in punitive damages, $305,291 in attorney's fees, and $9,575 in court costs. On June 30, 2006, the U.S. Court of Appeals affirmed the jury's decision, but ordered damages to be reduced since the battalion chief received pension benefits, 2006 106 1793268. In Florida, a federal jury awarded a firefighter with the City of Deerfield Beach $633,000 in back wages and attorney's fees in excess of $300,000. He was placed on forced FMLA sick leave after he made allegations of favoritism at a union meeting. The fire chief allegedly considered the comments as veiled threats and ordered a psychiatric exam. The psychiatrist told the fire department to put him on forced FMLA sick leave for possible "employee terrorism."

Ohio Deputy Sheriff Wins Arbitration—Denied Light Duty after Death of Her Police Officer Husband in Bank Robbery and Requested More Leave after Birth of Her Child

A deputy sheriff filed a grievance, and an arbitrator ordered that she be reinstated with back pay, finding that the sheriff had violated FMLA and wrongly denied the deputy's request for light duty, when she was four months pregnant, and her husband, a Columbus (OH) police officer, was killed in a bank robbery. The sheriff ordered her to continue her full-time duty patrol. After giving birth, the sheriff fired her when she did not return to work promptly.

Newspapers Investigating Alleged Sick Leave Abuse

Newspaper investigations into sick leave records, first in Boston, then Providence, and then Jacksonville (FL) allege widespread abuse. The *Jacksonville Times-Union* examined 69,000 leave requests in 2000–2003 and "estimated" that 30 percent appeared bogus, with sick rate doubling Friday to Sunday, particularly during football season, costing the city over $2 million in paid time off and $1 million in overtime for the replacement firefighters.

Arranging coverage for personnel on leave can be difficult. Fire Chief Rich Fletcher, Mason Fire Department, began an "incentive pay" program for part-time personnel who are able to regularly work daytime shifts. *Photo by author.*

This chapter addresses FMLA issues in the fire service, including efforts by employers to avoid inappropriate use of FMLA leave, including abuse of **intermittent leave**.

On February 5, 1993, President Clinton signed into law the Family and Medical Leave Act of 1993. The FMLA provides up to 12 weeks of unpaid, job-protected leave in a 12-month period to eligible employees for covered family and medical needs. The express purpose of the Act is to help balance the needs of the families with the demands of the workplace. 29 U.S.C. 2601(b).

The FMLA applies to all federal, state, and local government employers, regardless of the number of employees, and also to private employers employing 50 or more employees for each working day during 20 or more calendar workweeks in the current or preceding calendar year.

intermittent leave
Employees may take up to 12 weeks of leave, including leave in which the employee is off for only one day, returns to work, and then takes other days off intermittently.

◆ **DEPARTMENT OF LABOR GUIDE**

On May 22, 2006, the U.S. Department of Labor announced that it revised its "Employment Law Guide" (*www.dol.gov/compliance/guide/index.html*). The guide includes a helpful summary of the FMLA, 29 U.S.C. 2601 et seq., and the department's regulations, 29 CFR 825. The law covers every public agency, state and local, including fire departments, regardless of the number of employees. Private employers who engage in commerce with 50 or more employees are covered.

◆ **ELIGIBILITY UNDER FMLA**

To be eligible for FMLA leave, an employee must have worked at least 12 months (which do not have to be consecutive) for the employer and have worked at least 1,250 hours during the 12 months immediately before the FMLA leave begins.

The FMLA provides that eligible employees have a right to take up to 12 weeks of job-protected leave in any 12-month period for qualifying events without interference or restraint from their employers. In the fire service, such leave can be very disruptive to crews, so departments should encourage employees to provide as much advance notice as possible. A common requirement of many public agencies, including fire and EMS departments, and authorized under the U.S. Department of Labor regulations, is to require personnel first to use their sick leave, "comp" time, and vacation prior to going on unpaid FMLA leave.

Enforcement of the FMLA can be by a complaint filed by the employees with the U.S. Department of Labor, Wage and Hour Division, or by filing a private lawsuit in federal court.

◆ **NFPA STANDARDS**

◆ *NFPA 1582: Standard on Comprehensive Occupational Medical Program for Fire Departments* (2003 edition). This standard establishes a minimum medical condition for firefighters, which a department should consider when determining if a firefighter is fit for duty.

◆ KEY STATUTES

FAMILY AND MEDICAL LEAVE ACT OF 1993 (29 U.S.C. 2601)

Congress established a nationwide program allowing employees to take up to 12 weeks of unpaid leave annually for serious medical conditions for themselves or their immediate family members. Through a U.S. Supreme Court decision, the statute now also applies to state and local governments, including public fire departments.

◆ CASE STUDIES

CASE STUDY 10-1   Employee on Leave for Wife's Illness Decides Not to Retire; Previous Position Already Filled

Sorrell v. Rinker Materials Corporation, 359 F.3rd 332 (6th Cir. January 14, 2005); U.S. Court of Appeals for the 6th Circuit.

FACTS

Charles Sorrell has been an outside salesman for the company for the last 13 years, selling pre-cast concrete to the commercial construction industry. He lived in Dayton (OH), and his job required frequent travel, with a sales territory covering Cincinnati, Clermont County to the east, and northern Kentucky. Three days a week he was also at their Dayton facility.

In November 2000, he advised his supervisor that he planned to retire. Part of his motivation was to care for his wife, who had recently developed an eye disorder, requiring treatments using Prednisone daily. Because of possible dangerous side effects, the eye doctor wanted her to keep the dosage as low as possible. They decided to spend the winter in Florida where they owned a condominium.

Sorrel thought at the time that the FMLA covered only new parents, so he didn't even ask about taking a medical leave of absence. He and his supervisor discussed his coming back after the winter in a part-time assignment and the need to train a permanent replacement. They agreed his last day of work would be December 21, 2000, and with his accrued vacation, the actual termination date would be January 16, 2001.

His replacement was soon announced—Steven Jeffries, from the Indianapolis (IN) facility. During December, Mr. Sorrell trained Jeffries about the sales territory and its customers. On December 20, 2000, Sorrell signed retirement papers.

On December 21, 2000, Sorrell learned that the FMLA might cover this situation, and he informed his supervisor that perhaps he would take FMLA leave instead of retiring. He was told first to submit a doctor's note about his wife's condition. Dr. Oprecak wrote a note stating: "Due to decreasing Prednisone the body increases in stress, and decreases the immune system. Therefore a 3-month leave of absence would help the above."

The supervisor next told Sorrell that he would have to complete certain forms, and because it was so close to the holidays, he could submit the leave forms after the first of the year. The Sorrells left for Florida just after Christmas. In January, after some delays, the necessary forms were sent to them in Florida. Sorrell submitted the completed forms to his supervisor on January 22, 2001. Sorrell's FMLA leave was officially approved soon afterward.

On or about February 19, the supervisor completed a personnel change notification form, which indicated that Sorrell would not be retiring and reinstated him to active employee status, retroactive to his original hire date, so that he would be entitled to take FMLA leave.

Sorrell's FMLA leave was to expire on April 12, 2001. He and his wife returned to Ohio during the first week of April. On or about April 7, before the leave expired, Sorrell called his supervisor and informed him that he was back in town and ready to return to work. The supervisor told him that, unfortunately, senior management had imposed a company-wide "hiring freeze" and that there were no positions available.

Sorrell argued that a hiring freeze should not affect someone returning from leave taken pursuant to the FMLA. His supervisor said he would discuss this with the senior manager and get back to him. Two weeks later, the supervisor called to tell Sorrell that the company had "something" for him. They were willing to reinstate him to an outside sales position, but that he would have to accept a territory covering the southeast quadrant of Indiana.

This new sales territory is at least sixty miles from Sorrell's home and included areas of Indiana over 180 miles away from his home, that would take over three hours of driving time to reach. Sorrell was also told that he would be expected to spend two to three nights per week on the road, which he never had to do while working in his prior territory.

Sorrell met with company managers about this proposed arrangement but explicitly indicated that this was not the territory that he desired. The senior manager explained that his previous position, covering the Cincinnati territory, was not available because it was occupied by Jeffries. Sorrell suggested the possibility of working out a part-time arrangement, but the senior manager said that such an arrangement probably would not work because the company's customers expected salespeople to be available five days per week.

Sorrell declined the new sales territory in Indiana, stating that he wanted his old territory back. The senior manager immediately rejected this proposition but said that he would call Sorrell in a couple of days. No one from the company ever contacted him.

Sorrell filed a lawsuit in federal court in Cincinnati alleging a violation of his rights under the FMLA. In particular, Sorrell claims that the company violated his rights because he was not **restored to his prior position** or to an **equivalent position** upon his return from leave.

The District Court judge granted the company's motion for summary judgment on the grounds that Sorrell had relinquished his outside sales position prior to requesting FMLA leave and, therefore, was not entitled to that position, or an equivalent one, upon his return from leave. Sorrell filed an appeal to the U.S. Court of Appeals to the 6th Circuit.

YOU BE THE JUDGE

1. Was the wife's medical condition a serious health condition under the FMLA?
2. Is the company required to reinstate him immediately to his former sales territory?

HOLDING

Assuming that it is not a serious health condition, the employer having perhaps erroneously awarded FMLA leave must follow the law and reinstate the employee to the previous position or an equivalent position. The case was remanded to a trial judge to hold a hearing.

restored to prior position An employee who takes leave is "entitled, on return from such leave (A) to be restored by the employer to the position of employment held by the employee when the leave commenced; or (B) to be restored to an equivalent position."

equivalent position Under FMLA, when an employee returns from leave, the employee is entitled to his or her old job back immediately or a job that is equivalent.

The three-judge court wrote:

Sorrell's Family and Medical Leave Act claim is based upon what has been described as an "interference" theory. Such a theory, we have explained, arises from [29 U.S.C. Sec. 2615(a)(1)] which states that "[i]t shall be unlawful for any employer to interfere with, restrain, or deny the exercise of or the attempt to exercise, any right provided in this subchapter," and from § 2614(a)(1), which provides that "any eligible employee who takes leave . . . shall be entitled, on return from such leave (A) to be restored by the employer to the position of employment held by the employee when the leave commenced; or (B) to be restored to an equivalent position." In order to prevail on his claim, Sorrell must prove by a preponderance of the evidence that: (1) he was an eligible employee as defined in the Act; (2) Rinker [Materials Corporation] was an employer as defined in the Act; (3) he was entitled to leave for one of the reasons set forth in the Act; (4) he gave notice of his intention to take leave as required by the Act; and (5) Rinker [Materials Corporation] improperly denied him benefits under the Act.

The sole issue raised in Sorrell's appeal challenges the district court's holding that Sorrell was not entitled to his prior position or an equivalent one upon his return from leave because he had already relinquished his outside sales position prior to requesting leave. In its cross-appeal, [the company] argues, among other things, that Sorrell was not entitled to leave because his wife did not have a "serious health condition" as required by the Act and because Sorrell did not "care for" his wife within the meaning of the Act. Sorrell responds that principles of waiver or **estoppel** preclude Rinker from challenging his entitlement to leave. According to Sorrell, because Rinker granted his request for leave without informing him that Dr. Opremcak's medical certification was incomplete, without seeking to clarify the certification and without seeking a second opinion, "it cannot now challenge the sufficiency of the grounds for [his] request for leave." Although Sorrell raised this argument in his memorandum in opposition to Rinker's motion for summary judgment, the district court never explicitly considered or resolved it; instead, the court held as a matter of law that Sorrell was entitled to leave in order to care for his wife, who had a serious health condition.

The court said that on remand the trial judge should conduct a hearing and decide if the company is precluded from contesting Sorrell's eligibility for leave by virtue of its alleged failure to comply with or to avail itself of certain procedures under the Act.

On appeal, Sorrell identifies at least three such procedures. First, Sorrell argues that Rinker failed to advise him that it found Dr. Opremcak's medical certification incomplete or to provide him a reasonable opportunity to cure any deficiency in the certification, as it is required to do under 29 CFR Sec. 825.305(d), (providing that "[t]he employer shall advise an employee whenever the employer finds a certification incomplete, and provide the employee a reasonable opportunity to cure any such deficiency"). Second, Sorrell argues that Rinker failed to insist that a second medical opinion be submitted, as is its right under 29 U.S.C. Sec. 2613 (c)(1) (providing that "[i]n any case in which the employer has reason to doubt the validity" of a medical certification submitted by an employee, "the employer may require, at the expense of the employer, that the eligible employee obtain the opinion of a second health care provider designated or approved by the employer"). Finally, Sorrell argues that Rinker never sought clarification of Dr. Opremcak's medical certification, as it is permitted to do under 29 CFR Sec. 825.307(a), (providing that "a health care provider representing the employer may contact the employee's health care provider, with the employee's permission, for purposes of *clarification* and authenticity of

estoppel Under FMLA, if the employer has granted leave, even though the medical condition did not qualify as a serious health condition, the company may have waived its right to challenge the basis for the leave in court.

the medical certification") [emphasis in original]. According to Sorrell, the fact that Rinker approved his leave without complying with or availing itself of any of these procedures precludes Rinker from now contesting his entitlement to that leave.

Pursuant to 29 CFR Sec. 825.305(d), if Rinker found Dr. Opremcak's medical certification incomplete, it had a duty to advise Sorrell of that fact and to provide Sorrell a reasonable opportunity to cure any such deficiency in the certification. Although we have located no cases from this Court that provide any guidance with respect to this issue, several district courts in other jurisdictions have held that an employer may not assert incompleteness of a medical certification as grounds for disciplining an employee where the employer never notified the employee of the problem or gave him an opportunity to cure it.

Although we offer the foregoing observations as guidance, the applicability and impact of 29 CFR 825.305(d)—along with the other provisions referenced above—is a matter for the district court to consider in the first instance on remand.

LEGAL LESSONS LEARNED

Fire departments, like other employers, must carefully review medical certifications and other "notes from the doctor" and, if they are deficient, advise the employee that the FMLA leave will not be granted without adequate medical documentation.

CASE STUDY 10-2   FMLA—Application to State and Local Governments/U.S. Supreme Court

Nevada Department of Human Resources v. Hibbs, 538 U.S. 1 (2003).

FACTS

Williams Hibbs worked for the Welfare Division of the State of Nevada's Department of Human Resources. In April and May 1997, he sought leave under the FMLA to care for his ailing wife, who was recovering from a car accident and neck surgery. The department granted his request for the full 12 weeks of FMLA leave and authorized him to use the leave intermittently as needed between May and December 1997. Hibbs did so until August 5, 1997, after which he did not return to work. In October 1997, the department informed the respondent that he had exhausted his FMLA leave, that no further leave would be granted, and that he must report to work by November 12, 1997. Hibbs failed to do so and was terminated.

He filed a lawsuit in federal court under the FMLA provisions that allow "eligible employees" to take up to 12 workweeks of unpaid leave annually for any of several reasons, including the onset of a "serious health condition" in an employee's spouse, child, or parent. The Act creates a private right of action to seek both equitable relief and money damages "against any employer (including a public agency) in any Federal or State court of competent jurisdiction," 29 U.S.C. 2617 (a)(2), "should that employer interfere with, restrain, or deny the exercise of FMLA rights."

The District Court judge granted the State of Nevada's motion for summary judgment on the grounds that the FMLA claim was barred by the Eleventh Amendment to the U.S. Constitution. Hibbs appealed to the U.S. Court of Appeals, and the U.S. Department of Justice joined his appeal. A three-judge panel of the U.S. Court of Appeals reversed the trial judge and held that Mr. Hibbs was entitled to proceed with his FMLA lawsuit. The State of Nevada asked the U.S. Supreme Court to hear their appeal, and the court agreed to hear the case.

YOU BE THE JUDGE

Should Mr. Hibbs be entitled to sue the State of Nevada, or does the Eleventh Amendment protect the state?

HOLDING

Yes.

The Supreme Court opinion was written by Chief Justice Rehnquist: "We hold that employees of the State of Nevada may recover money damages in the event of the State's failure to comply with the family-care provision of the Act."

The Chief Justice explained that there has been "a split among the Courts of Appeals on the question whether an individual may sue a State for money damages in federal court for violation of the FMLA."

> For over a century now, we have made clear that the Constitution does not provide for federal jurisdiction over suits against nonconsenting States.
>
> Congress may, however, abrogate such immunity in federal court if it makes its intention to abrogate unmistakably clear in the language of the statute and acts pursuant to a valid exercise of its power under Sec. 5 of the Fourteenth Amendment. The clarity of Congress' intent here is not fairly debatable. The Act enables employees to seek damages "against any employer (including a public agency) in any Federal or State court of competent jurisdiction," 29 U.S.C. 2617 (a)(2) and Congress has defined "public agency" to include both "the government of a State or political subdivision thereof" and "any agency of . . . a State, or a political subdivision of a State." Sec 203 (x), Sec. 2611(4)(A)(iii).
>
> This case turns, then, on whether Congress acted within its constitutional authority when it sought to abrogate the States' immunity for purposes of the FMLA's family-leave provision. In enacting the FMLA, Congress relied on two of the powers vested in it by the Constitution: its Article I commerce power and its power under Sec. 5 of the Fourteenth Amendment to enforce that Amendment's guarantees. Congress may not abrogate the States' sovereign immunity pursuant to its Article I power over commerce. Congress may, however, abrogate States' sovereign immunity through a valid exercise of its Sec. 5 power, for the Eleventh Amendment, and the principle of state sovereignty which it embodies, are necessarily limited by the enforcement provisions of Sec. 5 of the Fourteenth Amendment.
>
> The text of the Act makes this clear. Congress found that, "due to the nature of the roles of men and women in our society, the primary responsibility for family caretaking often falls on women, and such responsibility affects the working lives of women more than it affects the working lives of men." 29 U.S.C. Sec. 2601 (a)(5). In response to this finding, Congress sought "to accomplish the [Act's other] purposes . . . in a manner that . . . minimizes the potential for employment discrimination *on the basis of sex* by ensuring generally that leave is available . . . *on a gender-neutral basis*[,] and to promote the goal of equal employment opportunity for women and men." Sec. 2601 (b)(4) and (5) [emphasis added].
>
> As the FMLA's legislative record reflects, a 1990 Bureau of Labor Statistics (BLS) survey stated that 37 percent of surveyed private-sector employees were covered by maternity leave policies, while only 18 percent were covered by paternity leave policies. S. Rep. No. 103-3, pp. 14-15 (1993), U.S. Code Cong. & Admin. News 1993, p. 3. The corresponding numbers from a similar BLS survey the previous year were 33 percent and 16 percent, respectively. *Ibid.* While these data show an increase in the percentage of employees eligible for such leave, they also show a widening of the gender gap during the same period. Thus, stereotype-based beliefs about the allocation of family duties

remained firmly rooted, and employers' reliance on them in establishing discriminatory leave policies remained widespread.

Congress also heard testimony that "[p]arental leave for fathers . . . is rare. Even . . . [w]here child-care leave policies do exist, men, *both in the public and private sectors,* receive notoriously discriminatory treatment in their requests for such leave." Joint Hearing 147 (Washington Council of Lawyers) [emphasis added]. Many States offered women extended "maternity" leave that far exceeded the typical 4- to 8-week period of physical disability due to pregnancy and childbirth, but very few States granted men a parallel benefit: Fifteen States provided women up to one year of extended maternity leave, while only four provided men with the same. M. Lord & M. King, The State Reference Guide to Work-Family Programs for State Employees 30 (1991). This and other differential leave policies were not attributable to any differential physical needs of men and women, but rather to the pervasive sex-role stereotype that caring for family members is women's work.

LEGAL LESSONS LEARNED

The FMLA applies to fire departments and other emergency responders. Violation of the law can result in costly litigation, including damages and attorney fees.

CASE STUDY 10-3   FMLA—Returning Employee Entitled to Prompt Reinstatement to Previous Job

Hoge v. Honda of America Mfg. Inc., No. 03-3452 (6th Cir. September 16, 2004); U.S. Court of Appeals in Cincinnati, OH.

FACTS

This case concerns the timing and nature of an employee's right to job restoration under the FMLA.

In November 1995, Plaintiff Lori Hoge, a production associate at Honda's East Liberty (OH) plant, sustained a back injury in a non-work-related car accident. She was hospitalized, took an extended leave of absence from her job, and returned to work in March 1996. Her injury, a fracture of a lumbar vertebra, imposed several permanent physical restrictions on her work activities. Plaintiff's permanent work restrictions included: no jumping in or out of cars; no lower back extensions in excess of 15 degrees; no lower back flexion in excess of 30 degrees; no pushing or pulling liner racks; no lifting of more than 15 pounds; and a 40-hour workweek limitation. After her back injury, Hoge returned to work on the "door line," a position that accommodated her physical restrictions. She worked on the door line, taking intermittent FMLA leave for her back injury, until she took the approved FMLA leave leading to the instant dispute.

On April 20, 2000, Honda approved Hoge's request for continuous FMLA leave from May 11 until June 12, 2000 for abdominal surgery unrelated to her back injury. On or about June 12, 2000, Hoge telephoned Honda to request an extension of her FMLA leave, informing Honda that she would need additional time to recover from her abdominal surgery.

Although the parties agree that Honda approved two requested extensions of FMLA leave beyond Hoge's original June 12 expected return date, they dispute the date of her anticipated return. She appeared and attempted to return to work on June 27, 2000, but Honda had no position open for her until July 31, 2000.

The evidence of her expected return date is in dispute. In a letter dated June 28, 2000, Honda approved a continuous FMLA leave extension "beginning on 6/12/00 and ending on 7/19/00." The letter stated that the company "expected [Plaintiff] to return to work at the beginning of [her] shift on 7/20/00." However, another letter dated June 30, signed by a different representative of Honda's Leave Coordination Department, Mr. Lippencott, approved continuous FMLA leave for Hoge from June 26, 2000 (the day before she attempted to return to work) until July 12, 2000, with her expected return on July 13, 2000.

Mr. Lippencott signed a third letter sent to Hoge, also dated June 30, 2000, which approved continuous "medical leave" from July 13 until December 31, 2000. These documents reveal that Honda approved Hoge's absence from work as FMLA leave for the period between June 12 and June 27, 2000. The letters also suggest that Honda did not expect her to return to work on the morning of June 27.

On the other hand, Hoge's affidavit that she filed with the trial judge states that she did not request FMLA leave beyond June 26, 2000. Further, a leave of absence extension request, dated June 19, 2000, and approved by Honda on June 28, 2000, establishes June 26, 2000 as the return date for Hoge.

During her FMLA leave, Honda continued instituting a "new model changeover" that included multiple engineering and stylistic changes for the production of its year 2000 models. The model changeover directly affected Honda's assembly department where Hoge worked and was gradually implemented between February 8 and August 15, 2000. After obtaining a release from her treating physician, Dr. Ronald Spier, Hoge appeared for work on June 27, 2000, expecting to return to her door line position. Upon her return, she presented to Honda's medical department a "Physician's Permit," which stated that she was able to return to her previous position on the door line. Hoge returned with the same physical restrictions associated with her back injury that she had before taking leave. She expected to be placed in a position that accommodated those restrictions. The medical department contacted Brett Strine, the person responsible for placing Hoge. Mr. Strine considered possible placements in light of the ongoing model changeover and staffing levels but informed Hoge that no positions were available.

Honda then conducted a placement review but did not find a suitable position for Hoge until July 26, 2000. She eventually returned to a position on the engine line on July 31, 2000. The company claims that the delay in finding an equivalent position was reasonable and was caused by several factors including Hoge's unexpected return and the time required to locate an equivalent position to accommodate her physical restrictions in light of the substantial changes made to its production processes. Her restoration to a position on the engine line was accomplished in accordance with a Gradual Return to Work (GRTW) program. Under this program, Hoge worked less than full time, and her weekly hours increased over six weeks. Honda did not restore Hoge to a full-time work schedule of 40 hours per week until September 18, 2000.

She filed a lawsuit in federal district court, and the trial judge held that she was entitled to be restored to her former position or an equivalent position with Honda within *one day* of her notifying them of her availability to return to work, June 28, 2000. The trial judge ruled that Honda violated the FMLA when it failed to return Hoge to an equivalent position until July 31, 2000. The judge subsequently awarded Hoge monetary damages, attorneys' fees, and court costs.

Honda has filed an appeal to the U.S. Court of Appeals. They argue that the FMLA required Honda to reinstate Hoge to her employment position or an equivalent position only within a "reasonable time," not immediately, and that there is no issue of fact that Honda did so in this case. Honda claims that Hoge's physical limitations, her unanticipated return, and the significant changes made by Honda to its production processes during a model changeover reasonably prevented Honda from restoring her to work until July 31, 2000.

YOU BE THE JUDGE

Can an employer delay reinstatement for a reasonable period of time because they are not yet ready for the employee to return?

HOLDING

An employer must reinstate the employee within one day to her previous job or an equivalent position.

The three-judge decision began with a review of the FMLA statute:

> 29 U.S.C. § 2614(a) describes the FMLA restoration right. It provides, in relevant part:
> (a) Restoration to position
> (1) In general
> Except as provided in subsection (b) of this section, any eligible employee who takes leave under section 2612 of this title for the intended purpose of the leave shall be entitled, *on return from such leave—*
> (A) to be restored by the employer to the position of employment held by the employee when the leave commenced; or
> (B) to be restored to an equivalent position with equivalent employment benefits, pay, and other terms and conditions of employment.
>
> (3) Limitations
> Nothing in this section shall be construed to entitle any restored employee to—
>
> (B) any right, benefit, or position of employment other than any right, benefit, or position to which the employee would have been entitled had the employee not taken the leave. [emphasis added].

The 6th Circuit then reviewed the definition of equivalent position in the U.S. Department of Labor regulations. An equivalent position is "one that is virtually identical to the employee's former position in terms of pay, benefits and working conditions, including privileges, perquisites and status. It must involve the same or substantially similar duties and responsibilities, which must entail substantially equivalent skill, effort, responsibility, and authority." 29 CFR 825.215(a).

> The Secretary of Labor has also promulgated a regulation describing employee rights on returning from FMLA leave. 29 CFR 825.214. It provides:
> (a) On return from FMLA leave, an employee is entitled to be returned to the same position the employee held when leave commenced, *or to an equivalent position with equivalent benefits, pay, and other terms and conditions of employment. An employee is entitled to such reinstatement even if the employee has been replaced or his or her position has been restructured to accommodate the employee's absence.* See also § 825.106(e) for the obligations of joint employers.
> (b) If the employee *is unable to perform an essential function of the position because of a physical or mental condition, including the continuation of a serious health condition, the employee has no right to restoration to another position under the FMLA.* However, the employer's obligations may be governed by the Americans with Disabilities Act (ADA). See § 825.702.
> *Id.* [emphases added].

The court concluded with the following,

The right to restoration arises when the employee is able to perform the essential functions of the position he left or an equivalent. If an employee can do so and has provided medical certification (if required by a uniform policy), the employer cannot simply delay restoration while it takes a reasonable amount of time to find a suitable position. Again, a "reasonable" delay in restoration after reasonable notice is given would force the employee to take more FMLA leave than is required and would interfere with an employee's exercise of FMLA rights.

LEGAL LESSONS LEARNED

In the fire service, when the firefighter is ready to return to work, the returning employee should be returned to their previous station and their previous shift.

Chapter Review Questions

1. In Case Study 10-1, *Sorrell v. Rinker Materials Corporation,* the court addressed the issue of reinstating the employee to his previous position. Discuss what a fire department should do when a female firefighter/paramedic, who has been on the department for only two years, returns to work after FMLA leave for pregnancy. Because of the recent retirement of her very experienced crew partner on "C shift" and his replacement by a probationary member, there is no current opening for her on the "C shift."

2. In Case Study 10-2, *Nevada Department of Human Resources v. Hibbs,* the court held that the FMLA applies to state and local governments. Discuss how many employees must be with a private company before the FMLA applies to that private employer (see 29 CFR 825.104; *www.gpoaccess.gov/cfr/index.html*) and discuss whether that minimum number of employees applies to public fire departments.

3. In Case Study 10-3, *Hoge v. Honda of America Mfg. Inc.,* the court addressed how quickly an employer must reinstate an employee returning from FMLA leave to his or her previous position. In the fire service, it is particularly important that the department knows when an employee is returning from leave so that there is always coverage for that position, and the minimum manning requirements are maintained. Discuss whether a department should have an SOP or written policy requiring employees to provide notice of their expected date of return. Obtain a copy of such a policy from a fire department in your state.

Expand Your Learning

Read and complete the individual student or group assignment, as directed by your instructor.

1. Many employers have viewed the FMLA as being overused and even abused by employees, particularly the intermittent leave provisions. Some public employers, including fire departments, require employees who take FMLA leave to use their sick leave or vacation first. Only when these days are exhausted is the employee allowed to go on unpaid FMLA. The U.S. Department of Labor, Wage and Hour Division,

administers the FMLA and has issued regulations that are published in the Code of Federal Regulations. (Go to *www.dol.gov* and click on "Family Medical Leave Act.")

Discuss whether fire departments should require employees to use their sick time, "comp" time, and vacation first prior to going on unpaid FMLA leave. Identify the CFR regulation that allows this practice and share with the class an SOP from a department in your state that addresses this issue.

2. The U.S. Department of Labor, Wage and Hour Division, has issued "model" forms for employers and employees (go to *www.dol.gov* and click on "Family and Medical Leave Act"). See Form WH-380 (March 1995), "Certificate of Health Care Provider," and Form WH-381 (March 1995), "Employer Response to Employee Request for FMLA Leave," 29 CFR 825 (Appendix B). Appendix C is the "Notice to Employee of Rights under FMLA," which should be posted by every employer. Discuss why it is useful for fire departments, in addition to having posted the notice, to also require employees to use Form WH-380, or an equivalent, when requesting FMLA leave and for the department to use WH-381 or an equivalent when reviewing employee FMLA leave requests. See also the "Compliance Guide to the Family and Medical Leave Act." Discuss why it is important that the family members of the requesting employees complete a written application for leave, supported by a doctor's certificate (using a form similar to WH-380), and that the fire department provides a written response (using a form similar to WH-381).

See Appendix A for additional Expand Your Learning activities related to this chapter.

Module III: Operational and Managerial Issues

CHAPTER 11

Fair Labor Standards Act

Kings Island Disaster Drill—Mason Quint at King's Island Amusement Park. *Photo by author.*

Key Terms

bifurcated trial, p. 141   **class action lawsuit, p. 143**
good faith, p. 141   **amicus curiae, p. 145**

From the Headlines

Louisville Owes Millions

A Jefferson County (KY) judge has ruled that the Louisville Metro government owes firefighters millions of dollars in overtime because officials calculated the pay incorrectly for years. The suit filed on behalf of 700 current and former firefighters alleged that from 1995 until 2001 the officials failed to calculate overtime using their total salaries (including salary supplements and clothing stipends) but instead merely used their base pay. Plaintiffs' counsel estimates that $10 to $35 million is owed.

Houston—$79.5 Million Settlement

Houston City Council approved a $79.5 million settlement of 2,600 paramedics' nine-year legal battle to collect overtime. The city claimed that they were paid

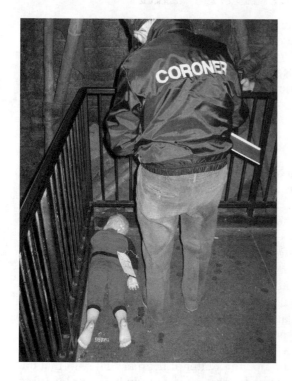

At the Kings Island drill, a triage unit was set up. The stretcher patients were removed from rides first and then the walking wounded. The coroner tags a "victim." *Photo by author.*

overtime the same as firefighters, but the paramedics successfully proved that they were entitled to overtime after a 40-hour workweek. The federal lawsuit was dismissed by the federal District Court judge but reversed by the U.S. Court of Appeals. (Congress in 1999 changed the FLSA, and paramedics/EMTs who are cross-trained in a fire department are now classified as firefighters under the FLSA.)

Pittsburgh—Some of the Highest Paid Employees Are Firefighters

A newspaper study found that 42 of the city's 50 highest paid employees are firefighters. In some cases, overtime doubled their base pay, so some firefighters were making in excess of $100,000 a year.

◆ INTRODUCTION

This chapter discusses the Fair Labor Standards Act (FLSA) and overtime pay for firefighters and EMS personnel. There has been substantial litigation in the fire service in recent years concerning computation of overtime pay, with millions of dollars being awarded.

◆ U.S. DEPARTMENT OF LABOR REGULATIONS

Employers must carefully read the U.S. Department of Labor regulations and interpretative opinion letters to avoid "willful" violations that subject the employer to three years of back pay plus interest, liquidated damages (doubling of back pay), and paying the plaintiffs' attorney's fees.

On May 22, 2006, the U.S. Department of Labor issued its revised "Employment Law Guide" (*www.dol.gov/compliance/guide/index.html*), including its helpful discussion of the FLSA of 1938, as amended (29 U.S.C. 201 et seq.) and enforcement regulations (29 CFR Parts 570 to 580). Refer to the sections on minimum wage and overtime pay and child labor (non-agricultural).

◆ SICK LEAVE BUY-BACK

Case law continues to expand the reach of the FLSA in the fire service. For example, on February 8, 2006, in *Chris N. Acton et al. v. City of Columbia, MO,* ___ F.3d___ (8th Cir. 2006), 2006 U.S. App. LEXIS 3005, the court held that the city's "sick leave buy-back program" should be included in determining the firefighter's "regular rate of pay" under 29 U.S.C. 207(e). Under the city's sick leave buy-back program, firefighters who work a 24-hour shift accumulate ten days of sick leave each year. They can "sell back" all unused sick leave days at 75 percent of their regular hourly pay, after they have worked enough years to have accumulated at least six months of sick leave reserve. The court concluded that this buy-back should be counted as remuneration and made reference to the Department of Labor regulation under 29 CFR 778.223 (monies paid to remain on-call, while not related to any specific hours of work, are nonetheless awarded as "compensation for performing a duty involved in the employee's job"). The court also referred to the Department of Labor Opinion Letter of February 24, 1986, that under 29 CFR 778.211, attendance bonuses and other bonuses to induce employees to work steadily or efficiently must be included in the calculation of an employee's regular rate of pay. The court denied, however, the firefighter's claim that the city's per diem meal allotment program should be included in calculating their regular rate of pay, because the Department of Labor regulations supports the city's exclusion of this item.

◆ STATE OVERTIME LAWS

State laws also impact fire departments. For example, based on a Texas statute, on August 23, 2005, in the *City of Houston v. Steve Williams et al.,* 183 S.W.3d 409, 2005 Texas App. LEXIS 7275, the Court of Appeals of Texas held that the city must pay numerous retired firefighters "their full salary for unused vacation or sick pay."

SALARIED/EXEMPT

Another area of litigation concerns classification of senior fire officers as "salaried, exempt" from receiving overtime pay. Under the FLSA, certain employees are exempt from both the minimum wage and overtime requirements, including "executive, administrative, and professional employees."

The Department of Labor issued new regulations on January 12, 2006, clarifying that senior executives in the public section may fall within the "General Rule for Executive Employers," 29 CFR 541.100, even if they do not have the power to hire and fire employees. In the public sector, the power to hire and fire is often reserved to elected officials, such as a mayor, city council, or township trustees. The new regulations broaden the definition of "exempt" executives to include public sector employees who "recommend" that employees be hired or fired. 29 CFR 541.105. The Department of Labor explained that they revised these regulations in response to the comments received on the draft rules from public sector employers "such as the Metropolitan Transport Authority, the New York State Police," which indicated that "the requirement in the proposal that an employee have the authority to hire or fire will cause many exempt employees to lose exempt status since employees in the public sector do not have authority to make such decisions." See *www.dol.gov/esa* (ESA Final Rule, Vol. 69, Number 79, pages 22121-22274).

Private sector lawsuits on "exempt" supervisors are also providing advance notice to fire departments. For example, in *Borda v. Sandusk Ltd.,* April 28, 2006, the Ohio Court of Appeals for the 6th District, 2006-Ohio-2112, affirmed a trial court decision finding an employer violated both Ohio law (Rev. Code 4111.03) and the FLSA. The suit was filed by two supervisors in a vinyl production plant, each managing about 30 employees, who were classified by their employer as "salaried" ($32,000 per year) and "exempt" from overtime pay of time-and-a-half. The company required them to work substantial overtime, for which they were paid straight time, based on an hourly rate computed from their $32,000 salary. The court held that because their average total compensation *far exceeded* their base salary, they were improperly classified as exempt. The court held that the employer should have known this was improper, by reading 29 CFR 541.118(a), and the applicable Wage and Hour Opinion Letters. The company, therefore, was ordered to pay the plaintiffs unpaid overtime (willful violations go back three years; non-willful go back two years), plus interest, "liquidated damages" under the FLSA of double the back pay, and attorney's fees.

◆ JUNIOR FIREFIGHTERS—CHILD LABOR LAWS

Another area of the law that fire departments with "junior firefighter" programs should review is the child labor provisions of the FLSA (*www.dol.gov/compliance*). For example, there are 17 hazardous occupations that the Secretary of Labor has declared are too dangerous for those under age 18 to perform. Employers are subject to

Richard L. Johnson teaching an Incident Safety Officer class. He is the former training supervisor at the Ohio Fire Academy. When full-time firefighters attended week-long classes at the Academy, it was common for their departments to switch them from their normal 24/48-hour schedule to a 40-hour workweek. *Photo by author.*

penalties of up to $11,000 per worker for each violation, criminal fines of up to $10,000 for willful violations, and imprisonment up to six months for two or more convictions. State child welfare laws should also be reviewed by fire departments.

Many states also have state statutes that enforce overtime laws, and lawsuits may be filed in state court alleging violation of both state law and the FLSA.

◆ KEY STATUTES

FAIR LABOR STANDARDS ACT OF 1938 (29 U.S.C. 201–219)

Congress passed the FLSA in 1938, during the Depression, to encourage private employers to hire more employees. The FLSA, as originally enacted, required employers to pay their employees a minimum wage of not less than 25 cents an hour and prohibited the employment of any person for workweeks in excess of 40 hours after the second year following the legislation "unless such employee receives compensation for his employment in excess of [40] hours . . . at a rate not less than one and one-half times the regular rate at which he is employed." In recent years, there has been substantial litigation in the fire service under this statute, including its "liquidated damages" provision under which the court can double the damages for unpaid overtime compensation under 29 U.S.C. 216(b). Significantly, the court can also award attorney's fees for the victorious plaintiff, to be paid by the defendant employer.

FLSA AMENDMENT APPLICABLE TO EMS [29 U.S.C. 203(y)] (DECEMBER 9, 1999)

In response to federal court decisions holding that paramedics and EMTs were entitled to overtime pay after 40 hours, Congress passed, and President Clinton signed into law, this amendment whereby paramedics and EMTs employed at a fire department will earn the same overtime as firefighters, if they are trained in fire suppression and employed by a fire department of a municipality or other political subdivision.

FLSA LAW ENFORCEMENT AND FIRE PROTECTION PROVISION [SECTION 207(k); 29 U.S.C. 207(k)]

This statute provides an exception for law enforcement and fire protection employees in which, instead of earning overtime after 40 hours in a workweek, they have special rules. For fire protection employees, overtime is earned after working 212 hours in a 28-day period if paid monthly, or 53 hours in a workweek if paid weekly.

◆ CASE STUDIES

CASE STUDY 11-1 Paramedics Who Are "Cross-Trained" but Do Not Perform Fire Suppression Duties

Cedrick Cleveland et al. v. City of Los Angeles, ____F.3d ___ (9th Cir. August 22, 2005); U.S. Court of Appeals for the 9th Circuit in San Francisco. On February 21, 2006, the U.S. Supreme Court declined the city's request to issue a writ of certiorari and hear their appeal, No. 05-851 (S. Ct. 2/21/06) (S. Ct. 2006).

FACTS

A lawsuit was filed claiming a violation of the FLSA by 119 cross-trained fire-fighter/paramedics employed by the City of Los Angeles, claiming they should be paid for overtime after 40 hours in a workweek when they are assigned to ambulances (not while assigned as firefighters). The lawsuit was filed in 1999 in federal District Court in Los Angeles, three months before Congress amended the FLSA on December 9, 1999, to include cross-trained EMS in the definition of fire suppression employees.

The parties to the lawsuit agreed to a bench trial (no jury) that was a **bifurcated trial** (the first trial phase is to determine if the city is liable; if so, then a second phase determines damages). The federal District Judge found that the cross-trained plaintiffs were not "employees in fire protection activities" because (1) they did not have the responsibility to engage in fire prevention, control, or extinguishment, and (2) they were not regularly dispatched to fire scenes, and (3) their nonexempt work as paramedics was not limited to less than 20 percent of their total work hours. During the liability phase of the trial, the federal judge determined that the city did not act reasonably or in **good faith** and ordered the City of Los Angeles to pay unpaid overtime, plus liquidated damages (doubled the unpaid overtime) for a total of $5,131,514.02, $116,500.00 in attorney's fees, and $11,389.12 in costs. The city filed an appeal to the U.S. Court of Appeals.

The trial judge's findings were based on the following evidence. Plaintiffs work a rotating schedule that results in nine scheduled 24-hour shifts every 27 days, for a total of 216 hours worked per pay period. The 119 cross-trained firefighter/paramedics were each assigned to work on advanced life support (ALS) "paramedic ambulances."

Some of the ambulances have two cross-trained paramedics, others have just one who is cross-trained, and others have two "single-function" paramedics. The paramedic ambulances are not dispatched to every fire call, only when there appears to be a need for ALS services. They do not assist in fire suppression.

Plaintiffs testified that they are sent to fire scenes an average of one or two times a year and only to perform emergency medical services. Dual-function paramedics can offer their services as firefighters at a fire scene, but there is no discipline for failure to do this.

The City of Los Angeles does not dispatch dual-trained ambulances for swift water rescue, cliff rescues, or underground tunnel rescues—such rescues are performed by fire suppression personnel assigned to rescue companies. Because dual-trained paramedics are so busy with motor vehicle accidents and normal EMS runs, they often miss required fire suppression training at their fire stations.

The City of Los Angeles pays their cross-trained paramedics for overtime, the same as firefighters, after 212 hours in a 28-day period. This lawsuit was filed prior to Congress amending the FLSA to include cross-trained EMS personnel in the definition of fire suppression employees.

The Department of Labor regulations in effect prior to the 1999 amendment to the FLSA included a four-part test of which employees were considered "fire suppression" personnel. Regulation 29 CFR 553.210(a) stated in pertinent part that a "fire suppression" employee was someone (1) employed by an organized fire department, (2) who has been trained in fire suppression, (3) who has the legal authority "and responsibility to engage in fire protection, control or extinguishment of a fire of any type," and (4) who performs activities concerning fire suppression and incidental functions. The regulation concludes with the following sentence, "The term would also include rescue and ambulance service personnel if such personnel form an integral part of the public agency's fire protection activities. See Sec. 553.215."

bifurcated trial   In lawsuits in federal court under the FLSA, the trial judge may ask the jury in the first phase of the trial to determine if the employer violated the FLSA. In the second phase, the jury will be asked to determine back pay and other damages for each plaintiff.

good faith   Under the FLSA, if the trial judge determines that the employer failed to review carefully the federal regulations and U.S. Department of Labor opinion letters on overtime pay, plaintiffs may also be awarded "liquidated damages" (the back pay award is doubled).

Section 553.215 states that ambulance and rescue personnel who work for a public agency that is not a fire department may also be treated as employees engaged in fire protection if they are "regularly engaged in fire protection."

The regulations also provide a 20 percent rule. Under 553.212, an employee who spends more than 20 percent of his or her time engaged in nonexempt activity ("activities not related to fire suppression") may not be considered to be an "employee engaged in fire protection." 29 CFR 553.212(a).

YOU BE THE JUDGE

1. Should the 119 cross-trained firefighters/paramedics be paid overtime after 40 hours?
2. Does the congressional amendment to the FLSA on December 9, 1999, Sec. 203(y), apply retroactively as clear congressional intent that cross-trained paramedics should be paid the same overtime as firefighters?

HOLDING

The plaintiffs keep the $5 million. The new definition is not retroactive.

The three-judge panel of the 9th Circuit Court wrote: "The FLSA is construed liberally in favor of employees; exemptions are to be narrowly construed against employers seeking to assert them."

The court said that when interpreting a statute or regulation, courts are to give plain meaning to undefined terms.

> Applying this ordinary, common-sense meaning, for Plaintiffs to have the "responsibility" to engage in fire suppression, they must have some real obligation or duty to do so. If a fire occurs, it must be their job to deal with it. The undisputed evidence shows that, among other things:
>
> (1) the paramedic ambulances do not carry fire-fighting equipment or breathing apparatuses;
> (2) a dispatcher does not know if he or she is sending single or dual function paramedics to a call;
> (3) paramedic ambulances are not regularly dispatched to fire scene and are dispatched only when there appears to be a need for advanced life support medical services;
> (4) dual function paramedics are not expected to wear protective gear;
> (5) dual function paramedics are dispatched to a variety of incidents (e.g., vehicle accidents and crime scene) at which they are expected to perform only medical services;
> (6) there is no evidence that a dual function paramedic has ever been ordered to perform fire suppression.

> The City contends that the definition of Sec. 203(y) supersedes any DOL language as the exclusive definition. Further, the City contends that the definition should be applied retroactively and prospectively because Congress' enactment of Sec. 203(y) constitutes "clarification" rather than amendment. However, because Plaintiffs do not have the "responsibility" to engage in fire suppression and consequently do not qualify under either definition, there is no need to discuss the temporal scope of Sec. 203(y).

LEGAL LESSONS LEARNED

Fire departments with cross-trained paramedics or EMTs must read 29 U.S.C. 203(y) carefully to ensure that they pay overtime consistent with the FLSA. Departments that provide EMS-only services are not exempted under the revised statute. They must pay their personnel overtime after 40 hours of work.

CASE STUDY 11-2   U.S. Supreme Court Decides When "Get Ready" Time Is on the Clock

IBP, Inc. v. Alvarez, 546 U.S. ___ (November 8, 2005); 03-1238 (S. Ct. 5/3/2004) (*www.supremecourtus.gov*).

FACTS

IBP, Inc., is a large producer of fresh beef, pork, and related meat products. At its plant in Pasco (WA) it employs approximately 178 workers in 113 job classifications in the slaughter division and 800 line workers in 145 job classifications in the processing division. All production workers in both divisions must wear outer garments, hardhats, hairnets, earplugs, gloves, sleeves, aprons, leggings, and boots. Many of them, particularly those who use knives, must also wear a variety of protective equipment for their hands, arms, torsos, and legs. This gear includes chain-link metal aprons, vests, Plexiglas armguards, and special gloves. IBP also requires its employees to store their equipment and tools in company locker rooms, where most of them don their protective gear.

Production workers' pay is based on the time spent cutting and bagging meat. Pay begins when the first piece of meat is cut and ends when the last piece of meat is bagged. Since 1998, however, IBP has also paid for four minutes of clothes-changing time.

In 1999, IBP employees filed a **class action lawsuit** in federal court to recover compensation for preproduction and postproduction work, including the time spent donning and doffing protective gear and walking between the locker rooms and the production floor before and after their assigned shifts.

class action lawsuit   This is a lawsuit filed on behalf of all similarly situated employees, such as all paramedics on a fire department who, prior to the 1999 change in the FLSA, did not receive overtime after 40 hours in a workweek.

The federal District Court judge, after a lengthy bench trial (no jury), held that donning and doffing of protective gear that was unique to the jobs at issue were compensable under the FLSA because they were integral and indispensable to the work of the employees who wore such equipment. Moreover, consistent with the continuous workday rule, the District Court concluded that, for those employees required to don and doff unique protective gear, the walking time between the locker room and the production floor was also compensable because it occurs during the workday. The trial judge did not, however, allow any recovery for ordinary clothes-changing and washing or for the donning and doffing of hardhats, ear plugs, safety glasses, boots, or hairnets.

The trial judge determined that employees in various job classifications were entitled to additional pay. For example, employees who were classified as "Processing Division knife users" were entitled to compensation for between 12 and 14 minutes of pre-production and post-production work, including 3.3 to 4.4 minutes of walking time.

The company filed an appeal to the U.S. Court of Appeals. A three-judge panel agreed with the District Court's ultimate conclusions on these issues but, in part, for different reasons. The Court of Appeals endorsed the distinction between the burdensome donning and doffing of elaborate protective gear, on the one hand, and the time spent donning and doffing non-unique gear such as hardhats and safety goggles, on the other. It did so because, in the context of this case, the time employees spent donning and doffing non-unique protective gear was "de minimis as a matter of law."

The company appealed to the U.S. Supreme Court, and the court agreed to take the case.

YOU BE THE JUDGE

When these employees put on protective gear such as chain-link metal aprons, vests, Plexiglas armguards, and special gloves, should they be considered "on the clock"?

HOLDING

Time spent putting on special protective gear is on the clock, as is their time walking to the production line. Time spent putting on normal protective gear, like hardhats, is not on the clock.

Justice Stevens wrote the opinion for a unanimous (nine to zero) court,

> IBP does not challenge the holding below that, in light of [prior case precedence] the donning and doffing of unique protective gear is [on the clock]. Thus, the only question for us to decide is whether the Court of Appeals correctly rejected IBP's contention that the walking between the locker rooms and the production areas is excluded from FLSA. . . .
>
> IBP argues that the [Portal to Portal Act of 1947, which exempted time employees walk from their locker to the job site from "on the clock" time], the history and purpose of its enactment, and the Department of Labor's interpretive guidance compel the conclusion that the Portal-to-Portal Act excludes this walking time from the scope of the FLSA. We find each of these arguments unpersuasive.
>
> IBP correctly points out that our decision in *Steiner* [*v. Mitchell,* 350 U.S. 247], held only that the donning and doffing of protective gear in that case were activities "integral and indispensable" to the workers' principal activity of making batteries. 350 U. S., at 256.
>
> In other words, IBP argues that, even though the court below concluded that donning and doffing of unique protective gear are "integral and indispensable" to the employees' principal activity, this means only that the donning and doffing of such gear are themselves covered by the FLSA. According to IBP, the donning is not a "principal activity" that starts the workday, and the walking that occurs immediately after donning and immediately before doffing is not compensable.
>
> The regulations adopted by the Secretary of Labor in 1947 support [the plaintiff employees'] view that when donning and doffing of protective gear are compensable activities, they may also define the outer limits of the workday. Under those regulations, the few minutes spent walking between the locker rooms and the production area are similar to the time spent walking between two different workplaces on the disassembly line. See 29 CFR §790.7(c) (2005) (explaining that the Portal-to-Portal Act does not affect the compensability of time spent traveling from the place of performance of one principal activity to that of another). See also §785.38 (explaining, in a later regulation interpreting the FLSA, that "[w]here an employee is required to report at a meeting place to receive instructions or to perform other work there, or to pick up and to carry tools, the travel from the designated place to the work place is part of the day's work, and must be counted as hours worked . . .").
>
> For the foregoing reasons, we hold that any activity that is "integral and indispensable" to a "principal activity" is itself a "principal activity" under §4(a) of the Portal-to-Portal Act. Moreover, during a continuous workday, any walking time that occurs after the beginning of the employee's first principal activity and before the end of the employee's last principal activity is excluded from the scope of that provision, and as a result is covered by the FLSA.

Barber Foods, Inc., operates a poultry processing plant in Portland (ME) that employs about 300 production workers. These employees operate six production lines and perform a variety of tasks that require different combinations of protective clothing. They are paid by the hour from the time they punch in to computerized time clocks located at the entrances to the production floor. They were sued by current and former employees who wanted to be paid for their time (a) donning and doffing required protective gear and (b) the attendant time walking and waiting. That case went to a jury, which found for the company. The Supreme Court said,

Before the case was submitted to the jury, the parties stipulated that four categories of workers—rotating, set-up, meat room, and shipping and receiving associates—were required to don protective gear at the beginning of their shifts and were required to doff this gear at the end of their shifts. The jury then made factual findings with regard to the amount of time reasonably required for each category of employees to don and doff such items; the jury concluded that such time was de minimis and therefore not compensable. The jury further concluded that two other categories of employees—maintenance and sanitation associates—were not required to don protective gear before starting their shifts. Accordingly, the jury ruled for Barber on all counts.

The plaintiff employees "argued in the Court of Appeals that the waiting time associated with the donning and doffing of clothes was compensable."

[The employees], supported by the United States as **amicus curiae**, maintain that the predonning waiting time is "integral and indispensable" to the "principal activity" of donning, and is therefore itself a principal activity. However, unlike the donning of certain types of protective gear, which is always essential if the worker is to do his job, the waiting may or may not be necessary in particular situations or for every employee. It is certainly not "integral and indispensable" in the same sense that the donning is. It does, however, always comfortably qualify as a "preliminary" activity.

> **amicus curiae** In cases before the U.S. Supreme Court, the court may allow others with an interest in a case, such as the U.S. Solicitor General or the American Civil Liberties Union (ACLU), to file "friend of the court" briefs.

We thus do not agree with [the employees] that the predonning waiting time at issue in this case is a "principal activity" under §4(a).

For example, walking from a time clock near the factory gate to a workstation is certainly necessary for employees to begin their work, but it is indisputable that the Portal-to-Portal Act evinces Congress' intent to repudiate *Anderson* [*v. Mt. Clemens Pottery,* 328 U.S. 680] holding that such walking time was compensable under the FLSA. We discern no limiting principle that would allow us to conclude that the waiting time in dispute here is a "principal activity" under §4(a), without also leading to the logical (but untenable) conclusion that the walking time at issue in *Anderson* would be a "principal activity" under §4(a) and would thus be unaffected by the Portal-to-Portal Act.

In short, we are not persuaded that such waiting—which in this case is two steps removed from the productive activity on the assembly line—is "integral and indispensable" to a "principal activity" that identifies the time when the continuous workday begins. Accordingly, we hold that §4(a)(2) excludes from the scope of the FLSA the time employees spend waiting to don the first piece of gear that marks the beginning of the continuous workday.

LEGAL LESSONS LEARNED

The Supreme Court has expanded the compensable workday for many employees. This case may be useful to off-duty personnel who must respond from home to get their gear (such as an off-duty arson investigator), because this response time would appear to be "on the clock" from the moment they leave home.

CASE STUDY 11-3 Comp Time—Public Employers Offering Instead of Overtime Pay/U.S. Supreme Court

Christensen v. Harris County, 529 U.S. 576 (2000).

FACTS

The plaintiffs are 127 deputy sheriffs, and they filed suit in federal court against their employer, Harris County (TX) and its sheriff, Tommy B. Thomas. Each of the deputies had individually agreed to accept compensatory time ("comp time"), in lieu of cash, as compensation for overtime.

As the deputies accumulated more and more comp time, Harris County became concerned that it lacked the resources to pay monetary compensation to employees who worked overtime after reaching the statutory cap on comp time accrual and to employees who left their jobs with sizable reserves of accrued time. As a result, the county began looking for a way to reduce accumulated comp time.

Harris County wrote to the Department of Labor's Wage and Hour Division, asking "whether the Sheriff may schedule non-exempt employees to use or take compensatory time." The acting administrator of the Wage and Hour Division replied:

> [I]t is our position that a public employer may schedule its nonexempt employees to use their accrued FLSA compensatory time as directed if the prior agreement specifically provides such a provision. . . . Absent such an agreement, it is our position that neither the statute nor the regulations permit an employer to require an employee to use accrued compensatory time.

Department of Labor, Wage and Hour Division (September 14, 1992), 1992 WL 845100 (Opinion Letter)

Harris County, after receiving the letter, implemented a policy under which the employees' supervisor sets a maximum number of comp hours that may be accumulated. When an employee's stock of hours approaches that maximum, the employee is advised of the maximum and is asked to take steps to reduce accumulated comp time. If the employee does not do so voluntarily, a supervisor may order the employee to use his comp time at specified times.

The deputies sued, claiming that the county's policy violates the FLSA because Sec. 207(o)(5)—which requires that an employer "reasonably accommodate" employee requests to use comp time—provides the exclusive means of using accrued time in the absence of an agreement or understanding permitting some other method. The District Court agreed, granting summary judgment for the deputy sheriffs and entering a declaratory judgment that the county's policy violated the FLSA.

The county appealed to the U.S. Court of Appeals for the 5th Circuit. A three-judge panel reversed the trial judge and held for Harris County, holding that the FLSA did not speak to the issue, and thus did not prohibit the county from implementing its compensatory time policy.

The deputy sheriffs asked the U.S. Supreme Court to hear their appeal, and the court agreed to take the case.

YOU BE THE JUDGE

Can the county tell its deputies they must use their comp time?

HOLDING

Yes.

Justice Thomas wrote the decision for the court:

Both parties and the United States as amicus curiae concede that nothing in the FLSA expressly prohibits a State or subdivision thereof from compelling employees to utilize accrued compensatory time. [Harris County] and the United States, however, contend that the FLSA implicitly prohibits such a practice in the absence of an agreement or understanding authorizing compelled use.

Title 29 U.S.C. Sec. 207(o)(5) provides: An employee . . . (A) who has accrued compensatory time off . . ., and (B) who has requested the use of such compensatory time, shall be permitted by the employee's employer to use such time within a reasonable period after making the request if the use of the compensatory time does not unduly disrupt the operations of the public agency.

Justice Thomas writes,

Sec. 207(o)(5) is better read not as setting forth the exclusive method by which compensatory time can be used, but as setting up a safeguard to ensure that an employee will receive timely compensation for working overtime. Section 207(o)(5) guarantees that, at the very minimum, an employee will get to use his compensatory time (i.e., take time off work with full pay) unless doing so would disrupt the employer's operations. And it is precisely this concern over ensuring that employees can timely "liquidate" compensatory time that the Secretary of Labor identified in her own regulations governing Sec. 207(o)(5):

> Compensatory time cannot be used as a means to avoid statutory overtime compensation. An employee has the right to use compensatory time earned, and must not be coerced to accept more compensatory time than an employer can realistically and in good faith expect to be able to grant within a reasonable period of his or her making a request for use of such time. 29 CFR Sec. 553.25(b) (1999).

At bottom, we think the better reading of Sec. 207(o)(5) is that it imposes a restriction upon an employer's efforts to prohibit the use of compensatory time when employees request to do so; that provision says nothing about restricting an employer's efforts to require employees to use compensatory time. Because the statute is silent on this issue and because Harris County's policy is entirely compatible with Sec. 207(o)(5), petitioners cannot, as they are required to do by 29 U.S.C. Sec. 216(b), prove that Harris County has violated Sec. 207.

As we have noted, no relevant statutory provision expressly or implicitly prohibits Harris County from pursuing its policy of forcing employees to utilize their compensatory time. In its opinion letter siding with the petitioners, the Department of Labor opined that "it is our position that neither the statute nor the regulations permit an employer to require an employee to use accrued compensatory time." Opinion Letter [emphasis added]. But this view is exactly backwards. Unless the FLSA prohibits respondents from adopting its policy, [the deputy sheriffs] cannot show that Harris County has violated the FLSA. And the FLSA contains no such prohibition.

LEGAL LESSONS LEARNED

Employers, including fire departments, may establish mandatory use of comp time procedures. These procedures should be set forth in writing and uniformly applied. There has been litigation concerning public employers refusing to grant comp time off because of budget shortfalls. It would be helpful to address this issue in an employee handbook or other written procedures.

■ ■

Chapter Review Questions

1. In Case Study 11-1, *Cedrick Cleveland v. City of Los Angeles,* we discussed overtime after 212 hours in a 28-day pay period. The FLSA authorizes overtime under other pay periods. Identify the various overtime periods used by fire departments in your state. Discuss how these departments keep track of the hours worked by part-time personnel for purposes of overtime pay.

2. In Case Study 11-2, *IBP, Inc. v. Alvarez,* the U.S. Supreme Court held that employees were "on the clock" for the time they were required to put on protective equipment and walked to the production area. Discuss whether a dispatcher should be considered "on the clock" when the city required her to submit to a fitness-for-duty evaluation for stress and then required her, on the doctor's recommendation, to attend weekly psychotherapy treatment, on her off-time, for the next six months. See U.S. Court of Appeals for the 7th Cir. Decision in *Sehie v. City of Aurora,* 2005 WL 3534472 (7th Cir. 2005).

3. In Case Study 11-3, *Christensen v. Harris County,* the court addressed the issue of comp time. Many fire departments have rules regarding the maximum comp time that employees may accumulate and also how comp time is paid on retirement or termination. Describe the rules for a large fire department in your state.

■ ■

Expand Your Learning

Read and complete the individual student or group assignment, as directed by your instructor.

1. On August 23, 2004, the Department of Labor's new regulations on FLSA went into effect, 29 CFR 541. See Fact Sheet #17J, "First Responders and the Part 541 Exemptions Under the Fair Labor Standards Act (FLSA)" at *www.dol.gov/esa.* The Department of Labor advises that firefighters, paramedics, EMTs, and rescue workers are *not* exempt and will continue to be paid overtime under Sec. 13(a)(1) of the new regulations. Describe under these new regulations when fire department executives, such as fire chiefs, assistant chiefs, and other senior fire officers, can be classified as "salaried/exempt" and therefore not receive overtime payment.

2. The FLSA regulations authorize fire departments to exclude "sleep time" when calculating overtime hours. Describe the Department of Labor rules for sleep time compensation. Review the payroll record-keeping practices of a fire department in your state that excludes sleep time hours.

See Appendix A for additional Expand Your Learning activities related to this chapter.

Drug-Free Workplace
Random Drug Testing and Firefighter DUIs

12 CHAPTER

Post-accident drug testing of firefighters injured on duty is a common requirement. In this photo, an officer driving a command vehicle to an alarm, with red lights and siren activated, was hit head-on by a young civilian driver who skidded on bald tires. Both survived the crash. *Photo supplied by Captain Tim Keene, Delhi (OH) Fire Department.*

Key Terms

medical review officer, p. 153   **last-chance agreement, p. 156**   **random drug testing, p. 157**

The air bag saved this officer's life. He was transported to University Hospital and, according to the fire department's drug-free workplace policy, he was tested for drugs and alcohol. The test was negative. *Photo supplied by Captain Tim Keene.*

FDNY Firefighter Files Suit against City

A firefighter, critically injured on New Year's Eve when hit in the face with a steel chair by a drunken firefighter, files a lawsuit against New York City.

IAFC Policy—No Alcohol within 8 Hours Prior to Duty

The International Association of Fire Chiefs (IAFC) recommends that fire departments adopt a written policy that prohibits firefighters from drinking any alcohol within eight hours of their next shift.

◆ INTRODUCTION

This chapter will address firefighter alcohol and other drug consumption issues. One of the most common issues concerns firefighters charged with driving under the influence. See Appendix E for the Cincinnati Fire Department's provision in their Labor-Management Agreement concerning firefighters' driving apparatus after being charged with driving under the influence.

The juvenile driver of this car was transported to the hospital, where he required facial surgery. Bald tires on the car were seized as evidence. *Photo supplied by Captain Tim Keene.*

Fire departments should establish a drug-free workplace policy. This policy should address post-accident and "reasonable suspicion" testing. Many fire departments have also established random drug-testing programs and Employee Assistance Programs (EAP).

NIOSH FIREFIGHTER FATALITY REPORT

Employee privacy rights are a concern with drug testing. On March 17, 2006, NIOSH published their Fire Fighter Fatality Investigation Report F2005-30 on the death of a 58-year-old New Jersey firefighter from sudden cardiac arrest while working out on duty (*www.cdc.gov/niosh*). Recommendation No. 6 was that the fire department discontinue routine testing for illegal drugs as part of the annual medical evaluation, unless there is cause to suspect the firefighter is using illegal drugs. This recommendation cites NFPA 1582, "Standard on Comprehensive Occupational Medical Program for Fire Departments" (2003), which does *not* recommend testing for illegal drugs as part of annual firefighter medical evaluations.

The NIOSH report addresses the employee privacy concern. "A comprehensive drug-free workplace approach includes five components—a policy, supervisor training, employee education, employee assistance, and drug testing. Such programs, especially when drug testing is included, must be reasonable and take into consideration employee privacy rights. Although not required by OSHA, drug-free workplace programs are natural complements to other initiatives that help ensure safe and healthy workplaces and add value to America's business and communities."

SPLIT SAMPLES

Fire departments that send personnel to be drug tested should ensure that a "split sample" is taken, in which the testing laboratory tests only a part of the urine. If the firefighter disagrees with a "positive" test result, the firefighter can pay for the remainder of the sample to be sent to another certified laboratory.

See, for example, *Morgan v. City of St. Louis,* 154 S.W.3d 6 (Mo. Ct. Appeals 2004). A captain with 24 years on the fire department took a drug test as part of a promotion process. He tested positive for marijuana. He asked that the split sample be retested. It also came back positive, and he was terminated. The city's Civil Service Commission upheld the termination. He appealed to court, which focused on the level of marijuana found in his urine. The city's policy stated that a concentration of 15 ng/ml was considered a positive test. The Captain's first test came back at 25 ng/ml. The second test reportedly came back positive, but the lab did not record the amount of ng/ml. The court therefore held that he did not test positive, and he was reinstated with full back pay.

STATE LAWS

Numerous states have adopted drug-free statutes, and readers are encouraged to review their state laws. For example, California tests drivers of school vehicles. The District of Columbia tests public-sector vehicle operators and those who work with children. Minnesota conducts random drug testing of professional athletes. North Carolina requires CDL drivers who test positive to inform the state. South Dakota tests public-sector job applicants. Tennessee tests drivers of childcare agencies. Texas requires reporting of positive drivers with a commercial driver's license (CDL). Washington also requires reporting of positive CDL drivers. See the Institute for a Drug-Free Workplace (*www.drugfreeworkplace.org*) for additional information on state laws.

◆ NFPA STANDARDS

- *NFPA 1071: Standard for Emergency Vehicle Technician Professional Qualifications* (2000 edition). This standard identifies the knowledge and skills required of vehicle technicians.
- *NFPA 1500: Standard on Fire Department Occupational Safety and Health Program* (2002 edition). This standard makes recommendations on the steps that a fire department can take to minimize hazards to firefighters.
- *NFPA 1582: Standard on Comprehensive Occupational Medical Program for Fire Departments* (2003 edition). This standard identifies medical conditions that possess a significant risk to the firefighter and other firefighters.

◆ KEY STATUTES

DRUG-FREE WORKPLACE ACT OF 1988

This federal statute, 41 U.S.C. 701–707, requires some federal contractors and all who receive federal grants to agree to provide drug-free workplaces as a condition of receiving the federal grant. Fire departments and others who receive grants are required to adopt a drug-free workplace program, including notifying employees that it is prohibited to have controlled substances in the workplace, requiring employees to report drug convictions within five days, and establishing a drug-free awareness program. The statute does not require that the employer implement employee drug testing.

OMNIBUS TRANSPORTATION EMPLOYEE TESTING ACT OF 1991

This statute requires drug and alcohol testing of safety-sensitive transportation employees in aviation, trucking, railroads, mass transit, pipelines, and other transportation industries (including CDL employees). The Department of Transportation (DOT) has published rules on random and suspicion drug testing at 49 CFR 40, covering about 12 million employees. In some states and municipalities, firefighters who drive emergency apparatus must obtain CDL licenses. This statute would mandate post-accident drug testing of these CDL employees.

◆ CASE STUDIES

CASE STUDY 12-1 Random Drug Testing—No History of Drug Problems on Fire Department

Petersen v. City of Mesa, 83 P.3d 35 (Ariz. 2004); Arizona Supreme Court. On October 4, 2006, the U.S. Supreme Court denied the city's petition for a writ of certiorari (declined to hear their appeal), 543 U.S. __, No. 03-1599.

FACTS

Craig Petersen is a firefighter for the City of Mesa (AZ). In 2001, after Petersen was hired, the City implemented a substance abuse program for the Mesa Fire Department. The program requires testing of firefighters (1) if the department has reasonable suspicion to believe that an individual firefighter has abused drugs or alcohol, (2) after a firefighter is involved in an accident on the job, (3) following a firefighter's return to duty or as a follow-up to "a determination that a covered member is in need of assistance," and (4) "on an unannounced and random basis spread reasonably throughout the calendar year."

The Mesa Fire Department used a computer program to select the firefighters to be tested. The department notifies firefighters of their selection for random testing immediately before, during, or after work. The firefighters are to be tested within 30 minutes of their notification, with allowance for travel time to the laboratory for collection. Once at the laboratory, firefighters are permitted to use private bathroom stalls when providing urine samples, which are then inspected by a monitor for the proper color and temperature.

The laboratory tests the sample for the presence of marijuana, cocaine, opiates, amphetamines, and phencyclidine. The laboratory initially tests the specimens by using an immunoassay test that meets the requirements of the Food and Drug Administration for commercial distribution. The laboratory then confirms all positive test results using the gas chromatography/mass spectrometry technique and reports positive results to a **medical review officer** (MRO), who has a "detailed knowledge of possible alternate medical explanations."

medical review officer
A physician or other medical expert licensed to confirm a "positive" drug test.

The MRO reviews the results before giving the information to the department's administrative official. Only confirmed tests are reported to the department as positive for a specific drug. Before verifying a positive result, however, the MRO must contact the firefighter on a confidential basis.

The Mesa Fire Department does not release information in a firefighter's drug testing record outside the department without the firefighter's consent. Twenty percent of those tested are selected for an alcohol breath test.

The department may discipline or terminate the employment of a firefighter who tests positive a second time or who refuses to submit to a required test.

The trial judge held that the random drug testing program violated the U.S. Constitution. The City of Mesa appealed to the Arizona Court of Appeals, which reversed the trial judge, holding that the department's random testing component is reasonable under both the Arizona and U.S. Constitutions. The court reasoned that the city's compelling need to discover specific but hidden conditions representing grave risks to the health and safety of the firefighters and the public outweighed Petersen's privacy interests. *Petersen v. City of Mesa*, 204 Ariz. 278, 286 ¶ 34, 63 P.3d 309, 317 (App. 2003).

YOU BE THE JUDGE

Should the Mesa Fire Department be allowed to conduct random drug testing of firefighters?

HOLDING

No.

The Arizona Supreme Court wrote,

[W]e hold that the Program's random testing component is unreasonable and therefore violates the Fourth Amendment to the United States Constitution.

The Fourth Amendment to the United States Constitution protects "[t]he right of the people to be secure in their persons, houses, papers, and effects, against unreasonable searches and seizures." U.S. Const. amend. IV. "The Amendment guarantees the privacy, dignity, and security of persons against certain arbitrary and invasive acts by officers of the Government or those acting at their direction." *Skinner v. Ry. Labor Executives' Ass'n*, 489 U.S. 602, 613–14 (1989).

In this case, the parties agree that the City's collection and testing of a firefighter's urine and breath constitutes a "search" under the Fourth Amendment.

As the language of the Fourth Amendment makes clear, "the ultimate measure of the constitutionality of a governmental search is 'reasonableness.'" To be reasonable, a search generally must be based upon some level of individualized suspicion of wrongdoing. *Skinner*, 489 U.S. at 624. The purpose of requiring individualized suspicion "is to protect privacy interests by assuring citizens subject to a search or seizure that such intrusions are not the random or arbitrary acts of government agents." *Id.* at 621–22.

The Supreme Court, however, has recognized limited exceptions to this general rule when special needs, beyond the normal need for law enforcement, make the warrant and probable-cause requirement impracticable.

The City concedes that its use of random, suspicionless testing is not based on any level of individualized suspicion. The City argues, however, that such testing is reasonable under the Fourth Amendment because the search "serves special governmental needs, beyond the normal need for law enforcement." *Nat'l Treasury Employees Union v. Von Raab*, 489 U.S. 656, 665 (1989).

Neither the [U.S.] Supreme Court nor this court has considered the reasonableness of random, suspicionless testing of city firefighters. The Supreme Court, however, has examined the constitutionality of suspicionless drug testing requirements analogous to the procedures Petersen challenges. *See Bd. of Educ. of Indep. Sch. Dist. v. Earls*, 536 U.S. 822 (2002) (high school students participating in competitive extracurricular activities); *Chandler v. Miller*, 520 U.S. 305 (1997) (candidates for political office); *Vernonia*, 515 U.S. 646 (high school students participating in interscholastic athletics); *Skinner*, 489 U.S. 602 (railway employees); *Von Raab*, 489 U.S. 656 (customs service agents); *see also Ferguson v. City of Charleston*, 532 U.S. 67 (2001) (holding unconstitutional a state hospital's drug testing of pregnant patients that involved hospital personnel notifying the police of patients who tested positive for cocaine).

As each of these decisions illustrates, when presented with an alleged "special need" in support of a particular Fourth Amendment intrusion, a court must weigh the individual's Fourth Amendment interests against the proffered governmental interests to determine whether the search in question "fit[s] within the closely guarded category of constitutionally permissible suspicionless searches." *Chandler*, 520 U.S. at 309.

Applying this "special needs" balancing test to the facts presented in this case, we begin by analyzing the City's proffered interests. Although the City need not present a "compelling" interest, the City's interest must be "important enough" to justify the government's intrusion into the firefighters' legitimate expectations of privacy. *Vernonia*, 515 U.S. at 661.

The City asserts that it has a "special need" to test firefighters because they occupy safety-sensitive positions. The City alleges that random testing furthers this interest by deterring "prohibited alcohol and controlled substance use" and detecting "prohibited use for the purpose of removing identified users from the safety-sensitive work force." We agree that the City has an interest in deterring and detecting prohibited alcohol and drug use among the City's firefighters.

Fourth Amendment analysis, however, requires that we do more than recognize that the City has an interest in deterring drug use among employees in safety-sensitive positions. In addition, we must look to the nature and immediacy of the City's concern. *Id.* at 660. That is, has the City identified a real and substantial risk? *Chandler*, 520 U.S. at 323. If so, will the City's proposed invasion of its firefighters' privacy interests further the City's interest in deterring and detecting drug use among its firefighters? *Skinner*, 489 U.S. at 624. Answering that question requires that we consider the efficacy of the Program in meeting the City's concern, *Vernonia*, 515 U.S. 11 at 660, and whether the invasion of privacy is calibrated to the defined risk. *Chandler*, 520 U.S. at 321–23.

The record before us provides little information about the City's reasons for adopting random testing and provides no evidence to explain the City's perceived need to conduct such testing. As the City conceded at oral argument, the record is devoid of any indication that the City has ever encountered any problem involving drug use by its firefighters. The record lacks not only evidence of even a single instance of drug use among the firefighters to be tested but also any evidence of accidents, fatalities, injuries, or property damage that can be attributed to drug or alcohol use by the City's firefighters. No evidence of record suggests that the firefighters asked for or consented to the testing policy, and the record includes not even an allegation or rumor that the City's firefighters used or abused drugs or alcohol.

Based on this record, we detect no real and substantial risk that the public safety is threatened by drug or alcohol use among the firefighters to be tested. The absence of evidence of drug use, at least as reflected in the record, provides no basis for us to conclude that random, suspicionless testing is calibrated to respond to any defined risk. At most, the Program's random testing component furthers only a generalized, substantiated interest in deterring and detecting a hypothetical drug abuse problem among the City's firefighters.

LEGAL LESSONS LEARNED

The majority of courts have upheld random drug testing of firefighters.

[Note: The Phoenix Fire Department has restarted drug testing in 2006, after a two-year suspension of their program.]

CASE STUDY 12-2 Last-Chance Agreement—Assistant Fire Chief in Drug Treatment Told He Must Sign

DePalma v. City of Lima, 155 Ohio App.3d 81, 799 N.E.2d 207 (3rd Dist. Court of Appeals, 2003). On December 8, 2004, the Ohio Supreme Court declined to hear the city's appeal. The court, however, ordered that the Court of Appeals decision "can not be cited as authority except by the parties inter se." 104 Ohio St. 3d 1201, 2004-Ohio-6401.

FACTS

For more than 20 years, Anthony DePalma was employed by the City of Lima Fire Department. During his career, he received very high scores on his exams and received numerous awards for valor and dutiful service. In 2000, he was promoted to assistant chief.

In December of 2000, he developed kidney stones and was prescribed various narcotic pain medications. He became addicted to these medications. When the medications were no longer available from his treating physicians, he began to purchase them illegally and eventually began taking heroin. Realizing he had a drug addiction, he spoke with his pastor and to another firefighter who was in recovery from an addiction to crack cocaine.

DePalma voluntarily checked himself into Shepherd Hill, a nationally known addiction-treatment center at the beginning of October 2001. On October 5, 2001, he was visited by the fire chief, who informed him that he had to sign a **last-chance agreement** (LCA) or his employment would be terminated. The LCA required him to (1) complete treatment at Shepherd Hill and (2) submit to random drug and alcohol testing. DePalma signed the LCA.

On March 17, 2002, he was taken to the hospital for another kidney stone. At the hospital, he was given a full 30-day prescription for Vicodin. His pain increased, and he returned to the hospital for surgery. He was given Demerol for two weeks while awaiting surgery.

He returned to work on April 1, 2002, and was required to submit to a drug test. He tested positive for pain killers and was terminated.

He appealed to the city's Civil Service Board, which reversed the termination. The board reinstated DePalma but suspended him without pay for 14 days. This was the same penalty received by the other firefighter who had relapsed for cocaine use.

On July 26, 2002, the City of Lima appealed the board's decision to the Court of Common Pleas. On January 17, 2003, the trial court, after reviewing the record of the Board's proceedings, reversed the decision of the board and affirmed the termination. DePalma appealed to the Ohio Court of Appeals.

YOU BE THE JUDGE

Should the assistant chief be terminated in accordance with the last-chance agreement?

HOLDING

No.

    The three-judge Court of Appeals wrote:

> In this case, the mayor and the fire chief both testified that prior to DePalma's entering the treatment program at Shepherd Hill, there was nothing to indicate that DePalma had violated any rules of work. His performance and behavior at work were excellent. However, once DePalma voluntarily entered the treatment program, the city became aware of his drug addiction and immediately changed the terms of DePalma's employment by having him sign the LCA. The city argues that this was a permissible action because the LCA was not discipline. The city claims that the LCA is not discipline because it did not adversely affect DePalma at the time he signed it. However, a written reprimand is discipline if it is placed in an employee's file and the implications of the writing continue beyond the placement in the file.
>
> We don't see any difference between the LCA and a written warning that is placed in one's file. Neither action adversely affects the subject at the time made. However, both actions can provide the basis for further action at a later time, including termination. Thus, the LCA is a form of discipline.
>
> The LCA was signed while DePalma was actively seeking treatment at a rehabilitation center and was no longer using the substance to which he is addicted. Thus, DePalma was a disabled individual under federal law when he was presented with the LCA. DePalma had been at Shepherd Hill for approximately one week when the fire chief, the

last-chance agreement
Occurs when an employee seeks treatment for alcohol or drug addiction and signs an agreement that if he or she tests positive upon return to work, employment will be terminated.

union representative, and a substance-abuse counselor arrived and told DePalma to either sign the document or be terminated. This, in effect, was a disciplinary action for being a drug addict. The mayor testified at the hearing that he viewed the status of being an addict the same as actively using drugs and therefore, he believed, equally subject to discipline. However, the city had no valid grounds for disciplining DePalma at this time as there was no violation of any workplace rule. The city had no authority to require the LCA at that time, as changing DePalma's terms of employment and disciplining him for being a drug addict violate federal law. Since the LCA could not be legally imposed while DePalma was in treatment at Shepherd Hill and no longer using drugs, the LCA is not valid, and the violation of its terms alone cannot form just cause for DePalma's termination. The first assignment of error is sustained.

LEGAL LESSONS LEARNED

Fire departments should adopt a drug-free workplace policy in which employees are "encouraged" to come forward for help before they are caught in a drug test. Employee Assistance Programs (EAP) should extend not only to the firefighter but also to the firefighter's spouse and children. Consider inviting the EAP counselor for alcoholism to the fire department for an annual presentation.

CASE STUDY 12-3 Random Drug Testing in Collective Bargaining Agreement—Firefighter Tests Positive for Marijuana

Washington v. Akron Civil Service Commission, 2002-Ohio-459 (Ohio 9th Dist.; February 6, 2002).

FACTS

The Akron Fire Department and IAFF Local 330 entered into a drug-testing agreement in January, 1996. The agreement called for drug screening of sworn personnel and was incorporated into the collective bargaining agreement. The fire department began **random drug testing** in July, 1996.

random drug testing
Employees are selected at random, often by computer search, for a drug test.

Jerry Washington was a firefighter/paramedic and was selected for random drug testing on July 27, 2000. He tested positive for marijuana. He was placed on indefinite suspension, and a pre-termination hearing was held on August 4, 2000, with the Deputy Mayor of Labor Relations, who recommended his discharge. He was fired by the mayor, effective August 23, 2000.

He filed an appeal to the city's Civil Service Commission, which held a hearing and upheld the termination. The firefighter then filed an appeal to the Ohio Court of Common Pleas. The trial judge reversed the Commission and ordered him reinstated. The city has filed an appeal to the Ohio Court of Appeals.

YOU BE THE JUDGE

Should the firefighter be reinstated?

HOLDING

No.

The three-judge panel wrote:

In Ohio, a member of a fire or police department may utilize either of two separate and distinct avenues of appeal to the court of common pleas from a decision of suspension, demotion or removal from office by a municipal civil service commission. Under Ohio Rev. Code 124.34, an appeal may be made on questions of law and fact, *i.e.,* a *de novo* hearing, or an appeal may be taken pursuant to Chapter 2506 of the Revised Code and the reviewing court will give deference to the judgment of the commission. However, when a collective bargaining agreement has been entered into pursuant to Chapter 4117 of the Revised Code, the provisions of the collective bargaining agreement prevail over other applicable statutes except for those laws specifically exempted by Ohio Rev. Code 4117.10(a).

Article V, Section C of the [Collective Bargaining] Agreement states: "Employees may appeal any formal disciplinary action to the Mayor and the Civil Service Commission subject to the conditions stated in Section 72 of the City Charter and the Akron Fire Department Rules and Regulations." Section 24.4 of the Akron Fire Department Rules and Regulations entitled "Akron City Charter-Section 72. Removal Of Officers And Employees" states, in pertinent part: "The employee or appointing authority may appeal the decision of the Civil Service Commission to the Court of Common Pleas pursuant to Ohio Revised Code Chapter 2506." This provision makes it clear that the only avenue for appeal that an officer or employee employed by the City of Akron's police or fire division has is an appeal pursuant to Chapter 2506 of the Revised Code.

Although the trial court does not state that it conducted its review pursuant to R.C. 124.34, a review of the trial court's decision reveals that the trial court did not give due deference to the decision of the Akron Civil Service Commission as required by Chapter 2506 of the Revised Code.

Accordingly, the City's first assignment of error is well taken.

LEGAL LESSONS LEARNED

Fire departments with collective bargaining agreements should attempt to negotiate a drug-free workplace provision in the Collective Bargaining Agreement (CBA). Some provisions, such as random drug testing, may be a "hard sell." Consult with a local legal counsel who is knowledgeable of your state labor laws regarding management's rights to implement a random drug-testing program.

■■■

Chapter Review Questions

1. In Case Study 12-1, *Petersen v. City of Mesa,* the Arizona Supreme Court held that random drug testing of a firefighter was not authorized because there had been no history of drug abuse presented. Most states have authorized drug testing of "safety-sensitive" positions. Discuss whether fire departments in your state have random drug-testing programs and whether these programs are included in the collective bargaining agreement with the union.

2. In Case Study 12-2, *DePalma v. City of Lima,* the court addressed "last-chance agreements." Describe how a fire department in your state handles a situation in which a firefighter tests positive for cocaine and then discloses he has an addiction. Does the policy allow him to enter into treatment? If so, can he be terminated for a second violation within a certain time period?

3. In Case Study 12-3, *Washington v. Akron Civil Service Commission,* the court addressed the issue of a firefighter terminated after his first positive test for marijuana. Some fire departments have a more lenient rule in their collective

bargaining agreements or employee handbooks whereby firefighters are not fired until their second positive drug test. Describe the policy at a fire department in your state. Discuss what rule you would advocate if you were the fire chief.

Expand Your Learning

Read and complete the individual student or group assignment, as directed by your instructor.

1. The U.S. Fire Administration, in its report, "Firefighter Fatalities in the United States in 2003," wrote the following about alcohol use in the fire service, "According to the United States Department of Health and Human Services, Drug and Alcohol Information Clearinghouse, approximately 14 million Americans—7.4 percent of the population—meet the diagnostic criteria for alcohol abuse or alcoholism. The fire service is not immune from these facts. Alcoholics live in every walk of life in our society, including the fire service. Through their jobs, firefighters see first-hand the effects of alcohol abuse. Firefighters often respond to alcohol-related vehicle crashes, domestic violence assaults, and other situations that are caused or made worse by alcohol."

Discuss what steps that a fire department can take to implement a "zero-tolerance policy" for on-duty personnel as well as those who respond from home. Identify the fire departments in your state that have adopted effective programs.

2. On January 10, 2006, the Federal Aviation Administration issued new regulations, effective April 10, 2006, which require anyone who performs "safety-sensitive" maintenance on aircraft, including subcontractors, be subject to random drug and alcohol testing; see 14 CFR 121. The FAA issued its alcohol-testing program eight years ago, and almost half of the 876 alcohol violations were attributable to maintenance employees. Discuss whether fire department random drug and alcohol testing should include civilian emergency vehicle maintenance employees.

See Appendix A for additional Expand Your Learning activities related to this chapter.

Emergency Medical Service (EMS) and the Health Insurance Portability and Accountability Act (HIPAA)

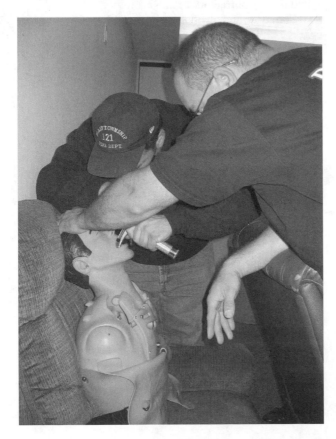

Firefighter/paramedics with the Plain Township (OH) Fire Department use a dummy to practice their cardiopulmonary resuscitation (CPR) and advanced life support (ALS) skills. *Photo by author.*

Key Terms

willful or wanton misconduct, p. 162   state-created danger, p. 168

dismissed without prejudice, p. 167   qualified immunity, p. 168

From the Headlines

First HIPAA Criminal Conviction

U.S. Attorney's Office in Seattle announces plea of guilty by Richard W. Gibson who, while working in a cancer treatment facility, obtained a patient's name, date of birth, and Social Security number and used this information to forge credit cards. He charged over $9,000 in the patient's name, including video games, home improvement supplies, jewelry, and gasoline. He faces a federal prison term of 10 to 16 months.

Patient Not Dead

A 51-year-old man was pronounced dead in his basement by Detroit EMS crews. While waiting for police, a family member went to the basement, and the patient opened his eyes and moved his arms. He was a diabetic and was taken to the hospital and treated.

Ride-Along Doctor Is a Fake

A 20-year-old repeat offender was arrested after he pretended to be a physician and went on numerous ride-alongs with the Cincinnati Fire Department ambulances. Investigation reveals he had previously pretended to be a police officer in Columbus and had also been convicted for making a false fire report.

◆ **INTRODUCTION**

This chapter will address litigation, exposure to infectious diseases, and other legal issues in the delivery of emergency medical services (EMS). Unfortunately, there has been increased litigation against paramedics, emergency medical technicians, and their employers.

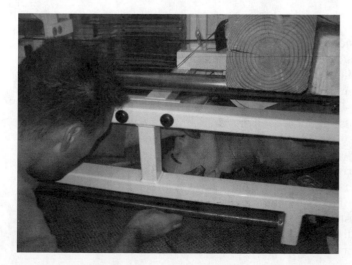

Some of the "patients" in this ALS training are in unusual positions, including under workout equipment. *Photo by author.*

willful or wanton misconduct
EMS conduct that amounts to
gross misconduct.

Many states have enacted statutes that protect EMS personnel from personal liability for their negligent conduct, but not from "gross negligence" or **willful or wanton misconduct.** Because there is no clear delineation for applying these standards in actual cases, many lawsuits are not dismissed until depositions have been taken during pre-trial discovery. In close cases, particularly when the EMS run reports are not very detailed, the insurance company defending the EMS personnel and their employer may decide to settle the case rather than risk a judge's denial of a motion for summary judgment and an adverse jury verdict.

◆ HEALTH INSURANCE PORTABILITY AND ACCOUNTABILITY ACT (HIPAA)

The HIPAA statute and increased attention to patient privacy have altered the manner in which EMS reports are stored and the requirements for release of patient information. Unfortunately, there has also been continuing litigation against EMS personnel.

The U.S. Department of Health and Human Services has issued detailed regulations on HIPAA. In 2003, the U.S. Department of Health and Human Services, Office of Civil Rights, issued a helpful summary of the HIPAA Medical Privacy Rule, April 11, 2003 (*www.hhs.gov/ocr/hipaa/privacy.html*).

These regulations provide that health care providers may not disclose protected health information except under certain exceptions provided under HIPAA. Section 164.502, Title 45, CFR. Protected health information is defined as individually identifiable health information that is transmitted or maintained in electronic media or any other form. Section 164.103, Title 45, CFR. In turn, "health information means any information, whether oral or recorded in any form or medium, that * * * [i]s created or received by a health care provider * * * and * * * [r]elates to the past, present, or future physical or mental health or condition of an individual; the provision of health care to an individual; or the past, present, or future payment for the provision of health care to an individual." Section 160.03, Title 45, CFR.

◆ DUI CONSENSUAL BLOOD DRAW

Courts are facing issues concerning HIPAA, including motions to suppress evidence from blood drawn at a hospital from a person suspected of driving under the influence (DUI) of alcohol or controlled substance or other defendant.

For example, in *State of Ohio v. Timothy B. Neely,* Slip Copy, 2005 WL 3610426 (Ohio App. 11 Dist.), 2005-Ohio-7045, December 29, 2005, the defendant was convicted of aggravated vehicular homicide, leaving the scene of an accident, and driving under the influence. Police were dispatched at 3:45 a.m. to the scene where a pedestrian had been struck by a vehicle. Police found a body by the road, and they interviewed the defendant, the driver of a Ford pickup that was no longer at the scene (it was later recovered at his father's house). The officer noticed the defendant's bloodshot eyes, slurred speech, and difficulty maintaining balance. He refused to take the field sobriety tests and was arrested. At the police station, he refused to take a Breathalyzer test. The police then drove him to Geneva Memorial Hospital, where a medical laboratory technician took a blood draw at 5:25 a.m. The technician testified he would never take a blood draw without the individual's consent. The blood tested at 0.23 percent for alcohol, plus cocaine; the deceased pedestrian tested at 0.128 percent. Prior to trial, the defendant filed a motion to suppress the blood drawn from him at the hospital. The trail judge denied the motion, and the defendant was subsequently convicted by a jury on all counts.

On appeal, the Ohio Court of Appeals held that the defendant had consented to the blood draw at the hospital, and that the blood did not violate HIPAA.

> Here, Geneva Memorial [Hospital] is a health care provider as contemplated by HIPAA. However, appellant's blood draw was not procured for the purpose of obtaining health care, diagnosis, or medical advice. To the contrary, appellant's blood was extracted for the sole purpose of investigating possible criminal activity. As a result, the blood draw does not relate to a health condition, provision of health care, or payment for the provision of health care, as established by HIPAA. Clearly, HIPAA was not intended to prohibit the disclosure of a blood draw test result that was taken as part of a criminal investigation, which included the offender's consent to the blood draw. Therefore, HIPAA's general prohibition against disclosure of "health information" [45 CFR 164.103] is not applicable to the case at bar.

◆ ENFORCEMENT ACCESS TO EMS RUN REPORTS

Fire and EMS personnel are often requested to provide immediate information to law enforcement. Occasionally, law enforcement will also request a copy of the completed EMS Run Report, which documents the details of the event. Specific federal regulations are now available on this issue [45 CFR 164.512(f)]. Fire departments should conduct training and develop a policy document concerning the release of patient information to law enforcement.

◆ HIV PATIENTS—U.S. DEPARTMENT OF JUSTICE LAWSUITS

Another area of litigation concerns EMS personnel and their alleged failure to treat or transport patients with human immunodeficiency virus (HIV). See the July 30, 2004, U.S. Department of Justice motion to intervene and join a patient's lawsuit against the City of Philadelphia for alleged failure of the EMS to treat an HIV patient properly, *John Gill Smith v. The City of Philadelphia,* Case No. 03-6494. See also March 18, 1994, Department of Justice settlement with the City of Philadelphia for failure to transport an HIV patient. (Go to *www.usdoj.gov* and search for "HIV.")

◆ EMS EXPOSURE TO INFECTIOUS DISEASE

Fire and EMS workers' exposure to infectious disease, including hepatitis C, has led to workers compensation claims and statutes, which establish a "rebutable presumption" that the firefighter or EMS personnel contracted the disease on duty.

For example, in *City of Philadelphia v. Workers' Compensation Board (Cospelich),* 2006 Pa. Commw. LEXIS 71 (Commonwealth Court of PA; February 15, 2006), Stephen Cospelich joined the Philadelphia Fire Department in 1985, and in 1999 he tested positive for hepatitis C. He immediately notified the department and, in 2002, filed a workers compensation claim for reimbursement of medical expenses. The city denied the claim, and it was litigated before the Workers Compensation Board. The board found his testimony to be credible and accepted the facts that he had responded to numerous shootings, stabbings, and childbirths and had performed CPR and many other duties that could expose him to blood and bodily fluids. The fire department did not distribute protective gloves until the 1980s, and he often found them to be too small—he is 6′ 5″ and weighs 260 pounds. The backboards used to transport patients were originally cleaned only with alcohol; the current practice is to use bleach. He admitted that during his off-duty days he worked for a medical transport service and for a nursing home, but he testified that he had little exposure to body fluids. He had never been exposed to a needle stick, undergone body piercing, used IV drugs, or engaged in homosexual activities.

The court found in favor of the firefighter, citing Pennsylvania workers compensation statutes that were amended in 2001 to cover this very situation. Further, Section 301(e) of the Act, 77 P.S. § 413, provides a rebuttable statutory presumption that an occupational disease is work-related. Section 301(e) provides:

> If it be shown that the employee, at or immediately before the date of disability, was employed in any occupation or industry in which the occupational disease is a hazard, it shall be presumed that the employee's occupational disease arose out of and in the course of his employment, but this presumption shall not be conclusive.

Firefighters have also lost court cases involving hepatitis C. For example, in *Robert Family v. City of Orlando,* February 23, 2006, Court of Appeals of Florida, 2006 Fla. App. LEXIS 2443, an Orlando firefighter was denied workers compensation coverage by the Judge of Compensation Claims who found that his hepatitis C was neither presumptively suffered in the line of duty pursuant to Florida statute 112.181 nor an occupational disease. The Court of Appeals agreed, writing,

> For Claimant to establish entitlement to recover on an occupational disease theory, the following four-part test must be met:
> (a) the disease must be actually caused by employment conditions that are characteristic of and peculiar to a particular occupation; (2) the disease must be actually contracted during employment in the particular occupation; (3) the occupation must present a particular hazard of the disease occurring so as to distinguish that occupation from usual occupations, or the incidence of the disease must be higher in the occupation than in usual occupations; and (4) if the disease is an ordinary disease of life, the incidence of such disease must be substantially higher in the particular occupation than in the general public.
>
> Here, the record supports the JCC's finding that Claimant failed to meet this test. Specifically, Claimant testified he could not recall any instances of exposure to anyone infected with hepatitis C. There was no evidence that Claimant contracted the disease during his employment or that he did not have the disease prior to his employment as a firefighter. Consequently, Claimant failed to establish the first and second prongs of the test: that his hepatitis C was actually contracted during his employment as a firefighter, and was caused by employment conditions characteristic of and peculiar to being a firefighter.
>
> Additionally, Dr. Rischitelli, an occupational disease specialist, testified he had conducted a study of firefighters, police officers, and correctional officers to determine whether there was an elevated prevalence of hepatitis C in those groups. The study revealed the rate of prevalence in those groups was essentially the same as that in the control group of voluntary blood donors. The record also indicates the Center for Disease Control published a report stating there is no higher incidence of hepatitis C in firefighters than in the general public. Dr. Rischitelli opined hepatitis C is an ordinary

disease of life, because it was reflected in the general population, and was a disease a physician would not be surprised to find in his practice.

Based on this evidence, Claimant also failed to establish the third and fourth prongs of the test: that being a firefighter presents a particular hazard of contracting hepatitis C, and the incidence of hepatitis C is substantially higher among firefighters than in the general public. Accordingly, the JCC's conclusion that Claimant's hepatitis C was not an occupational disease is affirmed.

[Fortunately for the firefighter, he had reached a private settlement agreement with the City of Orlando in 1996; the JCC had vacated it, but the Court of Appeals reversed the JCC and ordered that it be enforced.]

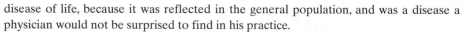

◆ **NFPA STANDARDS**

- *NFPA 1201: Standard for Providing Emergency Services to the Public* (2004 edition). This standard identifies requirements for establishing levels of emergency medical services—Basic Life Support (BLS) and Advanced Life Support (ALS).
- *NFPA 1581: Standard on Fire Department Infection Control Program* (2005 edition). This standard identifies standards to minimize the risk of infectious disease. The standard requires a written infectious disease policy addressing effective control procedures at scenes, in vehicles, and in facilities.

The Plain Township (OH) Fire Department's Dalmatian, "Blaze," watches the EMS training. The author needed a signed authorization to use each firefighter's photo, and Blaze's authorization came back "signed" with his inked paw marks. *Photo by author.*

CONTROLLED SUBSTANCES ACT OF 1970 (21 U.S.C. 301, 304, AND 1008)

This is the primary federal statute banning the illegal possession or sale of marijuana, cocaine, and other "controlled substances." The statute also applies to the storage and dispensing of controlled substances on ambulances. The U.S. Attorney General, acting through the Drug Enforcement Administration (DEA), may revoke or suspend a physician's certificate to dispense a controlled substance. The statute and DEA regulations also require prompt reporting to the DEA of any theft or loss of a controlled substance from an ambulance.

HEALTH INSURANCE PORTABILITY AND ACCOUNTABILITY ACT (HIPAA; 42 U.S.C. 1320)

This statute prohibits health care providers from disclosing "protected health information" except under certain limited exceptions. 45 CFR 164.502. Protected health information is defined as "individually identifiable health information that is transmitted or maintained in electronic media or any other form." 45 CFR 164.103. In turn, health information means "any information, whether oral or recorded in any form or medium, that * * * [i]s created or received by a health care provider * * * and * * * [r]elates to the past, present, or future physical or mental health or condition of an individual; the provision of health care to an individual; or the past, present, or future payment for the provision of health care to an individual." 45 CFR 160.03.

HIPAA—CRIMINAL PENALTIES UNDER THE HEALTH INSURANCE PORTABILITY AND ACCOUNTABILITY ACT (42 U.S.C. 1320d-6)

This statute requires "covered" medical service providers, including emergency EMS departments that bill Medicare or Medicaid patients for services, to protect the privacy of individually identifiable heath information. The statute includes civil and criminal provisions, with penalties of imprisonment for up to ten years, a fine of up to $250,000, and a period of supervision following release from prison of up to three years.

HIPAA—LAW ENFORCEMENT ACCESS TO EMS RECORDS [45 CFR 164.512(f)]

Under these regulations, disclosures to law enforcement are authorized for gun shots, child abuse, and public heath purposes. The statute also authorizes limited information to be provided to law enforcement "in response to a law enforcement official's request for such information for the purpose of identifying or locating a suspect, fugitive, material witness, or missing person," or "the law enforcement official represents that immediate law enforcement activity that depends upon the disclosure would be materially and adversely affected by waiting until the individual is able to agree to the disclosure."

BLOODBORNE PATHOGENS (29 CFR 1910.1030)

This OSHA regulation requires fire and EMS organizations to perform a risk assessment and appoint an infection control officer. Many states have adopted similar regulations, including Ohio Administrative Code, Section 4123: 1-21-07, that require fire and EMS departments to "operate an infection control program that actively attempts to identify and limit or prevent the exposure of employees to infection and contagious diseases in the performance of their assigned duties." Regarding hepatitis B, the regulation of the Ohio Bureau of Workers' Compensation, Division of Safety and Hygiene (614-466-7053), states: "The employer shall make available, at no cost to the employee, the hepatitis B vaccine and vaccination series to all employees who have potential occupational exposure."

EMS PEER REVIEW AND QUALITY IMPROVEMENT RECORDS ARE NOT PUBLIC RECORDS

Some states encourage peer reviews by protecting these communications from discovery in civil suits. For example, Ohio Revised Code 4765.12 (2000) requires that "each medical service organization in this state shall implement ongoing peer review and quality assurance programs designed to improve the availability and quality of the emergency services it provides. These quality review records are not public records, and are "not subject to discovery in a civil action and shall not be introduced into evidence in a civil action against the emergency medical service organization...."

◆ **CASE STUDIES**

CASE STUDY 13-1   EMS Sued for Alleged Failure to Transport Gunshot Victim

Patricia Jackson v. Schultz, 429 F.3d 586 (6th Cir. 2005); U.S. Court of Appeals for the 6th Circuit, located in Cincinnati.

FACTS

Patricia Jackson filed a lawsuit against two Detroit EMTs, alleging deprivation of the constitutional rights (42 U.S. Code 1983) of her deceased son, Alter Keith Jackson. According to the complaint, the decedent sustained gunshot wounds in a bar on September 16, 2000. The bullet that ended decedent's life was fired "indiscriminately" by another bar patron. In response to the shooting, someone called 9-1-1. The Detroit Fire Department dispatched the EMTs to the scene. Another team of Detroit Fire Department EMTs was also dispatched (they were not sued).

According to the lawsuit, when the EMTs arrived on the scene, the decedent was alive but bleeding profusely. The EMTs placed the decedent in their ambulance, where they later "watched him die." Jackson alleges that fire department policy and proper procedure mandates the use of life support in this situation and that the EMTs did not administer life support.

Fire department policy also required the EMTs to transport the decedent to a trauma enter. The lawsuit alleges that the EMTs failed to transport the decedent even though a center was located less than two miles from the scene. The decedent died while in the back of the ambulance.

On September 16, 2003, the lawsuit was filed in federal District Court in Detroit. Patricia Jackson had previously pursued litigation in Michigan state courts, but those claims were **dismissed without prejudice.**

The suit alleged that the EMTs violated the decedent's constitutional "substantive due process rights" by providing him no medical care while he was in custody. The

dismissed without prejudice
When a lawsuit has been dismissed by the trial judge, the plaintiff may later file an amended complaint if additional facts are discovered.

state-created danger EMS personnel may be liable if they increase the risk of danger to a patient.

lawsuit also alleges that the EMTs' conduct amounts to a **state-created danger,** which triggers a right to medical care; that by being placed in the back of the ambulance, the decedent was left in a worse position than if the EMTs had simply left him on the sidewalk bleeding. The plaintiff alleges that if the decedent had simply been left on the street, the second set of EMTs might have provided life support and transportation to a trauma center. The lawsuit does not allege that any private aid was attempted or that there was any meaningful private aid on the scene.

On April 1, 2004, the EMTs filed a motion to dismiss the lawsuit on the grounds of "claim preclusion" (case already dismissed in Michigan state courts). The plaintiff's response argued that previous Michigan decisions concerning the same events could not have claim-preclusive effect because the state court dismissed her complaint without prejudice. On September 3, 2004, the district court denied the motion to dismiss based on claim preclusion. Claim preclusion is not an issue raised on appeal.

qualified immunity EMS and other governmental officials will normally not be held personally liable for their negligence, as long as they did not act with willful or wanton misconduct.

In reply to the plaintiff's response, the EMTs added a second ground to their motion to dismiss, arguing that they were entitled to **qualified immunity** and that the case should therefore be dismissed. The EMTs argued that (1) they did not violate the decedent's constitutional rights, and (2) even if they did, such rights were not clearly established.

Jackson responded to this new ground for dismissal by contending that the EMTs' conduct violated the decedent's clearly established substantive due process rights. Jackson argued that the due process clause required the state to render competent medical care to the decedent because he was (1) in state custody and (2) in a situation of state-created danger.

On September 3, 2004, the federal judge denied the EMTs' motion to dismiss based on qualified immunity. The court held that the plaintiff alleged the violation of a clearly established constitutional right. The EMTs filed an appeal.

YOU BE THE JUDGE

Should the EMTs be dismissed from the case?

HOLDING

Yes—they enjoy qualified immunity.
The three-judge panel of the Court of Appeals wrote:

> It is not a constitutional violation for a state actor to render incompetent medical assistance or fail to rescue those in need. The two applicable exceptions to this general rule are (1) the custody exception and (2) the state-created danger exception. Neither of these exceptions applies under any set of facts that could be proven consistent with Jackson's allegations.
>
> *A. The Custody Exception: the decedent was never in custody:* The "custody exception" does not apply because the decedent was never in custody. The "custody exception" triggers a constitutional duty to provide adequate medical care to incarcerated prisoners, those involuntarily committed to mental institutions, foster children, pre-trial detainees, and those under "other similar restraint of personal liberty." See [U.S. Supreme Court decision in *DeShaney v. Winnebago County Dep't of Social Services,* 489 U.S. 189 (1989) at page] 200.
>
> As the district court noted, "when the State takes a person into its custody and holds him there against his will, the Constitution imposes upon it a corresponding duty to assume some responsibility for his safety." *DeShaney* at 199–200. The overarching prerequisite for

custody is an affirmative act by the state that restrains the ability of an individual to act on his own behalf.

The district court improperly held that moving an unconscious patient into an ambulance is custody. This court's precedent has made clear that *DeShaney's* concept of custody does not extend this far. This court has never held that one merely placed in an ambulance is in custody. The proper custody inquiry is whether the EMTs engaged in a "restraint of personal liberty" similar to the restraints mentioned in *DeShaney* at 200.

The EMTs did not engage in a "restraint of personal liberty" similar to the restraints mentioned in *DeShaney* at 200. The restraints of personal liberty mentioned in *DeShaney* all require some state action that applies force (or the threat of force) and show of authority made with the intent of acquiring physical control. Because the decedent was unconscious, the proper custody inquiry focuses on what the EMTs did to restrain him. This is not to say that an unconscious person cannot be taken into custody. For example, custody might be found where an unconscious drunk is handcuffed and transported to jail.

In the instant case, the EMTs did nothing to restrain decedent. The EMTs did not cause decedent to be shot nor did they render him unconscious. There is no allegation that the EMTs restrained or handcuffed the decedent. There is no allegation that the decedent was not free to leave the ambulance or be removed from the ambulance. Decedent's liberty was "constrained" by his incapacity, and his incapacity was in no way caused by the defendants. In sum, no set of facts consistent with the allegations shows that the EMTs did anything to restrain the decedent's liberty. Thus, no set of facts consistent with the allegations supports a finding that the EMTs took decedent into custody. Based on the facts alleged, there is no constitutional violation under the custody exception.

B. State-Created Danger: the EMTs did not increase the risk that decedent would be exposed to private harms. Even liberally construing Jackson's allegations, she has also not pled sufficient facts to show a constitutional violation based on the "state-created danger" exception. To show a state-created danger, Jackson must plead (1) an affirmative act by the EMTs that creates or increases a risk that the decedent would be exposed to "private acts of violence," (2) a special danger to the decedent such that the EMTs' acts placed the decedent specifically at risk, as distinguished from a risk that affects the public at large, and (3) that the EMTs knew or should have known that their actions specifically endangered the decedent. Because the first prong is not met it is unnecessary to consider the second and third prongs.

Jackson does not specify a theory of state-created danger. A liberal construction of Jackson's allegations permits two possible theories. First, the EMTs moved the decedent to a more dangerous area. Second, the EMTs moved the decedent to a place where it was less likely another state actor or private person would render aid. No set of facts consistent with the allegations supports a constitutional violation under either of these theories.

Risk of "private acts of violence" was not more probable inside the ambulance than outside the ambulance. Before being moved into the ambulance, the decedent was at a bar where a patron "indiscriminately" fired his weapon numerous times at bar patrons. After being moved, the decedent was in the back of an ambulance. No set of facts consistent with the allegations could support a finding that the risk of violence by a private party was more probable in the inside of the ambulance versus the scene of a bar where decedent was actually shot. Thus, the EMTs' action did not increase the risk that the decedent would be exposed to private acts of violence.

Jackson also does not state a constitutional claim that the EMTs hindered third party aid. Jackson cannot state a constitutional claim based on cutting off state aid because state actors do not have a general duty to render aid. See *DeShaney* at 196. But a constitutional claim may be premised on state action "cutting off" private sources of rescue without providing an adequate alternative.

Jackson alleges only one relevant fact to this inquiry: the EMTs put decedent in an ambulance. The EMTs did not discourage others from entering the ambulance. All evidence indicates decedent was free to leave (or be removed from) the ambulance. Furthermore, there is no evidence that any private rescue was available or attempted. No set of facts consistent with the allegations shows that the EMTs interfered with private aid. Thus, Jackson does not allege sufficient facts to support a claim for a constitutional violation based on cutting off private aid.

In sum, Jackson does not allege sufficient facts to support a constitutional violation.

No "Clearly Established" Constitutional Violation. Assuming that Jackson stated a viable constitutional claim, such a claim is not based on clearly established law. To defeat qualified immunity, Jackson must show a violation of a constitutional right, and that the constitutional right was clearly established. As noted above, there are no cases finding a constitutional right to medical care under these exact (or even vaguely similar) circumstances.

LEGAL LESSONS LEARNED

Thoroughly document your EMS run report, including time of arrival, vital signs, follow-up actions, and time of transport to a trauma center, so that the victim's family will have no doubt that you did all you could to save the patient.

CASE STUDY 13-2 Defibrillator Would Not Shock—Cardiac Arrest Patient

Mitchell v. Norwalk Area Health Services, 2005 WL 2415995 (Ohio App. 6 Dist; September 30, 2005), 2005-Ohio-5261.

FACTS

Deborah Mitchell, the surviving spouse and administrator of the estate of a cardiac arrest victim, filed suit against North Central EMS (NCEMS) and unnamed EMS personnel ["John Does" and "Jane Does" 1–10] for the death of her husband, alleging their willful and wanton misconduct. The trial court granted the defendants' motion for summary judgment, and Mrs. Mitchell filed an appeal to the Ohio Court of Appeals.

On August 5, 2001, Mrs. Mitchell was awakened from sleep by a noise in the living room where her husband was watching television. She went to investigate and found him sitting in a chair, gasping, and unable to breathe. After a "few moments," she called 9-1-1 and told them that her husband was "having trouble breathing."

Mrs. Mitchell woke her son, Thomas, who was 15 years old at the time. He, in turn, awoke neighbors in the building and asked if they knew CPR. No one did. Mrs. Mitchell called 9-1-1 a second time because she thought her husband had stopped breathing altogether. She and Thomas got Mr. Mitchell onto the floor, and she began chest compressions. She could not find her husband's pulse and described him as "stiff" and "foaming at the mouth."

Police officers arrived, and shortly after, the North Central EMS squad arrived. Mrs. Mitchell stated that the EMS personnel immediately put a bag over Mr. Mitchell's mouth. She also stated that she saw a defibrillator sitting there right away. She saw the EMS personnel doing chest compressions and placing defibrillator pads on Mr. Mitchell's chest. After the defibrillator was in place, she saw the EMS personnel "try" to defibrillate, but she saw nothing happen.

She remembered hearing one say "that they couldn't get a connection or something." Mrs. Mitchell then saw them try to deliver another shock with the defibrillator. Shortly after this second shock, a police officer offered to take her and Thomas to the hospital to wait for the ambulance, and she and Thomas left.

The life squad submitted a chronology of events occurring from the time Mrs. Mitchell first called 9-1-1, compiled from evidentiary materials including the NCEMS squad run report that the two squad members, paramedic Shannon Belcher and EMT Billie Jo Morrow, hand-wrote after transporting Mr. Mitchell to Bellvue Hospital:

"00:36—call received for difficulty breathing
00:42:53—Arrival of S. Belcher and B. J. Morrow
00:43:52—Second call comes in—patient stopped breathing
00:44—CPR started
00:45:41—M. Adamcio and D. Short dispatched
00:47—Shock attempt [with] 200 [joules] with Zoll
00:49:40—M. Adamcio and D. Short arrive
00:49:45—Retrieved the mobile life Physio/Lifepak
00:50—Shock attempt [with] 200 [joules] with Zoll
00:53—Shock attempt [with] 200 [joules] with Zoll—indicated and crossed out [on Belcher's run report]
00:56—1 mg Epinephrine/PEA indicated
00:57—1 mg Atropine/v-fib indicated
00:57—Attempted to shock with Zoll; patient in v-fib
00:59—Shocked @ 200 with Zoll, are noted—fine v-fib.
1:00—Attempted to shock 300 with Zoll monitor—unsuccessful
1:08—Physio Lifepak applied and turned on per Code Summary Sheet
1:11—Defib @ 360 with Lifepack [sic]
1:13—Defib at 360 with Lifepack
1:14—Defib at 360 with Lifepack
1:16—Defib 360 with Lifepack
1:20—Defib 360 with Lifepack
1:21—[Emergency room] arrival
1:22—Defib at 360 with Lifepack"

In her lawsuit, Mrs. Mitchell contends that the squad members' failure to *immediately* use the Lifepack defibrillator that the second squad brought with them, and their continuing defibrillator attempts with the non-functioning Zoll unit, constituted willful and wanton misconduct.

During her deposition, the EMT acknowledged that NCEMS protocol required all squads to maintain two sets of defibrillator pads in the squad truck. During Mr. Mitchell's run, only one set of pads was in the squad.

YOU BE THE JUDGE

Should the lawsuit be reinstated so a jury can decide if the EMS personnel were liable?

HOLDING

No.

The three-judge court (two to one decision) began by referencing the Ohio immunity statute, Ohio Rev. Code 4765.49(B). The court wrote that the statute "provides immunity to political subdivisions performing emergency medical services. The statute also blankets corporations and other business entities under contract with a political subdivision to provide such services."

The statute provides:

> A political subdivision, joint ambulance district, joint emergency medical services district, or other public agency, and any officer or employee of a public agency or of a private organization operating under contract or in joint agreement with one or more political subdivisions, that provides emergency medical services, or that enters into a joint agreement or a contract with the state, any political subdivision, joint ambulance district, or joint emergency medical services district for the provision of emergency medical services, is not liable in damages in a civil action for injury, death, or loss to person or property arising out of any actions taken by a first responder, EMT-basic, EMT-I, or paramedic working under the officer's or employee's jurisdiction, or for injury, death, or loss to person or property arising out of any actions of licensed medical personnel advising or assisting the first responder, EMT-basic, EMT-I, or paramedic, unless the services are provided in a manner that constitutes willful or wanton misconduct.

The court cited "expert testimony" by both parties. The life squad's expert witness, Dr. Barry R. Cover, M.D., testified similarly in deposition. However, his practice is "strictly limited" to internal medicine, and he stated that he had no training in cardiology, had not been an EMS or paramedic, and was not authorized to train, instruct, or test an EMT or paramedic. His work does include "advising" Port Clinton area EMT units by reviewing their EMS protocols. He defined a protocol as "lists of actions that are to be undertaken in specific medical cases"; they are devised and approved by physicians. Cover, as did the appellants' experts, also made legal conclusions as to whether the conduct of NCEMS personnel constituted willful or wanton conduct. However, he had no knowledge of what "specific policies or procedures" NCEMS should have regarding how many defibrillator pads to stock with machines or keep in squad vehicles. He did, however, opine about why the Zoll monitor failed to work:

> **[Testimony]:** "I think there was a problem with the pads themselves, that there was not proper patient—there was no proper contact of the pads. Obviously, there was enough contact so that they could gain a reading of the rhythm as they have from the rhythm strips, but without proper total pad contact, they weren't able to deliver the appropriate amount of energy."

NCEMS Testimony

Donald Ballah was the executive director of NCEMS and was responsible for administering policies and procedures. He had also been a paramedic since 1984. When asked what the "duty of the EMS crew" was, he answered: "The duty to administer the protocols as provided by our Medical Director." He stated that it was the policy of NCEMS to have each squad vehicle stocked with two sets of defibrillator pads. He also stated that it would surprise him if a squad did not have two sets of pads; even if a squad had used a set of pads on a run, they would be expected to restock the squad vehicle with a second set.

Ballah also stated that in the protocol with respect to defibrillating a patient in cardiac arrest, paramedics were to follow "current ACLS guidelines." He insisted that those protocols were followed. With respect to the crew's choice to continue using the Zoll instead of switching to the available Lifepack, he also acknowledged that (in response to a rather convoluted question regarding the circumstances of the event) he "probably would have asked for the second defibrillator"; however, he would not give an answer as to the point at which he would have switched machines.

Analysis

Although plaintiffs have pointed to issues of fact, we agree with the trial court that these facts are not material insofar as they fail to create a genuine issue of whether appellees' misconduct was willful or wanton. Each expert and appellees' witnesses testified to substantially the same standard of care for paramedics delivering emergency care to cardiac arrest patients. We emphasize here that the question is whether the acts in *breaching* that standard of care evidence willful or wanton misconduct. The trial court fittingly characterized the level of care delivered to Mr. Mitchell—particularly in the squad's "choice" to continue to use the Zoll instead of expediently switching to the Lifepack—as negligent and "inept." Further, we agree with the trial court that the mere piling up of negligent acts does not, by virtue of sheer volume, thereby convert negligence into willful or wanton acts.

The Court of Appeals explained,

A defendant might be guilty of the grossest negligence, and his acts might be fraught with the direst consequences, without having those elements of intent and purpose necessary to constitute willful tort.

Negligence and wanton are mutually exclusive terms, implying radically different mental states. In particular, in order to advance issues of wanton misconduct, there must be a disposition to perversity on the part of the tortfeasor and such perversity must be under such conditions that the actor must be conscious that his conduct will in all probability result in injury. Unlike a willful act, intent need not be present for an act to be wanton, but the actor must be conscious that the probability that harm will result from such failure is great.

Applying these legal standards to the facts, having construed those facts in a light most beneficial to appellants, we must conclude that the paramedics' acts could not have constituted wanton misconduct. Knowledge that a sudden cardiac arrest patient will, in all probability, suffer great harm from a failure to administer proper care will always exist for every trained paramedic and EMT; thus, this factor carries little weight in circumstances such as these. That is, while the harm may be of greater or lesser degrees in other professions, death is the only certain outcome for paramedics caring for a sudden cardiac arrest patient if *proper* care is not delivered; this situation demonstrates that even with proper care, the risk of serious injury or death is high; with negligent care, the risk is even greater. Even when all proper choices are made by a paramedic when delivering care in these circumstances, death is still a significant probability. Owing to the specific situations paramedics encounter, we find the trial court's analogies between choices made by an automobile driver to the choices faced by paramedics to be weak ones. Thus, in these circumstances, emphasis may be appropriately placed upon the element of absence of all care and failure to exercise any care. Hypothetically, future facts may be such that *some* care was rendered, yet the care may be wantonly applied; thus, a grant of summary judgment under this legal standard to these facts does not foreclose this future possibility; however, such is not the case here.

The court stated,

[The life squad] crew members asserted that they had delivered defibrillating shocks to the decedent within the first few minutes of their arrival on scene. [Plaintiffs] argued that because their key witness, a police officer on the scene, stated that he saw no paramedic from the first squad use a defibrillator, a genuine issue of fact arose as to whether the paramedics recklessly disregarded treatment protocols and the standard of care by failing to deliver timely shocks. Most notably, as here, there was an absence of rhythm strips from the first defibrillator supposedly used by the first squad; however, rhythm strips did exist from a second defibrillator which evidenced that the second squad delivered defibrillating shocks some 20 minutes "later." That court concluded, "even if the first team of paramedics failed to use or have a defibrillator, that is not per se evidence of willful or wanton conduct."

Factually, the only difference is that here, a working defibrillator was on-scene for approximately 20 minutes *after* the paramedics attempted to use the Zoll unit and it failed twice; hence, there is no dispute that a functioning unit was available.

As previously noted, *any* deviation from the standard of care for paramedics attending to a sudden cardiac arrest patient carries a high probability of significant harm. From this evidence, a court could correctly conclude that the inferences drawn in appellants' favor do not, as a matter of law, create any issues with respect to the legal standard of willful or wanton misconduct on the paramedics' part.

LEGAL LESSONS LEARNED

Governmental immunity of political subdivisions and qualified immunity for fire and EMS personnel is an extremely important legal defense; but to avoid litigation, EMS personnel should carry extra pads and extra batteries on their defibrillators. Fire and EMS departments should have thorough, daily procedures to inventory supplies and check for missing or malfunctioning equipment.

CASE STUDY 13-3 Revocation of EMT's Certificate—Crimes of "Moral Turpitude"

Holycross v. State Board of EMS, 837 N.E.2d 423, 163 Ohio App. 3d 213, 2005 Ohio 4598 (Ohio 2005).

FACTS

Disciplinary proceedings were initiated by the Ohio EMS Board against EMT-B Nathan Holycross, after he pled guilty to three misdemeanors: telephone harassment, attempted telecommunications harassment, and criminal trespass. The Ohio State Board of Emergency Medical Services revoked the EMT's license to practice as an emergency medical technician. The EMT filed an appeal to the Ohio Court of Common Pleas Court, which upheld the revocation, and he now appeals to the Ohio Court of Appeals.

The EMT contends that the trial court erred in holding that he had committed crimes involving "moral turpitude"—the Ohio EMS Board's basis for revoking his license—and that the decision violated his constitutional right to due process of law.

As a result of these convictions, Holycross was notified by the Ohio EMS Board that it was proposing to take disciplinary action against him and that he had the right to request a hearing. He did not request a hearing. The Ohio EMS Board heard no evidence but found that the misdemeanor offenses of which Holycross had been convicted met the definition of "moral turpitude" as defined in Ohio Administrative Code Chapter 4765-1-01 and permanently revoked his certificate to practice as an emergency medical technician "at any level."

Holycross became enamored of Ashley Erwin, the 15-year-old daughter of his coworker, Captain Greg Erwin. Erwin was upset by this and desired to end all contact between Holycross and his daughter. The offense of attempted telecommunications harassment is described in the Champaign County Sheriff's Office report as an attempt by Holycross to send an e-mail message to Ashley Erwin through her MSN instant messenger service. When Ashley Erwin saw that Holycross had initiated contact, she had the option to accept the message and accept Holycross on her "buddy list" or to block him from further contact. She notified her father and blocked contact. This occurred on September 16, 2002, after Ashley's father had told Holycross not to have any further contact with her. There is no indication what this attempted message contained or that Ashley ever received it.

There is no description of the telephone harassment offense of which Holycross was convicted. Other than the fact that he was convicted of the offense, no facts were laid before the Ohio EMS Board concerning this offense.

Regarding the criminal trespass offense, Ashley Erwin gave a written statement concerning this offense, which is worth quoting in full: "He write me notes talking about getting married. We got close over the summer and I liked to be there to talk to but he started writing letters telling him his feeling for me. He would sometimes come up to my room while I was asleep and kiss me on the cheek or forehead but then he would leave. At first, I kinda enjoyed the attention but after a while it got old and I tried to let him know to back off by telling him that dad was getting upset and he needed to stop saying stuff like we were going to get married. I also ignored him and tried to stay out of the same room as him but he kept giving me notes."

The police also obtained a statement, signed by Ashley Erwin, of her handwritten answers to questions written by the detective. Some of these answers are illuminating:

Q. "Did Nathan Holycross ever enter your home, at * * *, when Greg Erwin, your father was not at home? Did Nathan Holycross enter the house without being invited in?"
A. "Yes, a lot of times, most of the time he would knock and then would come in before you could even get to the door."

Q. "Did Nathan Holycross enter your house at * * * without permission, after your father, Greg Erwin, had notified him not to be in or around the house? When was this?"
A. "Yes, in November but I'm not for sure."

Q. "How many times did Nathan Holycross enter your home without permission?"
A. "A lot of times, probably about 10–12 times."

Q. "What did you, Ashley Erwin, do when Nathan Holycross entered your home without permission?"
A. "Sometimes we would just talk for a while which normally ended with a hug. Towards January when I saw him pull in I would go to my room and pretend to be asleep."

Q. "Did you, Ashley Erwin, ever let Nathan Holycross into your home when your father, Greg Erwin, was not present?"
A. "Yes, if I got to the door before he came in. Normally he would come in on his own."

She was later asked:

Q. "Has Holycross made contact or attempted to contact you by telephone, mail or any message since the date of his conviction for Criminal Trespass? If so, how?"
A. "No. He asked me for my email address after the conviction of trespass."

Q. "Has Holycross' recent attempts to contact or instances of contact caused you any distress, hardship or problems at home?"
A. "No."

YOU BE THE JUDGE

Should Holycross' EMT certificate to practice be revoked?

HOLDING

No.

The three-judge court wrote:

We can make several conclusory observations based upon our review of the entire record of the administrative proceedings. One is that there is no allegation that Holycross ever touched an erogenous zone or did anything with Ashley Erwin more intimate or familiar than kissing her cheek and forehead while she was asleep in bed. The other is that Ashley's father, Greg Erwin, was severely distressed, if not outraged, by Holycross's contacts with his daughter after Holycross was told to stay away from her.

The issue we are called upon to decide is whether it can reasonably be found, from the evidence before the board, that the criminal-trespass offense of which Holycross was convicted involved moral turpitude. In other words, did it involve baseness, vileness, or depravity? Again, this is a different question from whether the conduct was wrong; obviously, it was, and Holycross has appropriately been convicted of a criminal offense and subjected to a criminal sanction in consequence.

Under other circumstances, we would be prepared to say that the unauthorized entry into the bedroom of a woman of any age, but particularly of the age of 15, followed by kissing that woman on the cheek or forehead, could be an act involving baseness, vileness, or depravity. If the perpetrator had reason to believe that his conduct would be highly offensive to the woman, we would be prepared to say that a reasonable finder of fact could find that conduct to be base, vile, and depraved, since, even though the kiss is chaste enough, the intrusion into the private sanctum of the bedroom for that purpose is likely to be so offensive to the personal security of the victim as to constitute a gross violation of one's duty to another. We hasten to add that no special relationship between perpetrator and victim would be necessary to make that finding—the unauthorized entry of a stranger into a woman's bedroom for the purpose of stealing a kiss, again, no matter how chaste, could reasonably be deemed to be likely to be so offensive to the victim's sense of personal security as to be base, vile, or depraved.

In the case before us, however, all of the evidence in the record calls out for a conclusion that, unlike her father, Ashley Erwin was not likely to find Holycross's unauthorized entry into her bedroom and kissing of her cheek or forehead to be highly offensive. It clearly appears that she was not, in fact, highly offended, even if she may have been somewhat offended. The nature of the relationship between Holycross and Ashley Erwin that clearly appears from this record belies any likelihood that she would find his conduct highly offensive.

But what of the evident outrage that Holycross's conduct visited upon Greg Erwin, Ashley's father? This is perhaps the closest question in this case. We are not prepared to hold that persisting in giving one's attention to a 15-year-old girl, in the face of her father's express, and strongly voiced, disapproval is base, vile, or depraved, wrong though it may be. After all, notwithstanding the violent offense that he undoubtedly gave to Lord Capulet by persisting in pursuing Capulet's daughter, who, if memory serves, was even younger than 15, who can find it in their hearts to declare Shakespeare's Romeo vile, depraved, or base?

We emphasize that in reviewing the evidence in the record, we have credited all of the evidence upon which the board could possibly rely, and have ignored contrary evidence. (Holycross told the investigating officers, for example, that he kissed Ashley Erwin while she was asleep on a downstairs couch, not in her bedroom.) Furthermore, we have given what we regard as due deference to the trial court. Nevertheless, we find no reasonable basis in the record of this administrative proceeding for a finding that the misdemeanor offenses of which Holycross was convicted involve moral turpitude.

LEGAL LESSONS LEARNED

EMS and fire department employees should use extreme caution when pleading guilty to offenses that may result in loss of their certificates to practice, because the courts normally give great deference to the decisions of administrative agencies such as the Ohio EMS Board. Fire departments may also have an obligation to report convictions of personnel to their state EMS boards.

■■■

Chapter Review Questions

1. In Case Study 13-1, *Patricia Jackson v. Schultz,* the court discussed "qualified immunity" for the EMTs when treating the gunshot victim who died in their ambulance. Discuss whether EMS personnel in your state purchase insurance to cover the risk of lawsuits, and whether such insurance is helpful when EMS personnel want to have private legal counsel in addition to the city's attorney.

2. In Case Study 13-2, *Mitchell v. Norwalk Area Health Services,* the court discussed liability for failure to carry a second set of pads for the defibrillator. Describe the inventory control policies at a fire and EMS department in your state and how those inventory check sheets can be extremely helpful in litigation.

3. In Case Study 13-3, *Holycross v. State Board of EMS,* the court decided that the administrative record did not support a finding of "moral turpitude." Describe what criminal offenses must be reported to your state board. Identify those misdemeanor (non-felony) convictions that must be reported.

■■■

Expand Your Learning

Read and complete the individual student or group assignment, as directed by your instructor.

1. In *Davidson v. City of Jacksonville, FL,* 395 F.Supp. 1291 (M.D. Fla. 2005), a patient filed a lawsuit in federal District Court, alleging his constitutional rights were deprived when Jacksonville Fire and Rescue personnel responded to a 9-1-1 call by Mr. Davidson's wife. The paramedics were advised that her husband was a diabetic and that he had been experiencing back pain recently. He was unresponsive and had an altered state of consciousness. The paramedics forcibly held him down in bed to set up an IV in his arm, which he then pulled out. It was eventually re-inserted, and they administered dextrose. The paramedics called for assistance, and a police officer handcuffed him. His ankles were bound with bandages, and he was "hog

tied" in order to transport him to the hospital. The hospital determined he had a herniated disc in his back and had suffered a permanent disability. The federal trial judge dismissed the lawsuit, because the paramedics did not intend to harm him.

Describe when it is appropriate to have police handcuff a patient and what procedures should be followed when the handcuffed patient is transported to the hospital.

2. In *Betnacur v. City of New York,* 782 N.Y.S. 2d 446 (N.Y. App. Div. 2004), a group of EMS personnel filed suit against the City of New York for injuries suffered while at a labor demonstration by off-duty correctional officers outside a city correctional facility. The injured EMS workers claim that city police officers directed them into a staging area for the demonstration. When they waded into the mob to rescue a coworker, they were attacked, and police failed to respond and protect them. The Court of Appeals held that the city was not liable, because the police have only a general duty to the public as a whole and there was no specific duty to assist EMS personnel.

When EMS are dispatched to a police scene, describe how incident command should be coordinated with police officials so there is an integrated command response. Describe any police/EMS joint training that has been held by fire departments in your state.

See Appendix A for additional Expand Your Learning activities related to this chapter.

Physical Fitness

14 CHAPTER

Annual physicals are given by Dr. William Lovett, who tows his medical van to the fire station to provide exams for all three shifts of on-duty personnel at Deerfield Township (OH) Fire and Rescue Department. *Photo by author.*

Key Terms

irreparable harm, p. 186   **special need showing, p. 187**

From the Headlines

U.S. Fire Administration—Cardiac Arrests

Study reveals 43.7% of the firefighters who died while on duty between 1995 and 2004 had cardiac arrests (440 out of 1006 deaths). Many had pre-existing cardiac problems (*www.usfa.fema.gov*).

Firefighter Dies of Cardiac Arrest—Fire Department Has No Physical Ability/Agility Testing

On November 29, 2005, a NIOSH Fire Fighter Fatality Investigation report, "Volunteer Fire Fighter Suffers Cardiac Arrest While Battling A Structure Fire—New

Dr. William Lovett's medical testing van is equipped with all of the necessary devices. *Photo by author.*

York," was issued concerning a May 13, 2004, line of duty death. The report's Recommendation 5 states, "Conduct pre-placement and annual physical performance (physical ability) evaluations on FFs to ensure that they are physically capable of performing the essential job functions of structural fire fighting." At present, the department does not require periodic physical ability/agility testing. No fitness equipment is currently available at the fire station and no formal wellness program has been established (*www.cdc.gov/niosh/face200432.html*).

Job Stress—Firefighter Wins Workers Comp for Heart Attack

The Alabama Court of Civil Appeals decided in favor of a retired firefighter, who won worker's comp coverage for the heart attack he suffered in 2000, finding it was caused by job stress.

◆ INTRODUCTION

This chapter addresses the issue of physical fitness and increased management attention to firefighter wellness programs, including the International Association of Fire Fighters/International Association of Fire Chiefs joint labor management wellness/fitness initiative (2000). See the Cincinnati Fire Department's wellness agreement in their Labor-Management Agreement, in Appendix F.

Newark (OH) Fire Department's new Station 3, with a complete workout facility in the basement. *Photo by author.*

◆ NIOSH RECOMMENDATIONS ON FIREFIGHTER FITNESS

On May 17, 2006, NIOSH issued Firefighter Fatality Investigation Report F2005-30 concerning the death of a 58-year-old New Jersey firefighter, who suffered sudden cardiac death on May 31, 2005, while exercising at his fire station (*www.cdc.gov/ niosh/fire/reports/face200530.html*). The firefighter was found collapsed on the rowing machine. An autopsy revealed a severe narrowing (75 percent) of the left anterior descending coronary artery. The fire department has 252 uniformed personnel in six fire stations, with a voluntary wellness/fitness program but no mandatory annual medical evaluations except for hazardous materials technicians. The NIOSH recommendations should get the attention of every fire department, including the following:

- Annual medical evaluation for *all* firefighters (see NFPA 1582; IAFF/IAFC wellness/ fitness initiative (2000)
- SCBA—Use OSHA respiratory standard medical questionnaire and clearance (see 29 CFR 1910.134).
- *Mandatory* wellness/fitness program for firefighters (see NFPA 1583; IAFF/IAFC Fire Service Joint Labor Management Wellness/Fitness Initiative of 1997)
- Annual physical ability evaluation (see NFPA 1500)

◆ LITIGATION: MANDATORY FITNESS AND OFF-DUTY ACTIVITIES

There has been litigation challenging fire department wellness/fitness programs that contain "mandatory" requirements (see Case Study 14-1). There has also been litigation concerning workers compensation coverage for firefighters injured in off-duty activities "encouraged" by the fire department. For example, in *David Hardt v. Town of Watertown,* 95 Conn. App. 52, 2006 Conn. App. LEXIS 190 (April 25, 2006), the Appellate Court affirmed the denial of workers compensation coverage for a volunteer deputy fire chief who injured his knee playing in the department's basketball program while off-duty. The fire department did not have a structured wellness/fitness program. Instead, senior officers encouraged a weekly basketball program.

The majority of the court (two to one decision) held that the basketball game was not "training" and therefore not covered by workers compensation. The court wrote, "On April 23, 2001, the plaintiff injured his knee while playing basketball in the department's basketball program. The chief of the department described the plaintiff's injury as having occurred while he was participating in the voluntary fire department sponsored open gym (physical fitness program)."

The court held that the basketball game was not "training."

In addition to the grammatical structure of the phrase, the dictionary definition of the word "training" supports the board's conclusion that the term refers to training directly related to firefighting rather than to general physical fitness. The primary definition of training is "the education, instruction, or discipline of a person . . . that is being trained." Random House Webster's Unabridged Dictionary, supra. When the phrase "in training for" is read as modifying "volunteer fire duty," we understand it to mean the education, instruction or discipline of a person that is being trained in fire duties, which are defined more specifically in *General Statutes § 7-314 (a)*. That subsection provides in relevant part that fire duties include "duties performed while at fires, while answering alarms of fire, while answering calls for mutual aid assistance, while returning from calls for mutual aid assistance, while directly returning from fires, while at fire drills or parades, while going directly to or returning directly from fire drills or parades, while at tests or trials of any apparatus or equipment normally used by the fire department, while going directly to or returning directly from such tests or trials, while instructing or being instructed in fire duties, while answering or returning from ambulance calls where the ambulance service is part of the fire service, while answering or returning from fire department emergency calls and any other duty ordered to be performed by a superior or commanding officer in the fire department. . . ." *General Statutes § 7-314 (a)*. Nothing in this definition persuades us that training for fire duties means training for the general physical demands of the position, as opposed to learning about and practicing the skills associated with fighting fires. Although we acknowledge that firefighting requires a certain degree of physical fitness, we are unable to conclude that members of volunteer fire departments are entitled to workers' compensation for injuries sustained while they are engaged in purely voluntary physical fitness activities.

The dissenting judge wrote,

I respectfully disagree with the conclusion of the majority that the participation by the plaintiff, David Hardt, in the open gymnasium program (program) arranged by the Watertown volunteer fire department (department) for the exclusive use of its members did not constitute training for volunteer fire duty under *General Statutes § 7-314 (a)*.

It thus is axiomatic that physical fitness is a prerequisite to adequate performance of fire duties. For that reason, volunteer firefighters in Watertown, like those in other municipalities, are required to pass annual physical examinations. Accordingly, I would conclude that, in certain circumstances, physical fitness programs may constitute training for volunteer fire duty. In my mind, the present case is such an instance.

The record reveals that the program was exclusive to department members. The department organized and regularly promoted the program, and it encouraged participation therein by its members. The chief of the department characterized the program as a "physical fitness program which is also recreational," and the commissioner found that the program was sponsored by the department "to promote physical fitness among

[its] members. . . ." The commissioner further found that "the major purpose of the program was not recreational." Under these circumstances, I would conclude that participation in this particular physical fitness program constitutes training for volunteer fire duty.

◆ **NFPA STANDARDS**

- *NFPA 1500: Standard on Fire Department Occupational Safety and Health Program* (2002 edition). This standard identifies actions that a fire department can take to minimize the risks to firefighters, including a comprehensive physical fitness program.
- *NFPA 1582: Standard on Comprehensive Occupational Medical Program for Fire Departments* (2003 edition). NFPA Standard 1582 calls for annual physical exams. This standard identifies medical conditions that can pose significant risks to a firefighter. It calls for a pre-employment (baseline) physical exam, as well as annual physical examinations.
- *NFPA 1583: Standard on Health-Related Fitness Programs for Fire Fighters* (2000 edition). This standard identifies the necessary steps in establishing a physical fitness program, including the involvement of the physician, baseline measurements, and evaluating the impact of the fitness program.

Newark (OH) workout equipment. *Photo by author.*

PUBLIC SAFETY OFFICER DEATH BENEFITS ACT

This statute, 42 USC 3796-3796c, was amended in 2003 by the Hometown Heroes Survivors Benefits Act of 2003 to extend public safety officer death benefits to include those who die within 24 hours of strenuous on-duty activity.

U.S. DEPARTMENT OF JUSTICE REGULATIONS

The U.S. Department of Justice administers the death benefits program through its Bureau of Justice Assistance (BJA). Their regulations broadly define line of duty deaths of a public safety officer to include "any action the officer is so obligated or authorized to perform." 28 CFR 32.2 (c)(1).

EEOC REGULATIONS

EEOC regulations prohibit employers from imposing physical fitness requirements that have a disparate impact on older employees unless "reasonably necessary for the specific work to be performed." 29 CFR 860.103(f)(1)(i) (1970).

FIREFIGHTER HEALTH AND WELLNESS PROGRAMS

Some states have issued regulations requiring wellness programs. For example, in Ohio, the Bureau of Workers' Compensation, Division of Safety and Hygiene, issued Ohio Administrative Code 4123:1-21-07, requiring fire departments to "establish and provide a health-related fitness and wellness program that enables members to develop and maintain a level of health and fitness to safely perform their assigned functions. This program will not be punitive, as the purpose of the program is to improve the health and well being of the individual."

PUBLIC SAFETY OFFICERS' BENEFIT ACT OF 1976

This statute established federal death benefits for families of firefighters and other public safety officers who die in the line of duty.

◆ **CASE STUDIES**

CASE STUDY 14-1 Physical Fitness and Wellness Program—Annual Medical Exam

Weimer v. City of Baton Rouge, 915 So. 2d 875 (La. App. May 6, 2005).

FACTS

In July, 2003, the City of Baton Rouge (LA) Fire Department implemented a "Wellness and Fitness Program" calling for the periodic physical fitness testing of firefighters, an exercise regimen, and annual medical examinations. The program was launched without the approval of the Municipal Fire and Police Civil Service Board or the State Examiner, which establishes statewide civil service requirements.

Firefighter Gregory J. Weimer, three other firefighters, and IAFF Local 557 filed a petition in District Court for East Baton Rouge Parish, alleging the program violated the "classification plans" approved by the state and local civil service commission. The parties agreed to a joint stipulation of uncontested facts that was submitted to the trial judge:

1. Firefighter Weimer and the other plaintiffs are classified employees of the fire department serving with permanent status.
2. IAFF Local 557 is the collective bargaining agent for the firefighters.
3. The fire department and Fire Chief Ed Smith implemented the wellness and fitness program on July 31, 2003.
4. Firefighters were ordered to complete a detailed medical history.
5. Each firefighter must undergo an annual physical examination.
6. The program was implemented without approval of IAFF Local 557.
7. The program was not presented to or adopted by the Municipal Fire and Civil Service Board, nor by the State Examiner.
8. The firefighters and union would testify that, in their opinion, the program goes beyond what is reasonably necessary to test the relative capacity and fitness of a firefighter, and the program presents confidentiality and invasion of privacy issues.
9. Chief Ed Smith would testify that he implemented the program in conjunction with the International Association of Firefighters and International Association of Fire Chiefs Joint Labor Management Wellness and Fitness Initiative.
10. State Examiner Melinda Livingston would testify that since the program does not affect firefighter hiring or promotion, neither her office nor the Municipal Fire and Police Civil Service Board has any duties or responsibilities concerning the program.

The trial judge issued an injunction, finding that it violated Louisiana Statute 33:2483, and prohibited the fire department from implementing portions of the program. The fire department filed an appeal to the Court of Appeals of Louisiana.

YOU BE THE JUDGE

Should the program be enjoined?

HOLDING

No. The three-judge Court of Appeals wrote: "The record does not reveal anything in the Program that could affect the firefighters' classification plan."

The IAFF/IAFC program outlines a comprehensive system to improve the wellness of the fire department uniformed personnel. "It is to be implemented as a nonpunitive, positive, individualized program."

The major areas of the program implemented by the Baton Rouge Fire Department are the annual medical evaluation, fitness regimen and evaluation, rehabilitation provisions, and a cardiac stress test every five years. As part of the program, each fire substation was provided with fitness/strength equipment and treadmills. Firefighters are required to participate in physical fitness training for up to one hour and 30 minutes for 24-hour shifts, and 45 minutes for straight day (40-hour) personnel.

It appears that the firefighters' main objections relate to the extensive medical questionnaire and the lack of written guarantees that an employee who fails to pass an annual physical or fitness evaluation cannot be disciplined or denied promotions. Regarding the medical questionnaire, the Court of Appeals wrote,

> The Program provides that fitness evaluations are to be conducted by contracted professional physical therapy technicians, who will communicate the results of the evaluation only to the firefighter and to the doctors who are contracted to provide the Program's medical evaluations.

Regarding the non-disciplinary issue, the court wrote,

> (T)he only evidence presented on this issue was the statement of Chief Ed Smith that the Program is not designed to test the firefighter's fitness for entrance or promotion and that the results will not be used for discipline or to deny promotions. This statement is supported by the [IAFF/IAFC Program] documents, and there was no evidence to refute it.

The court also noted that the firefighters and the union did not contest portions of the program that have significant monetary value.

> Under the Program, unless a firefighter chooses to use and pay his own physician for the annual medical evaluation, an option which is allowed, the valuable health benefits of the comprehensive annual medical examination and the five-year cardiac stress test are provided during working hours at no cost to the firefighter. The firefighters do not seek removal of the exercise equipment from the substations, nor did they ask to be relieved from being paid to use this equipment during work hours. Similar training could only be obtained by making individual purchases of such equipment or by investing in costly memberships at health and fitness clubs.

irreparable harm For plaintiffs seeking an injunction, such as firefighters challenging a wellness and physical fitness program, there must be evidence that the plaintiff will be permanently harmed unless the program is enjoined.

In conclusion, there was no basis for the injunction. There "was no showing that the firefighters or the local union will suffer any harm, much less **irreparable harm,** if the Program is fully implemented. . . . All costs of this appeal are assessed against the firefighters and the local union.

LEGAL LESSONS LEARNED

Fire departments intending to launch a fitness/wellness program should prepare a written program description, which stresses the non-compulsory nature of the program, including confidentiality of all medical and fitness information concerning the firefighters participating. If there is a collective bargaining agreement in place, the program should be reviewed with the union.

CASE STUDY 14-2 Blood Draws, Invasion of Privacy, and Physical Fitness and Wellness Program

Jason Anderson, et al. v. City of Taylor, Case No. 04-74345, _____ F. Supp. _____ (E.D. Mich. August 11, 2005; Judge Paul D. Borman, U.S. District Court in Detroit). The unpublished decision can be read at Administrative Office of U.S. Courts PACER free written opinions (*http://www.pacer.uscourts.gov*) Case 2:04-cv-74345-PDB-RSW.

FACTS

The Taylor Fire Department applied for and received a $105,400 FEMA grant for its wellness program. The program included mandatory health appraisals by a local hospital, including a mandatory blood draw for the Lipid Profile test to determine the levels of triglycerides and HDL, VLDL, and LDL cholesterol.

A group of firefighters filed suit in federal District Court, claiming the blood draws violate their constitutional rights. The city filed a motion for summary judgment, arguing that the U.S. Supreme Court in *National Treasury Employees Union v. Van Raab,* 489 U.S. 656 (1989), the court upheld drug testing of U.S. customs agents, because the government had a "compelling interest" in confirming their agents were drug-free. The city of Taylor (MI) likewise argued that they have a compelling interest in making sure that firefighters are physically fit to perform the job.

YOU BE THE JUDGE

Should the city be allowed to take mandatory blood draws?

HOLDING

No.

The federal trial judge held that the blood draws violate the right of privacy of the firefighters. The judge said that the Supreme Court's decision in *National Treasury Employees Union v. Raab* was not applicable. The trial judge wrote,

> In certain circumstances, a search or seizure unsupported by probable cause may be constitutional when special needs, beyond the normal need for law enforcement, make the warrant and probable cause requirements impracticable. Only if we can say that the government has made that **special need showing** do we then inquire into the relative strengths of the competing private and public interests to settle whether the testing requirement is reasonable under the Fourth Amendment. If the government has not made its special-need showing, then the inquiry is complete, and the testing program must be struck down as unconstitutional.

special need showing   When a fire department seeks to justify blood draws and other invasions of privacy, a public employer must demonstrate a special need to protect the public.

The trial judge concluded,

> [The City] failed to establish that there is a special need for the blood draws. The Court finds that there has been no showing by the City that there has been a problem with physically unfit fire and emergency response workers. Other courts have set forth, in the context of drug testing, the importance of determining whether the targeted group exhibits pronounced drug problems. Analogizing the mandatory blood test to drug testing procedures, the Court finds that there is no evidence that firefighters experience pronounced fitness problems. Nor is there a high degree of harm to the public from high cholesterol readings from the Plaintiffs.
>
> The City has not shown any indication of a concrete danger to public safety demanding departure from the Fourth Amendment's main rule. The instant blood draws were used to determine employee's cholesterol level. A cholesterol reading, while an important health barometer, cannot accurately determine the overall fitness of an employee as it related to how the employee is able to respond in an emergency situation. Further, any risk from high cholesterol is likely to take years to manifest. The Court finds that this is not a situation involving a high risk of harm to the public such as an intoxicated employee with a firearm or a drug-impaired train conductor. Thus, the Court finds that the City has failed to articulate a special need for the blood draws.

LEGAL LESSONS LEARNED

The City of Taylor may decide to file an appeal to the 6th Circuit Court of Appeals. The case points out the importance of including the firefighter's bargaining unit in the process of establishing the health and wellness program, in which mutually acceptable language may address the issue of blood draws and the handling of this personal medical information.

CASE STUDY 14-3   Physical Ability Testing, Adverse Impact on Female Firefighters

Victoria Pietras v. Board of Fire Commissioners of the Farmingville Fire District, 180 F. 3d 468 (2nd Cir. 1999).

FACTS

The Farmingville (NY) volunteer fire department, with about 100 firefighters, requires all probationary members to pass a physical ability test (PAT). The applicants had to pass nine tests, within specified times, including a "charged hose drag" weighing about 280 pounds, dragged over 150 feet.

In order to set the time limits, the fire department in 1993 gave the test to 44 test-takers, including one female, full-time firefighter. They determined that the tests should take four minutes each (the males averaged 3:30, and the female 5:30). Plaintiff Victoria Pietras took 5:21.

Pietras took the test twice, and each time failed the four-minute test. Six other probationary female firefighters took the test, and four out of seven passed it. Most of the male probationary volunteers passed it (63 out of 66). Pietras was terminated after she failed the PAT test for a second time.

She filed a charge of sex discrimination with the New York Division of Human Rights, and it was forwarded to federal EEOC. Pietras filed suit in federal court, claiming the PAT four-minute test had a "disparate impact" on females.

She supported this claim with the testimony of Dr. Robert Otto, an expert exercise physiologist. He conducted an extensive review of the physical agility tests administered by various volunteer and paid fire departments and concluded (a) the four-minute limit in the Farmingville Fire Department test had a disparate impact on woman, and (b) it was not job related.

The federal trial judge held a bench trial (no jury), took testimony, and then ruled for the plaintiff, noting that the male passage rate of the PAT was 95 percent (63 out of 66), whereas the female pass rate was only 57 percent (4 out of 7). Relying on the "four-fifths rule" (or 80 percent) in the EEOC Guidelines, the trial judge concluded that "a pass rate for women which is less [than] four-fifths (or 80%) of the pass rate for men typically signifies disparate impact sufficient to establish a prima facie case."

The city filed an appeal to the U.S. Court of Appeals for the 2nd Circuit.

YOU BE THE JUDGE

Should the fire department be ordered to reinstate the plaintiff as a volunteer firefighter?

HOLDING

Yes.

The three-judge Court of Appeals wrote,

The statistical analysis of the PAT results were designed to evaluate the impact of the test on women generally, and not merely on women who were probationary firefighters.

We have previously considered the EEOC's four-fifth's rule for determining whether statistical evidence gives rise to an inference of disparate impact.... Under this rule, a pass rate for women that is less than 80% of the pass rate for men constitutes prima facie evidence of disparate impact. See 29 CFR 1607.4(D)(1988). In the instant case, as the District Court found, the PAT pass rate for women (57%) was less than 80% of the pass rate for men (95%). The court therefore drew an inference of disparate impact.

In the case before us, Pietras presented more than just statistics. After conducting an exhaustive analysis of the practices of other fire departments, Dr. Otto provided expert testimony on the disparate impact of Farmingville's PAT. This expert testimony, combined with the statistics Pietras did present, comfortably tips the scales in favor of the district court's finding of disparate impact.

LEGAL LESSONS LEARNED

Physical ability tests that have a statistical "disparate impact" will be struck down unless there is clear expert testimony that the test is a fair measure of real job skills. The U.S. Department of Justice continues to file lawsuits challenging tests that have a disparate impact on women; see *United States v. City of Erie,* 2005 WL 3610687 (W.D. Pa. 2005), in which a federal district judge held that the city failed to prove that its police officer agility test was "job related." The test required 13 push-ups and the climbing of a six-foot-high wall or fence, with men passing at 71 percent and females at 12.9 percent.

■ ■

Chapter Review Questions

1. In Case Study 14-1, *Weimer v. City of Baton Rouge,* the court addressed the issue of health and wellness programs. Describe the process used by a fire department in your state to adopt such a program and the provisions in the program when a firefighter declines to participate and declines to workout with the rest of his or her crew.

2. In Case Study 14-2, *Anderson v. City of Taylor,* the court addressed the issue of blood draws in a health and wellness program. Describe the provisions you would include in a program, in which the blood draw information would not be disclosed to the fire department unless there was clear indication of a medical problem rendering the firefighter unfit for duty.

3. In Case Study 14-3, *Victoria Pietras v. Board of Fire Commissioners of the Farmingville Fire District,* the court addressed the issue of physical ability testing and its disparate impact on women. Describe the physical ability testing done by a fire department in your state and whether the various requirements of the test are each job-related to the "real world" requirements of firefighters.

■ ■

Expand Your Learning

Read and complete the individual student or group assignment, as directed by your instructor.

1. The National Fire Protection Association (NFPA) conducted a study in 2005 of firefighters who died of cardiac arrest from 1995 until 2004 and 43.7 percent (440 firefighters) who experienced sudden cardiac death typically triggered by stress or exertion (*www.nfpa.org*). Describe what can be done nationwide, and in your fire department, to change this tragic situation.

2. In 2003, City Council members in Clifton (NJ), concerned about workers compensation claims from firefighters, ordered all exercise equipment removed from the city's firehouses. Describe what can be done to reduce workers' compensation claims by firefighters.

See Appendix A for additional Expand Your Learning activities related to this chapter.

Critical Incident Stress Management

CHAPTER 15

Three students died in this off-campus home at Miami University, Oxford (OH). The fire chief advised the press that he called out the Southwest Ohio Critical Incident Stress Management (CISM) team for a debriefing of fire, EMS, and police officers from several departments. *Photo by author.*

Key Terms

retaliation, p. 194   failure to train, p. 198   public records, p. 200

From the Headlines

Miami University—Careless Smoking and Alcohol

Investigators have concluded that the fire, which killed three students on April 10, 2005, in an off-campus house in Oxford (OH) was caused by careless smoking that ignited a living room couch. All three students had high blood-alcohol levels. The Butler County Coroner is quoted, "Once you're intoxicated, all of your senses don't work as well. You can't hear as well, you can't see as well, and maybe you can't be aroused as easily."

Taking Care of Our Own

National Interagency Fire Center personnel needed counselors after Hurricane Katrina relief efforts. Crisis teams had also responded to the 9/11 attacks in New York City and the space shuttle crash in Texas (*www.RedCross.org*).

World Trade Center Rescue Workers

Release of audio tapes and interviews of rescue workers from the World Trade Center, with command systems shattered, showed the need for crisis intervention. These workers also face the pressure of worrying about health issues. Since 9/11, cancer has been diagnosed in 283 rescuers, and 33 of them have died, including paramedic Deborah Reeve (age 41; mesothelioma), Port Authority EMT Stephen "Rak" Yurek (age 46; brain cancer), and NYPD officer Ronald Weintraub (age 43; bile-duct cancer), *www.firehouse.com* (June 11, 2006).

Miami University students placed flowers and posted notes about their friends who died in the structure fire. *Photo by author.*

◆ INTRODUCTION

This chapter discusses the need for activation of Critical Incident Stress Management (CISM) teams and the legal issues concerning dealing with firefighters under stress. An important legal issue is the protection of the privacy rights of individuals who participate in a CISM debriefing or defusing.

Lawsuits for "invasion of privacy" can be avoided by having clear CISM team policies that confirm "what is said here, stays here." In addition, statutory protection of these disclosures can be very helpful, such as the new Ohio statute discussed in the next section.

◆ OHIO ENACTS CISM PRIVILEGE LAW

The author of this textbook asked Ohio State Senator Bob Schuler about getting a law passed that would enforce the CISM cautionary remarks when we begin a debriefing: "What is said here, stays here." The author is pleased to report that Senator Schuler and his legislative assistant, Tony Seegers, drafted such a bill, and it was enacted into law in 2006. It establishes a legal privilege in which CISM team members and others are prohibited from discussing what is said in the CISM gathering, with some exceptions such as threats to commit a crime.

◆ NFPA STANDARDS

- *NFPA 1584: Recommended Practice on the Rehabilitation of Members Operating at Incident Scene Operations and Training Exercises (2003 edition).* The standard suggests personnel be evaluated for "emotional and mental stress" at incident scenes.

◆ KEY STATUTES

OHIO ENACTS A CISM "TESTIMONIAL PRIVILEGE" LAW

This law generally protects CISM team members from testifying about information received from an individual who receives crisis response services (Sub. Senate Bill 10, effective 1/27/06); *www.legislature.state.ohio.us/*. It is similar to laws protecting the attorney–client privilege. The testimonial privilege does not apply if any of the following are true: (1) the communication or advice indicates "clear and present danger" to the individual who receives crisis response services or to other persons; (2) the individual who received such services gives express consent to the testimony; (3) if the individual who received such services is deceased, the surviving spouse or the executor or administrator of the estate of the deceased individual gives express consent; (4) the individual who received such services voluntarily testifies, in which case the team member may be compelled to testify on the same subject; (5) the court *in camera* determines that the information communicated by the individual who received such services is not germane to the relationship between the individual and the team member; or (6) the communication or advice pertains or is related to a criminal act.

Linda King-Edrington, Program Director, Southwest Ohio CISM Team (retired paramedic/firefighter), thanks Tony Seegers, Esq., Legislative Assistant to Ohio State Senator Bob Schuler, for helping to get the new Ohio "Testimonial Privilege" law passed. *Photo by author.*

U.S.C. 1983

This federal statute provides individuals with the right to sue federal, state, or local officials, and the governmental entity, for alleged violation of federal constitutional rights. It has been used to sue police officers and fire officers.

The statute provides:

> Every person who, under color of any statute, ordinance, regulation, custom, or usage, of any State ... subjects, or causes to be subjected, any citizen of the United States or other person within the jurisdiction thereof to the deprivation of any rights, privileges, or immunities secured by the Constitution and laws, shall be liable to the party injured in an action at law, suit in equity, or other proper proceeding for redress....

◆ **CASE STUDIES**

CASE STUDY 15-1   Threats by Firefighter; Warning Posters Posted in Fire Stations

Pflanz v. Cincinnati, 149 Ohio App.3d 743, 778 N.E.2d 1073 (Ohio App. 1st 2002).

FACTS

[Note: We reviewed this case in Chapter 9 on ADA; now let's look at the employee threats.]

Pflanz was a Cincinnati firefighter/EMT who injured his back while on an EMS run when the cot gave way and he tried catching the patient. He was medically separated from his employment after turning down a light-duty position in the police department property room at a salary $10,000 less than his firefighter's annual average income. He filed an ADA lawsuit in the Ohio Court of Common Pleas, and the trial judge granted summary judgment to the city on all claims. The Ohio Court of Appeals agreed with the trial judge that he was not "disabled" because lifting and living with pain are not "major life activities" under the state handicap-discrimination statute's definition of "disability."

Pflanz filed a lawsuit that claimed the city posted "hazard posters" about him in fire stations in July 1996 in **retaliation** for having filed an EEOC complaint (in June 1995) concerning the ADA (Americans with Disabilities Act). He further claimed in the lawsuit that this retaliation violated both federal and Ohio laws protecting individuals who file a complaint with the EEOC and the Ohio Civil Rights Commission (in Ohio, a complaint filed with one agency is automatically dual-filed with the other agency).

The hazard posters were posted after Pflanz sent this e-mail message to his fellow firefighters:

> Perhaps [Fire] Chief [Thomas E.] Steidel and the mayor [Roxanne Qualls] should read the response and opinion of the IAFF regarding the MS firefighters. Considering the fire department permanently crippled me by placing defective equipment on line; separated me without benefits; deprived my family of medical insurance and income and I doubt that I'm the only one—the very same conditions exist right here in the Queen City. A man can only be expected to take so much and July 7 will be my one year anniversary of separation. And it is also when my worker's compensation benefits run out leaving nothing but welfare as an alternative.

The fire chief and other city officials were alarmed by the threats in Planz's e-mail, and the fire chief requested that the Cincinnati police issue a "hazard poster" on Pflanz for posting throughout the fire department. The hazard poster stated that Pflanz "had made disparaging public statements about the division and city Administration." The poster also stated that Pflanz was "trained as a bomb tech" and had "written comments supporting radical right wing causes." It urged all members of the fire division to "use caution when dealing with suspicious mail or packages."

On September 29, 1998, the EEOC found probable cause to believe that the city's issuance of the hazard poster was retaliatory. Sometime thereafter, Pflanz applied for and was granted workers' compensation benefits for a psychological condition stemming from his injury and the city's publication of the hazard poster.

The trial judge dismissed his retaliation claim, and Pflanz filed an appeal to the Ohio Court of Appeals.

YOU BE THE JUDGE

Should the Court of Appeals give Pflanz the right to prove his allegations of retaliation to a jury, or should they sustain the trial judge's dismissal?

HOLDING

The trial judge properly dismissed this claim.

The three-judge court wrote,

> The test for establishing a claim of retaliation under federal law is basically the same one that we applied to Pflanz's state retaliation claim. Consequently, we analyze Pflanz's state and federal retaliations claims together.
>
> To establish his retaliation claim under the ADA and Ohio Rev. Code 4112.02(I), Pflanz had to show that (1) he engaged in protected activity; (2) his employer knew about

retaliation Employee who files an EEOC complaint is protected from discipline, if he or she can prove there is a causal connection between the filing of the EEOC complaint and the subsequent discipline.

the protected activity; (3) the employer took an adverse action; and (4) there was a causal connection between the protected activity and the adverse action. To show the requisite casual link, Pflanz had to present evidence sufficient to raise the inference that [his] protected activity was the likely reason for the adverse action. A "causal connection may be demonstrated by evidence of circumstances that justify an inference of retaliatory motive, such as protected conduct closely followed by adverse action." But the mere fact that an adverse employment action occurs subsequent to the protected activity does not alone support an inference of retaliation. If Pflanz established a prima facie case, the burden then shifted to the city to articulate a legitimate, nondiscriminatory reason for the adverse employment action.

Neither party disputes that Pflanz was engaged in a protected activity when he filed a charge of disability discrimination with the EEOC in June 1995, or that the city knew he was engaged in that protected activity. The parties do dispute, however, whether Pflanz established the third and fourth elements necessary for a prima facie case of retaliation.

The trial court ruled that Pflanz had not established either element. With respect to the third element, the trial court held that the anti-retaliation provisions of the ADA and Ohio Rev. Code 4112.02(I) did not extend to adverse actions taken against former employees. The trial court reasoned that because Pflanz was no longer employed by the city when the hazard poster was published, he could not establish that he had been subjected to an adverse employment action. The trial court further found that Pflanz could not establish the fourth element of a prima facie case because he could not show a causal link between his EEOC charge and the hazard poster.

Pflanz argues that there were genuine issues of material fact as to whether the city's publication of the hazard poster was an "adverse employment action." Relying on the United States Supreme Court's decision in *Robinson v. Shell Oil Co,* [519 U.S. 337 (1997)]… Pflanz contends that, even though he was a former employee, he could still pursue an action for retaliatory discrimination under the ADA. Pflanz further contends that there were genuine issues of material fact concerning the existence of a causal connection between the publication of the hazard poster and his initial charge of discrimination because the EEOC's investigation of his charge was ongoing at the time the poster was published. The trial court concluded that this evidence alone was insufficient to create a causal connection. We agree.

Timing alone is not sufficient to establish a causal connection, especially in this case where there was a significant gap in time between the time that Pflanz filed his charge of discrimination and the time that the hazard poster was published. The mere fact that the EEOC's investigation was ongoing was not enough to establish a causal connection. Even if we were to assume arguendo that Pflanz had set forth a prima facie case of retaliation, his retaliation claims still failed because he did not demonstrate that the reason articulated by the city for publishing and distributing the hazard poster was a mere pretext for discrimination.

The city asserts that its decision to publish the hazard poster was not based on retaliation. The city contends that it distributed the hazard poster as a reasonable precaution to help secure the safety of the employees and management personnel in the Fire Division. Pflanz argues that the city's reason was pretextual because no one in the Fire Division truly believed that he was a threat. In support of his argument, Pflanz points to the fact that Assistant Fire Chief Worley, who was designated on the hazard poster as the individual to contact if there were questions concerning Pflanz, testified that he did not recall the events leading up to the poster or his involvement in the poster. We fail to see how this created a genuine issue of material fact, especially when there is evidence in the record that another assistant fire chief, Daniel Kuhn, and Chief Steidel took the lead in publishing and distributing the hazard poster.

Pflanz also implies that because Worley claimed that he did not know about the hazard poster, he did not view Pflanz as a threat. Having reviewed Worley's testimony, we cannot agree with Pflanz's assertion. When asked whether he was "personally worried about Pflanz as a result of the posting," Worley replied: "Well, the one thing this says here that I concur with is if he [Police Officer Richardson] talked to me and I showed concern about what Mr. Pflanz might be capable of, that's absolutely correct. I would be worried about a person that's experienced in bombs and bomb technology. You would have to be insane not to be worried about it or concerned about what he's capable of. Did I think he would really do it? I'm not going to answer that."

As for Pflanz's reliance on affidavits from three former co-workers, each stating that Pflanz was not a threat, this evidence was similarly insufficient to create a genuine issue of material fact, especially when Pflanz's e-mail threats were not leveled at co-workers, but at those in management positions.

Similarly, Pflanz's reliance on a sentence in a memorandum from Chief Steidel to Betty Baker, Director of Personnel, dated August 2, 1996, in which Steidel stated that he did not live his life in fear of Pflanz, did not create a genuine issue of material fact. The memorandum, which was dated over a month after the hazard poster had been distributed, did not rebut the city's argument that Steidel and other management personnel viewed Pflanz's June e-mail as a threat to the Fire Division, nor did it evidence any retaliatory motive for the city's publication of the hazard poster. Furthermore, the fact that the Fire Division had published and distributed a hazard poster for only one other firefighter, who had a criminal record, similarly did not create an inference of discrimination. Pflanz argues that because he did not have a history of violence, the city should have investigated the threats in his e-mail prior to its publication of the hazard poster, and that because it did not, the city's actions were retaliatory.

Pflanz's argument fails for two reasons. First, we are not required to determine whether the city's decision to publish the poster without further investigation was a good or bad decision, because the legal provisions against retaliation are not violated by an exercise of erroneous or even illogical business judgment. "An employer's business judgment is relevant only insofar as it relates to the motivation of the employer with respect to the allegedly illegal conduct." And, second, there is ample evidence in the record to show that the city's publication of the hazard poster was not retaliatory. Thus, the city was entitled to judgment as a matter of law on Pflanz's federal and state retaliation claims. Consequently, we overrule the third assignment of error.

LEGAL LESSONS LEARNED

Fire departments should take threats very seriously. The courts will generally support strong management response to perceived threats. Employees should be very careful what they write, even on an off-site e-mail message board.

CASE STUDY 15-2   9-1-1 Calls Ignored—Dispatcher Never Sent Police to Reports of Drunk Driver Weaving Down Highway—Head-On Crash into Innocent Driver

Daniel J. Svette v. Jacob A. Caplinger, et al., Case No. 2:04-cv-780, ___ F.Supp. ___ (S.D. Ohio; May 18, 2005, Judge Gregory L. Frost); unpublished decision can be read on Administrative Office of U.S. Courts PACER free written opinions (*http://www. pacer.uscourts.gov*), Case 2:04-cv-00780-GLK-NMK.

FACTS

A motorist who was severely injured when struck head-on by a drunk driver sued after he learned that another motorist had called the 9-1-1 dispatcher to report the drunk driver, but the dispatcher never sent a deputy sheriff to investigate. The injured motorist is suing Ross County (OH) Sheriff Ronald L. Nichols and Sergeant Nancy Haggard, claiming deprivation of constitutional rights under 42 U.S.C. 1983, for failure to train the dispatcher adequately.

On October 30, 2002, Sergeant Nancy Haggard was answering calls received over the Ross County 9-1-1 emergency telephone line. Haggard received four cellular calls from a passenger in an automobile that was following another automobile. The caller indicated that the lead automobile (driven by Jacob Caplinger who was intoxicated) was traveling quite erratically, going left of center, running red lights, striking the guard rail, entering a field along the road, and varying in speed from 5 mph to over 75 mph.

Although Sergeant Haggard informed the caller that a sheriff's car would be sent, no car was dispatched. Eventually, Caplinger's automobile again crossed the center line and hit head-on with an oncoming automobile driven by the plaintiff, Daniel J. Svette. Svette sustained serious injuries involving several hundred thousands of dollars in medical bills and four hospitalizations.

Svette subsequently filed an action against the Ross County defendants in the Ross County Court of Common Pleas on July 15, 2003. After Svette filed a Second Amended Complaint on July 30, 2004, that included a claim under 42 U.S.C. 1983, the Ross County defendants then "removed" (if sued for violation of federal law, defendants can seek to have the case tried in federal court) the state lawsuit to federal District Court on August 19, 2004.

Svette's sole federal cause of action is under 42 U.S.C. 1983. He must show that, while acting under color of state law, the Ross County defendants deprived him of a right secured by the federal Constitution or laws of the United States.

YOU BE THE JUDGE

Should the sergeant, the sheriff, and the county stand trial in federal court?

HOLDING

No.

A U.S. Magistrate wrote the following opinion:

> Thus, to satisfy his burden on his failure to train/failure to supervise theory, Plaintiff must demonstrate the existence of and impropriety of an involved policy. This is because the Sixth Circuit has explained: Municipalities are not ... liable for every misdeed of their employees and agents. "Instead, it is when execution of a government's policy or custom, whether made by its lawmakers or by those whose edicts or acts may fairly be said to represent official policy, inflicts the injury that the government as an entity is responsible under [42 U.S.C. Sec.] 1983."
>
> A plaintiff must "identify the policy, connect the policy to the city itself and show that the particular injury was incurred because of the execution of that policy."
>
> The United States Supreme Court has held: "[T]he inadequacy of police training may serve as the basis for Sec. 1983 liability only where the failure to train amounts to deliberate indifference to the rights of persons with whom the police come into contact.... Only where a municipality's failure to train its employees in a relevant respect evidences

a 'deliberate indifference' to the rights of its inhabitants can such a shortcoming be properly thought of as a city 'policy or custom' that is actionable under Sec. 1983." City of Canton v. Harris, 489 U.S. 378 (1989).

failure to train Federal lawsuits may be filed, under 42 U.S.C. 1983, for failure to train emergency responders and dispatchers.

A city may also be liable, in narrow circumstances, for **failure to train** its officials, if that failure gives rise to a clearly foreseeable violation of constitutional rights reflecting deliberate indifference to them: It may happen that in light of the duties assigned to specific officers or employees the need for more or different training is so obvious, and the inadequacy so likely to result in the violation of constitutional rights, that the policymakers of the city can reasonably be said to have been deliberately indifferent to the need. In that event, the failure to provide proper training may fairly be said to represent a policy for which the city is responsible, and for which the city may be held liable if it actually causes injury.

The ultimate focus '[i]n resolving the issue of a [municipality's] liability, ... must be on adequacy of the training program in relation to the tasks the particular officers must perform.'

What makes the instant case unusual is that unlike most Sec. 1983 failure to train/failure to supervise cases, this Court cannot reach the ultimate issue here of whether a state actor's policy is responsible for the underlying constitutional deprivation—because here there *was* no constitutional deprivation. Plaintiff's unique theory of his case is that inadequate training and supervision of dispatchers led to a failure to respond to the 9-1-1 calls, which permitted Caplinger to injure him, which resulted in four hospitalizations constituting a deprivation of Plaintiff's liberty in contravention of his *procedural* due process rights under the Fourteenth Amendment. Citing that same constitutional protection, he also reasons that a property deprivation has occurred in that he has incurred over $440,000 in medical treatment, over $25,000 of which he has paid himself.

[T]he Sixth Circuit addressed a situation in which the decedent had called 9-1-1 three times in an attempt to get help during a domestic dispute that resulted in her death. The district court in that case had denied a police officer qualified immunity, resulting in an interlocutory appeal. On appeal, the Sixth Circuit held that no *substantive* due process violation existed based on the lack of both a custodial-type setting and a general 'special relationship' between the defendant and the victim. Rather, specific facts related to the officer's direct involvement *at the scene of the crime* defeated qualified immunity. On remand, another judge of this Court subsequently granted summary judgment in favor of the county defendants, reasoning that there was no constitutional right to a 9-1-1 service, that the failure to protect a citizen against private violence was not a constitutional violation, and that they had neither *created* nor *increased* the risk to the decedent. That a failure to act to protect a victim's liberty interests against harms inflicted by a third party is not a constitutional deprivation of due process.

These cases, although substantive due process cases, inform the instant analysis in that they reason that the existence of a 9-1-1 service, even an improperly functioning one, does not establish state actor liability for harm inflicted by a third-party private actor. Likewise, they therefore support the proposition that state action is needed for there to be a violation of the Due Process Clause. But as the Ross County Defendants correctly argue, here there was no governmental decision leading to Plaintiff's injuries at Caplinger's hands. In the procedural due process context, then, there is no governmental process to attack because the government did not act to harm Plaintiff, but at best failed to act to keep him safe—a duty that is not constitutionally compelled. Given these facts, the basis for Plaintiff's procedural due process claim is unclear. It is illogical to characterize his claim as a complaint that he was not afforded notice of state action or an opportunity to be heard as one would find in the traditional procedural due process case. Plaintiff is confusingly linking an allegedly flawed dispatching procedure to the deprivation of his asserted rights. Although creative, this is an attempt to bootstrap a third-party's actions into a deprivation of due process claim by conflating a failure to act with reasoned state actor deliberation. Plaintiff impermissibly mixes notions of who is an actor and equates private harm with state decision making.

The Court must conclude that Plaintiff has failed to point to any evidence supporting the existence of a constitutional violation committed by the Ross County Defendants. In the absence of such a predicate violation, the Court need not and can not reach the issue of whether there is insufficient training or supervision of the dispatchers, because even assuming that there was, such a policy lacks a connection to a constitutional deprivation so as to create liability.

Further, there is simply no evidence of a history of dispatching errors resulting in constitutional violations. The lack of the deprivation of a constitutional right informs the Court's qualified immunity analysis. To evade liability, the individual defendants invoke this affirmative defense, which is meant to safeguard an official's proper decision making process and offers that party potential relief from frivolous suits.

The Sixth Circuit has explained that qualified immunity shields government officials from liability for civil damages insofar as their conduct does not violate clearly established statutory or constitutional rights of which a reasonable person would have known.

In addition to shielding officials from liability, qualified immunity may entitle the official to not stand trial or face the other burdens of litigation. This principle directs courts to make a ruling on the issue of qualified immunity early in the proceedings, so that the costs and expenses of trial are avoided where the defense is dispositive. Of import here is that the Supreme Court has instructed lower courts to use a distinct analysis to determine whether summary judgment based on qualified immunity is warranted. In addressing the potential applicability of qualified immunity, a court follows a sequential inquiry: First, the court considers whether, on the plaintiff's facts, there has been a violation. Second, the court considers whether that violation involved 'clearly established constitutional rights of which a reasonable person would have known.'

The doctrine of qualified immunity recognizes that an officer can be found to have violated the Constitution, but be granted immunity for *reasonable* mistakes as to the legality of his or her action. The reasonableness of such mistakes is inherently dependent upon the clarity of the legal constraints governing particular police conduct. The contours of the right must be sufficiently clear that a reasonable official would understand that what he is doing violates that right.

Cognizant of this mandated inquiry, the Court finds that Plaintiff's case breaks down at the first stage of the sequential inquiry, because, as noted, even necessarily construing the facts and all reasonable inferences in Plaintiff's favor, there has been no violation of Plaintiff's constitutional rights. Absent such a predicate showing, Plaintiff has failed to present the Court with the threshold circumstances necessitating further inquiry as to whether the doctrine applies. Were the Court to engage in such inquiry, however, the Court notes that Plaintiff would face another insurmountable hurdle in demonstrating that his essentially confusing claim to a *procedural* due process right here constituted a clearly established constitutional right of which a reasonable person would have known. A failure to train or supervise is not actionable in and of itself. Rather, it must be linked to a predicate deprivation of a constitutional right. Because no evidence exists that suggests even a reasonable inference that the Ross County Defendants deprived Plaintiff of a right secured by the Federal Constitution or laws of the United

States, the Ross County Defendants are entitled to judgment as a matter of law on Plaintiff's Sec. 1980 claim.

LEGAL LESSONS LEARNED

It is important that dispatchers are trained to rapidly advise law enforcement of an intoxicated motorist, or similar life hazard, and that training is documented. In the event of litigation, cases will normally be dismissed if there was merely a mistake in following the training provided.

CASE STUDY 15-3 9-1-1 Audio Tapes: Prosecutor Offers Transcript— Newspaper Demands Audio Tape

The State ex rel. Dispatch Printing Company v. Morrow County Prosecutor's Office, 105 Ohio St.3d 172, 824 N.E.2d 64 (Ohio Sup. Ct., February 24, 2005).

FACTS

The *Columbus Dispatch* newspaper filed a lawsuit under the Ohio Public Records Act to compel the county, county prosecuting attorney, and county prosecuting attorney's office to immediately release a copy of an audiotape of 9-1-1 emergency calls relating to two homicides and sought a peremptory *writ of mandamus.*

In January 2005, the newspaper requested that Morrow County Prosecuting Attorney Charles S. Howland provide it with a copy of the 9-1-1 tapes relating to the homicides of Diana Cooper and Cameron Bateman. Howland permitted the newspaper to listen to the 9-1-1 tapes and offered to transcribe them, but he refused to provide the newspaper with a copy of the tape or allow the newspaper to record it.

On January 28, 2005, the newspaper filed a lawsuit under the Ohio Public Records Act, Ohio Rev. Code 149.43, to compel the respondents—Howland, the Morrow County Prosecutor's office, and Morrow County—to *immediately* produce a copy of the requested 9-1-1 tapes. The newspaper also moved for an immediate injunction and requested its costs and expenses, including attorney's fees. In its motion, the newspaper states, "Given the time sensitivity of this matter, it is requested that the relief be granted forthwith."

On February 16, 2005, the prosecutor filed an answer confirming the pertinent facts.

YOU BE THE JUDGE

Should the newspaper be entitled to a copy of the audiotapes?

HOLDING

Yes. The Ohio Supreme Court (seven to zero) wrote, "The Dispatch is entitled to the requested writ. Nine-one-one tapes in general * * * are public records which are not exempt from disclosure and must be immediately released upon request."

In *Cincinnati Enquirer v. Hamilton County,* 75 Ohio St.3d 374, 662 N.E. 2d 334 (Ohio Sup. Ct. 1996), the court decided:

public records 9-1-1 audio tapes of callers may be considered records open to the public and TV and radio stations for public broadcast.

[We] reasoned as follows in holding that all 911 tapes are **public records** subject to immediate release upon request: "Basic 911 systems * * * are systems 'in which a caller provides information on the nature of and location of an emergency, and the personnel receiving the call must determine the appropriate emergency service provider to respond at that location.' * * * 911 operators simply compile information and do not investigate. The 911 tapes are not made in order to preserve evidence for criminal prosecution. 911 calls that are received * * * are always initiated by the callers. * * *

From the foregoing, it is evident that 911 tapes are not prepared by attorneys or other law enforcement officials. Instead, 911 calls are routinely recorded without any specific investigatory purpose in mind. There is no expectation of privacy when a person makes a 911 call. Instead, there is an expectation that the information provided will be recorded and disclosed to the public. Moreover, because 911 calls generally precede offense or incident form reports completed by the police, they are even further removed from the initiation of the criminal investigation than the form reports themselves.

The moment the tapes were made as a result of the calls (in these cases—and in all other 911 call cases) to the 911 number, the tapes became public records. * * * Thus, any inquiry as to the release of records should have been immediately at an end, and the tapes should have been, and should now and henceforth always be, released.

The particular content of the 911 tapes is irrelevant. * * *

In addition, the fact that the tapes in question subsequently came into the possession and/or control of a prosecutor, other law enforcement officials, or even the grand jury has no significance. Once clothed with the public records cloak, the records cannot be defrocked of their status.

The County Prosecutor concedes "that they keep the requested record in audiotape format and that despite the *Dispatch*'s requesting a copy of the tape in this format, respondents refused to release copies of the tape and did not allow the *Dispatch* to copy it. Under R.C. 149.43(B)(2), they had a duty to provide the *Dispatch* with a copy of the 911 tape in that same format.

Similarly, the *Dispatch* is entitled to a copy of the 911 tape at cost. Moreover, because 911 tapes "must be immediately released upon request," we grant an *immediate* peremptory writ.

Attorney Fees

The *Dispatch* is also entitled to attorney fees. It has established a sufficient public benefit, and respondents failed to comply with the records requests for invalid reasons. We order the *Dispatch's* counsel to submit a bill and documentation of evidence in support of its request for attorney fees.

PFEIFER, J., concurs separately.

"I concur in the decision of the majority. The law is clear. However, as I wrote in [the *Cincinnati Enquirer* decision], the General Assembly should consider changing the law. The public's right to scrutinize the workings of the government should be balanced against an individual citizen's right to privacy. A person should be able to summon the help of police officers or firefighters without having his plea broadcast on the evening news. A transcript of a 911 call would convey the necessary information without transforming a personal tragedy into a public spectacle."

LEGAL LESSONS LEARNED

In Ohio, the audio tapes of emergency calls to dispatchers are public records, unless the legislature enacts a statute to protect the privacy of individual callers. Likewise, tapes of incident command and tactical channels at fire scenes are probably considered public records. Check the law in your state.

■ ■

Chapter Review Questions

1. In Case Study 15-1, *Pflanz v. Cincinnati,* the court discussed the issue of threats in the workplace. We have all read about tragic cases in U.S. Post Offices ("going postal") and other workplaces. Describe what steps that a fire department should take when it has been reported that a firefighter has a problem with anger management.
2. In Case Study 15-2, *Daniel J. Svette v. Caplinger,* the court addressed the issue of negligent conduct by a dispatcher. Discuss what steps might be taken to increase the knowledge of dispatchers concerning the information that firefighters need when responding to emergency calls (for example, structure fire calls in which a teenage caller advises that there is a grease fire in the kitchen and that he is alone and leaving the house).
3. In Case Study 15-3, *The State ex rel. Dispatch Printing Company v. Morrow County Prosecutor's Office,* the court addressed the issue of disclosure of 9-1-1 tapes. In the fire service, many states have passed laws that prohibit the public disclosure of "familiar" information about firefighters, EMS, and police, including home addresses, telephone numbers, and names of spouses and children. Describe whether such information is protected from disclosure in your state.

■ ■

Expand Your Learning

Read and complete the individual or student assignment, as directed by your instructor.

1. A structure fire in which children have died can be extremely stressful. A helpful practice after every major fire is to conduct a thorough debriefing of the fire ground actions and a discussion about improvements that can be made in tactics, communications, and so on. Several states have laws that specifically protect these types of critiques from disclosure to third parties, and some states protect EMS quality reviews (see Ohio Rev. Code 4765.12, "Information gathered solely for use in a peer review or quality assurance program conducted on behalf of an emergency medical service organization is not a public record under Section 149.43 of the Revised Code," and "is not subject to discovery in a civil action...." *www.legislature. state.ohio.us/*). Describe what laws exist in your state that protect such post-fire quality improvement reviews and reports. Locate additional laws from other states that you would recommend be adopted.
2. Some mental health professionals have publicly challenged the CISM process. Describe what programs are used by the U.S. military and how the CISM process might be further improved in your state.

See Appendix A for additional Expand Your Learning activities related to this chapter.

Module IV: The Fire Official as Rule-Maker and Enforcer

Discipline
Misconduct On Duty and Off Duty and Ethical Decision Making

16 CHAPTER

Sitka (AK) Fire Department has a new and impressive fire station. During summer months, many cruise ships dock in Sitka. With very limited on-duty staff, the department must rely on off-duty personnel to respond (and they must follow the same standards regarding alcohol and drug use and ability to perform their tasks).

Key Terms

coerced, p. 209

self-incrimination, p. 209

equal protection, p. 211

balancing test, p. 214

From the Headlines

Six Firefighters, Including Two Lieutenants Fired—Lewd Act On Duty

The Milwaukee Fire Department investigated a lieutenant who used a computer while on duty to chat with someone he thought was a woman and allegedly masturbated on camera. The lieutenant

Phoenix(AZ) new fire station. *Photo by author.*

did not know that he was actually chatting with a group of firefighters in another station who were given the woman's password.

Two Firefighters Indicted/Dumping Gasoline from Fireboat

Two White Plains (NY) firefighters were indicted for reportedly pouring gasoline from a fireboat, after the gas had been contaminated with water, down a sink at fire headquarters. The gasoline flowed into the sewer system to a public housing project, where vapors were ignited and badly burned two maintenance workers using an electric pump.

On-Duty Group Sex—One Firefighter Resigns, Three Others Sent Notices of Intent to Terminate

The Sacramento (CA) Fire Department investigated alleged on-duty group sex at a firehouse, including a female ten-year veteran, a captain, and two other firefighters. This is the latest in a string of scandals involving discipline of 24 firefighters (including two resignations and termination of two Captains) for drinking on duty, giving joy rides to women on fire trucks, and attending a "Porn Star Costume Ball."

Art work at a Phoenix fire station, including a metal tree in the backyard. *Photo by author.*

◆ INTRODUCTION

This chapter deals with employee discipline and investigation of firefighter misconduct. Departments that have well-defined disciplinary processes, including thorough investigations, "due process" for the accused, well-documented discipline, and clear chain-of-command appeal procedures can avoid frivolous litigation and unsupported grievances.

◆ ETHICAL DECISION MAKING

Ethical decision making should be a topic addressed in every officer development class and in fire recruit class. A helpful book on this topic is *Managing by Values, How to Put Your Values into Action for Extraordinary Results,* by Ken Blanchard and Michael O'Connor, Berrett-Koehler Publishers, Inc. (1997).

◆ OFF-DUTY CONDUCT—TERMINATION

Unfortunately, the law books are full of cases involving misconduct, both on duty and off duty. For example, in *Charles Lawrence v. City of Texarkana,* Case No. 05-310 (Ark. Supreme Court, January 5, 2006), an off-duty firefighter was terminated after fleeing from police in excess of 100 mph as police investigated a domestic dispute. In another example, *Ogden City Corporation v. Harmon,* 116 P.3d 973 (Utah Ct. App. 2005), the court upheld the termination of a fire captain, for arranging for topless female entertainers at a union fund-raising event and engaging in other prior inappropriate activities, including urinating on a firefighter taking a shower.

◆ INTERNAL INVESTIGATIONS—FIFTH AMENDMENT RIGHTS

Investigations of employee misconduct must be conducted with great care and with an understanding of the rights of both the employee and management. There are many legal issues concerning employee discipline. A "word to the wise" is to consult with knowledgeable legal counsel if questions arise, whether you are conducting an investigation or are the subject of an investigation. For example, an issue that often arises in internal investigations is whether the firefighter being interviewed about the conduct of another firefighter can be "compelled" to answer questions, over his Fifth Amendment rights not to incriminate himself. (See Case 16–1 in this chapter.)

◆ JURY VERDICTS AND THE FIRST AMENDMENT RIGHT OF FREE SPEECH

Jury verdicts against public employers, including fire departments, can be very steep. For example, a federal jury in October 2002 reportedly awarded a Springdale (AR) firefighter $350,000, after he was terminated because he made public comments about the fire department being understaffed.

The U.S. Supreme Court's decision (five to four vote) on May 30, 2006, in *Garcetti v. Ceballos,* 547 U.S.__, 126 S. Ct. 1951 (5/30/06) (*www.supremecourtus.gov*), is a "must read" decision. The majority of the court held that public employers have the right to discipline employees for their official speech, including memos that they write and investigations that they conduct. "When a public employee speaks pursuant to employment responsibilities, however, there is no relevant analogue to speech by citizens who are not government employees." See also the dissenting opinions, including Justice Stevens, who characterized the majority as "misguided."

◆ KEY STATUTES

EMPLOYEE POLYGRAPH PROTECTION ACT OF 1998 (29 U.S.C. 2002)

This federal statute generally prohibits private employers from requiring, requesting, suggesting, or causing any employee or applicant to take a lie detector test. The statute does not apply "with respect to the United States Government, or any state or local government, or any political subdivision...." The statute is enforced by the U.S. Department of Labor. The statute and regulations make it clear it does not apply to most government employers. 29 U.S.C. 2006(a); 29 CFR 801.10. In the fire service, use of lie detector tests may be restricted by state statutes or by collective bargaining agreements.

SARBANES-OXLEY ACT OF 2002

Congress enacted this law after the accounting scandals at Enron Corporation in Houston, and other companies, in order to protect shareholders from fraud. Section 406 requires covered entities to adopt a "code of ethics" for senior financial officers. Although most of these provisions do not apply to the fire service, the law does have federal criminal penalties that apply to *all organizations* for altering documents, retaliation against whistleblowers, or otherwise obstructing government investigations into fraud.

RIGHTS OF FIREFIGHTERS IN INVESTIGATIONS

Some states have enacted legislation that establishes the rights of firefighters and other public employees. For example, according to Florida's Section 112.82 , "Rights of Firefighters," whenever a firefighter is subjected to an interrogation, such interrogation shall be conducted pursuant to the following terms:

1. The interrogation shall take place at the facility where the investigating officer is assigned or at the facility that has jurisdiction over the place where the incident under investigation allegedly occurred, as designated by the investigating officer.
2. No firefighter shall be subjected to interrogation without first receiving written notice of sufficient detail of the investigation in order to reasonably apprise the firefighter of the nature of the investigation. The firefighter shall be informed beforehand of the names of all complainants.
3. All interrogations shall be conducted at a reasonable time of day, preferably when the firefighter is on duty, unless the importance of the interrogation or investigation is of such a nature that immediate action is required.
4. The firefighter under investigation shall be informed of the name, rank, and unit or command of the officer in charge of the investigation, the interrogators, and all persons present during any interrogation.

5. Interrogation sessions shall be of reasonable duration, and the firefighter shall be permitted reasonable periods for rest and personal necessities.

6. The firefighter being interrogated shall not be subjected to offensive language or offered any incentive as an inducement to answer any questions.

7. A complete record of any interrogation shall be made, and if a transcript of such interrogation is made, the firefighter under investigation shall be entitled to a copy without charge. Such record may be electronically recorded.

8. An employee or officer of an employing agency may represent the agency, and an employee organization may represent any member of a bargaining unit desiring such representation in any proceeding to which this part applies. If a collective bargaining agreement provides for the presence of a representative of the collective bargaining unit during investigations or interrogations, such representative shall be allowed to be present.

9. No firefighter shall be discharged, disciplined, demoted, denied promotion or seniority, transferred, reassigned, or otherwise disciplined or discriminated against in regard to his or her employment, or be threatened with any such treatment as retaliation for or by reason solely of his or her exercise of any of the rights granted or protected by this part.

OHIO STATUTE PROTECTING CAREER PERSONNEL FROM TERMINATION EXCEPT FOR SERIOUS OFFENSES

Many states have statutes that protect career firefighters from removal except upon proof of serious misconduct. In Ohio, for example, Ohio Rev. Code 733.35 states that delinquent department heads or officers may be removed if "guilty, in the performance of his official duty, or bribery, misfeasance, malfeasance, nonfeasance, misconduct in office, gross neglect of duty, gross immorality, or habitual drunkenness."

◆ **CASE STUDIES**

CASE STUDY 16-1   Internal Investigations—Police Officers Compelled to Answer Questions or Lose Jobs—Fifth Amendment Rights/U.S. Supreme Court

Garrity v. New Jersey, 385 U.S. 493 (1967).

FACTS

Police officers were accused of fixing tickets. During an investigation, they gave statements to New Jersey State Police that were later used against them in a criminal case.

The policemen worked in the New Jersey boroughs of Bellmawr and Barrington. Police Chief Garrity and Police Officer Virtue were from Bellmawr. Officer Holroyd, Officer Elwell, and Officer Murray were police officers in Barrington. Another defendant was Mrs. Naglee, the clerk of Bellmawr's municipal court.

In June 1961, the New Jersey Supreme Court received complaints of ticket fixing, and the court directed the state's attorney general to investigate reports of traffic ticket fixing in Bellmawr and Barrington. Subsequent investigations produced evidence that these police officers had falsified municipal court records, altered traffic tickets, and diverted moneys produced from bail and fines to unauthorized purposes.

During these investigations, the State of New Jersey obtained two sworn statements from each of the police officers. The first statements were taken from the second in November 1961.

The statements of the Bellmawr police officers were taken in a room in the local firehouse, for which Police Chief Garrity himself had made arrangements. None of the police officers were in custody before or after the depositions were taken. Each apparently continued to pursue his ordinary duties as a public official of the community. The statements were recorded by a court stenographer, who testified that he witnessed no indications of unwillingness or even significant hesitation on the part of any of the petitioners. The Bellmawr police officers did not have counsel present, but the deputy attorney general testified without contradiction that Police Chief Garrity had informed him as they strolled between Garrity's office and the firehouse that he had arranged for counsel, but he thought that none would be required at that stage. The interrogations were not excessively lengthy, and reasonable efforts were made to assure the physical comfort of the witnesses. Mrs. Naglee, the clerk of the Bellmawr municipal court, who was known to suffer from a heart ailment, was assured that questioning would cease if she felt any discomfort.

As interrogation commenced, each of the police officers was sworn, carefully informed that he need not give any information, reminded that any information given might be used in a subsequent criminal prosecution, and warned that as a police officer *he was subject to a proceeding under New Jersey law to discharge him if he failed to provide information relevant to his public responsibilities.*

Mrs. Naglee was not told that she could be removed from her position at the court if she failed to give information pertinent to the discharge of her duties. All of the police officers consented to give statements, none displayed any significant hesitation, and none suggested that the decision to offer information was motivated by the possibility of discharge.

Portions of those statements were admitted at their criminal trials. At the trial, the court excused the jury and conducted a hearing to determine whether the statements were voluntarily given. The State of New Jersey offered witnesses who testified as to the manner in which the statements were taken; the police officers did not testify at that hearing. The trial judge held the statements to be voluntary. The trial judge noted that there was "no physical coercion, no overbearing tactics of psychological persuasion, no lengthy incommunicado detention, or efforts to humiliate or ridicule the defendants." The trial judge found no evidence that any of the petitioners were reluctant to offer statements and concluded that the interrogations were conducted with a "high degree of civility and restraint."

The police officers were convicted in two separate trials of conspiracy to obstruct the proper administration of the state motor traffic laws. The Supreme Court of New Jersey affirmed all the convictions. The police officers asked the U.S. Supreme Court to hear their appeal, and the court agreed to hear their appeal.

YOU BE THE JUDGE

Should their statements have been suppressed?

HOLDING

Yes—convictions reversed.

Justice Douglas wrote the decision for the majority (six to three decision):

Before being questioned, each appellant was warned (1) that anything he said might be used against him in any state criminal proceeding; (2) that he had the privilege to refuse to answer if the disclosure would tend to incriminate him; but (3) that if he refused to answer he would be subject to removal from office.

[The New Jersey statute provides] that "Any person holding or who has held any elective or appointive public office, position or employment (whether State, county or municipal), who refuses to testify upon matters relating to the office, position or employment in any criminal proceeding wherein he is a defendant or is called as a witness on behalf of the prosecution, upon the ground that his answer may tend to incriminate him or compel him to be a witness against himself or refuses to waive immunity when called by a grand jury to testify thereon or who willfully refuses or fails to appear before any court, commission or body of this state which has the right to inquire under oath upon matters relating to the office, position or employment of such person or who, having been sworn, refuses to testify or to answer any material question upon the ground that his answer may tend to incriminate him or compel him to be a witness against himself, shall, if holding elective or public office, position or employment, be removed therefrom or shall thereby forfeit his office, position or employment and any vested or future right of tenure or pension granted to him by any law of this State provided the inquiry relates to a matter which occurred or arose within the preceding five years. Any person so forfeiting his office, position or employment shall not thereafter be eligible for election or appointment to any public office, position or employment in this State." N.J. Rev. Stat. 2A: 81–17.1 (Supp. 1965).

The police officers answered the questions. No immunity was granted, as there is no immunity statute applicable in these circumstances. Over their objections, some of the answers given were used in subsequent prosecutions for conspiracy to obstruct the administration of the traffic laws. Appellants were convicted and their convictions were sustained over their protests that their statements were **coerced**, by reason of the fact that, if they refused to answer, they could lose their positions with the police department.

[W]hether, valid or not, the fear of being discharged under [the New Jersey statute] for refusal to answer on the one hand and the fear of self-incrimination on the other was "a choice between the rock and the whirlpool" which made the statements products of coercion in violation of the Fourteenth Amendment.

We agree with the New Jersey Supreme Court that the forfeiture-of-office statute is relevant here only for the bearing it has on the voluntary character of the statements used to convict petitioners in their criminal prosecutions.

The choice imposed on petitioners was one between **self-incrimination** or job forfeiture. Coercion … can be "mental as well as physical"; "the blood of the accused is not the only hallmark of an unconstitutional inquisition." Subtle pressures may be as telling as coarse and vulgar ones. The question is whether the accused was deprived of his "free choice to admit, to deny, or to refuse to answer."

The choice given petitioners was either to forfeit their jobs or to incriminate themselves. The option to lose their means of livelihood or to pay the penalty of self-incrimination is the antithesis of free choice to speak out or to remain silent. That practice, like the interrogation practices that we reviewed in *Miranda v. State of Arizona,* 384 U.S. 436 is 'likely to exert such pressure upon an individual as to disable him from making a free and rational choice.' We think the statements were infected by the coercion inherent in this scheme of questioning and cannot be sustained as voluntary under our prior decisions.

The question in this case, however, is not cognizable in those terms. Our question is whether a State, contrary to the requirement of the Fourteenth Amendment, can use the threat of discharge to secure incriminatory evidence against an employee.

coerced Statements obtained from an employee, who is subject to being fired for refusal to answer, may be considered inadmissible in a criminal trial unless the employee is granted immunity from criminal prosecution.

self-incrimination The fifth Amendment protects firefighters and others from being compelled to answer questions that can incriminate them in a crime; there is no such protection against administrative discipline for their conduct.

"We conclude that policemen, like teachers and lawyers, are not relegated to a watered-down version of constitutional rights.

Reversed.

LEGAL LESSONS LEARNED

"Garrity Warnings" are now used in internal investigations to compel police and firefighters to answer internal investigation questions, after granting them immunity from criminal prosecution. The employees can still be disciplined administratively for their conduct. These "Garrity Warnings" should be used only if the local prosecutor concurs in the grant of immunity.

CASE STUDY 16-2   Fire Chief Did Not File Fire Reports with State Fire Marshal—Termination

Dennis v. Berne Twp. Board of Trustees, 2005-Ohio-5951 (5th Dist. October 31, 2005). After his demotion, Dennis filed an appeal to the Ohio Court of Common Pleas. He also filed a lawsuit against the Township alleging violation of his federal constitutional rights under 42 U.S.C. 1983. The Township moved that case from state to federal court. A federal district judge granted summary judgment to the Township on the 1983 claim and remanded his wrongful discharge claim to state court. Case No. C2-04-CV-1185, 2006 U.S. Dist. LEXIS 2450 (January 24, 2006).

FACTS

The state fire marshal sent a letter to the township trustees advising they were not receiving National Fire Incident Reporting System (NFIRS) reports on fires in the township. The trustees retained a private investigator who, on October 31, 2002, filed a complaint with the township trustees that Fire Chief Mark Dennis had failed to file reports of fires timely with the fire marshal's office. The trustees suspended the chief on November 5, 2002. On November 19, the three township trustees held a hearing, and they voted (two to one) to demote him from chief to fire lieutenant.

On December 6, 2002, the appellant filed an appeal with the Court of Common Pleas for Fairfield County. A hearing was held on November 14, 2003. By entry filed July 13, 2004, the trial court affirmed the trustee's decision.

YOU BE THE JUDGE

Did the trustees properly remove the fire chief?

HOLDING

Yes.

The three-judge court (vote of two to one) wrote,

Appellant claims the trial court erred in finding appellant's procedural and substantive due process rights had not been violated. We disagree.

Specifically, appellant argues Trustee Carmichael participated in the investigation and fact-finding process prior to the hearing, and he lacked the authority to handover the matter to a private investigator, Mr. Tatman. Also, appellant argues Trustee Bailey based his decision on other grounds.

Ohio Rev. Code 505.38 governs removal of fire chief. Subsection (A) provides in part, "To initiate removal proceedings, and for that purpose, the board shall designate the fire chief or a private citizen to investigate the conduct and prepare the necessary charges in conformity with those sections."

From the record, we conclude appellant's assertions as to Trustee Carmichael lack merit for the following reasons. Mr. Tatman testified he was formally hired on August 22, 2002 by Trustee Carmichael to investigate a letter from the State Fire Marshall regarding appellant's failure to file reports. Mr. Tatman stated the letter was passed on to him to investigate the situation. Although Mr. Tatman updated Trustee Carmichael on the progress of the investigation, he did not discuss "conclusions or facts of what the investigation revealed." During Mr. Tatman's meetings with the trustees, he was under the impression that they were "upset that there was problems occurring within the Fire Department and that they wanted the problems resolved. And it didn't have anything to do with Mark Dennis, it was just generally towards a dislike that there were problems within the Fire Department." Trustee Carmichael admitted to verifying the authenticity of the State Fire Marshal's letter but nothing further. The trustees agreed Trustee Carmichael was in charge of fire department issues.

We find these facts do not taint the due process requirements on the removal issue.

Appellant also argues Trustee Bailey relied on other issues in removing him. Although Trustee Bailey's testimony is inconsistent on this issue, we find no lack of due process to appellant. There is no contradictory evidence to appellant failing to file the statutory mandated reports pursuant to Rev. Code 3737.24. What is evident from Trustee Bailey's testimony is that appellant's failure to timely file reports was a factor for removal. Trustee Bailey felt appellant failed to do the job and the complaint was based on nonfeasance in office.

Upon review, we conclude there was no violation of appellant's rights to procedural and substantive due process.

Appellant also claims the trial court erred in finding appellee afforded him **equal protection** under the law. We disagree.

equal protection Employees facing discipline can assert under the Fourteenth Amendment that the employer is treating them more harshly than others for similar misdeeds.

Specifically, appellant argues because the previous fire chiefs failed to timely file reports, he was given disparate treatment. In support of his argument, appellant cites the rational basis test which states, "laws must be based upon fundamentally reasonable classifications, and must have the capability of being applied reasonably and fairly among all to whom such laws pertain." In addition, "equal protection means only that persons in the same classification must be treated alike and that reasonable grounds which further a legitimate governmental interest exist in making the distinction between those persons who fall within the class and those who fall outside the class."

Appellant argues previous chiefs failed to file reports from March to May of 2001 and for 1995 and 1999, and appellee failed to investigate and/or discipline the fire chiefs involved. Appellant's Brief at 18. Although appellee's investigation started with March of 2001, there is no proof in the record that the 1995 and 1999 dates were brought to appellee's attention at the time. Further, appellee did not have the power to discipline previous chiefs after Mr. Tatman reported the deficiencies during appellee's hearing. See, November 26, 2002 Public Hearing T. at 27-28. As the record indicates, Trustee Carmichael did not become a trustee until January 1, 2002. T. at 109.

Appellant also argues fulfilling the statutory requirements of reporting to the State Fire Marshal is not a legitimate governmental interest. We find a violation regarding a state mandated report is in the interest of good township government.

Ohio Rev. Code 505.38 provides for the appointment of a fire chief by the board of township trustees, and "shall continue in office until removed from office as provided by section 733.35 to 733.39 of the Revised Code."

As noted earlier, there was no dispute that the required reports were not filed by appellant and had not been filed up to appellee's hearing date. What appellant argues now are mitigating factors he believes should be considered in finding his removal by appellee to be unlawful. As Trustee Bailey stated, appellant "just didn't" file the required reports and get the job done when the deficiencies were brought to light. Also, appellant did not take advantage of additional training during the pending investigation. Appellant made one attempt at requesting a secretary for help, but the request was tabled and appellee "asked him to bring to us something in writing so that we could look at it and see why he needed a secretary." Appellant did not pursue the issue at subsequent meetings.

Upon review, we conclude there was in fact a failure by appellant to perform his statutory duties and despite the persuasiveness of the mitigating factors, the removal was not arbitrary, unreasonable or contrary to law.

LEGAL LESSONS LEARNED

Fire chiefs must comply with the statutory requirement to file fire reports with the state fire marshall. The fact that predecessors violated the law will have little weight with the current township trustees or the courts. Many states offer training programs for new fire chiefs; we all are familiar with the statement, "Ignorance of the law is not a defense."

[Note: the former fire chief also filed a lawsuit in U.S. District Court, alleging deprivation of federal constitutional rights under 42 U.S.C. 1983; that court dismissed the suit on January 24, 2006; 2006 WL 181989 (S.D. Ohio 2006).]

CASE STUDY 16-3   Home-Made Sex Videos—Police Officer Off Duty, But in Uniform/U.S. Supreme Court

City of San Diego v. John Roe, 543 U.S. 77 (2004).

FACTS

A police officer was fired for offering his home-made, sexually explicit videos for sale on an online auction site. He sued the city for violation of his First Amendment rights, claiming his off-duty, non-work-related activities should not be a basis for discipline. The federal District Court dismissed his suit. The 9th Circuit Court of Appeals reversed, and the city filed an appeal with the U.S. Supreme Court.

John Roe was a San Diego police officer, who made a video showing himself stripping off a police uniform and masturbating. He sold the video on the adults-only section of eBay, the popular online auction site. His user name was "Codestud3@aol .com," a wordplay on a high-priority police radio call.

The uniform apparently was not the specific uniform worn by the San Diego police department, but it was clearly identifiable as a police uniform. Roe also sold custom videos, as well as police equipment, including official uniforms of the San Diego Police Department (SDPD) and various other items such as men's underwear.

His eBay user profile identified him as employed in the field of law enforcement. Roe's supervisor, a police sergeant, discovered Roe's activities when, while on eBay, he came across an official SDPD police uniform for sale offered by an individual with the username "Codestud3@aol.com." He searched for other items Codestud3 offered and discovered listings for Roe's videos depicting the objectionable material. Recognizing Roe's picture, the sergeant printed images of certain of Roe's offerings and shared them with others in Roe's chain of command, including a police captain.

The captain notified the SDPD's internal affairs department, which began an investigation. In response to a request by an undercover officer, Roe produced a custom video. It showed Roe, again in police uniform, issuing a traffic citation but revoking it after undoing the uniform and masturbating.

The investigation revealed that Roe's conduct violated specific SDPD policies, including conduct unbecoming of an officer, outside employment, and immoral conduct. When confronted, he admitted to selling the videos and police paraphernalia. The SDPD ordered Roe to "cease displaying, manufacturing, distributing or selling any sexually explicit materials or engaging in any similar behaviors, via the internet, U.S. Mail, commercial vendors or distributors, or any other medium available to the public."

Although Roe removed some of the items he had offered for sale, he did not change his seller's profile, which described the first two videos he had produced and listed their prices as well as the prices for custom videos. After discovering Roe's failure to follow its orders, the SDPD—citing Roe for the added violation of disobedience of lawful orders—began termination proceedings. The proceedings resulted in Roe's dismissal from the police force.

Roe brought suit in the District Court under 42 U.S.C. 1983, claiming that the employment termination violated his First Amendment right to free speech. In granting summary judgment to the city, the District Court decided that Roe had not demonstrated that selling official police uniforms and producing, marketing, and selling sexually explicit videos for profit qualified as expression relating to a matter of "public concern" under this court's decision in *Connick v. Myers,* 461 U.S. 138 (1983).

The Court of Appeals for the 9th Circuit reversed and held Roe's conduct fell within the protected category of citizen commentary on matters of public concern. The city filed an appeal to the U.S. Supreme Court.

YOU BE THE JUDGE

Was the police officer properly discharged?

HOLDING

Yes.

The court wrote:

A government employee does not relinquish all First Amendment rights otherwise enjoyed by citizens just by reason of his or her employment. On the other hand, a governmental employer may impose certain restraints on the speech of its employees, restraints that would be unconstitutional if applied to the general public. The Court has recognized the right of employees to speak on matters of public concern, typically matters concerning government policies that are of interest to the public at large, a subject on which public employees are uniquely qualified to comment. Outside of this category, the Court has held that when government employees speak or write on their own time on topics unrelated to their employment, the speech can have First Amendment protection, absent some governmental justification "far stronger than mere speculation" in regulating it. We have little difficulty in concluding that the City was not barred from terminating Roe under either line of cases.

A

In concluding that Roe's activities qualified as a matter of public concern, the Court of Appeals relied heavily on the Court's decision in NTEU, [where] it was established that the speech was unrelated to the employment and had no effect on the mission and purpose of the employer. The question was whether the Federal Government could impose certain monetary limitations on outside earnings from speaking or writing on a class of federal employees. The Court held that, within the particular classification of employment, the Government had shown no justification for the outside salary limitations. The First Amendment right of the employees sufficed to invalidate the restrictions on the outside earnings for such activities. The Court noted that throughout history public employees who undertook to write or to speak in their spare time had made substantial contributions to literature and art, and observed that none of the speech at issue "even arguably [had] any adverse impact" on the employer.

The Court of Appeals' reliance on NTEU was seriously misplaced. Although Roe's activities took place outside the workplace and purported to be about subjects not related to his employment, the SDPD demonstrated legitimate and substantial interests of its own that were compromised by his speech. Far from confining his activities to speech unrelated to his employment, Roe took deliberate steps to link his videos and other wares to his police work, all in a way injurious to his employer. The use of the uniform, the law enforcement reference in the Web site, the listing of the speaker as "in the field of law enforcement," and the debased parody of an officer performing indecent acts while in the course of official duties brought the mission of the employer and the professionalism of its officers into serious disrepute.

The Court of Appeals noted the City conceded Roe's activities were "unrelated" to his employment. In the context of the pleadings and arguments, the proper interpretation of the City's statement is simply to underscore the obvious proposition that Roe's speech was not a comment on the workings or functioning of the SDPD. It is quite a different question whether the speech was detrimental to the SDPD. On that score the City's consistent position has been that the speech is contrary to its regulations and harmful to the proper functioning of the police force.

B

To reconcile the employee's right to engage in speech and the government employer's right to protect its own legitimate interests in performing its mission, the Pickering Court adopted a **balancing test**. It requires a court evaluating restraints on a public employee's speech to balance the interests of the [employee], as a citizen, in commenting upon matters of public concern and the interest of the State, as an employer, in promoting the efficiency of the public services it performs through its employees.

Underlying the decision in Pickering is the recognition that public employees are often the members of the community who are likely to have informed opinions as to the operations of their public employers, operations which are of substantial concern to the public. Were they not able to speak on these matters, the community would be deprived of informed opinions on important public issues... The interest at stake is as much the public's interest in receiving informed opinion as it is the employee's own right to disseminate it.

In *Connick,* an assistant district attorney, unhappy with her supervisor's decision to transfer her to another division, circulated an intraoffice questionnaire. The document solicited her co-workers' views on, *inter alia,* office transfer policy, office morale, the need for grievance committees, the level of confidence in supervisors, and whether employees felt pressured to work in political campaigns.

Finding that—with the exception of the final question—the questionnaire touched not on matters of public concern but on internal workplace grievances, the Court held no Pickering balancing was required. To conclude otherwise would ignore the "common-sense realization that government offices could not function if every employment decision became a constitutional matter." Connick held that a public employee's speech is entitled to Pickering balancing only when the employee speaks "as a citizen upon matters of public concern" rather than "as an employee upon matters only of personal interest."

balancing test Employees such as firefighters and police officers have a restricted right of free speech, balanced against the government employer right to avoid undue public concern.

Although the boundaries of the public concern test are not well-defined, Connick provides some guidance. It directs courts to examine the "content, form, and context of a given statement, as revealed by the whole record" in assessing whether an employee's speech addresses a matter of public concern. In addition, it notes that the standard for determining whether expression is of public concern is the same standard used to determine whether a common-law action for invasion of privacy is present.

These cases make clear that public concern is something that is a subject of legitimate news interest; that is, a subject of general interest and of value and concern to the public at the time of publication. The Court has also recognized that certain private remarks, such as negative comments about the President of the United States, touch on matters of public concern and should thus be subject to Pickering balancing.

Applying these principles to the instant case, there is no difficulty in concluding that Roe's expression does not qualify as a matter of public concern under any view of the public concern test. He fails the threshold test and Pickering balancing does not come into play.

Connick is controlling precedent, but to show why this is not a close case it is instructive to note that even under the view expressed by the dissent in Connick from four Members of the Court, the speech here would not come within the definition of a matter of public concern. The dissent in Connick would have held that the entirety of the questionnaire circulated by the employee "discussed subjects that could reasonably be expected to be of interest to persons seeking to develop informed opinions about the manner in which... an elected official charged with managing a vital governmental agency, discharges his responsibilities." No similar purpose could be attributed to the employee's speech in the present case. Roe's activities did nothing to inform the public about any aspect of the SDPD's functioning or operation. Nor were Roe's activities anything like the private remarks at issue in Rankin where one co-worker commented to another co-worker on an item of political news. Roe's expression was widely broadcast, linked to his official status as a police officer, and designed to exploit his employer's image.

The speech in question was detrimental to the mission and functions of the employer. There is no basis for finding that it was of concern to the community as the Court's cases have understood that term in the context of restrictions by governmental entities on the speech of their employees.

The judgment of the Court of Appeals is Reversed.

LEGAL LESSONS LEARNED

Fire departments and other governmental employers have a right to reasonably restrict the "free speech" of firefighters when that speech has a detrimental effect on the department.

■ ■

Chapter Review Questions

1. In Case Study 16-1, *Garrity v. New Jersey,* the court addressed the issue of internal investigations. If you were directed by the fire chief to investigate the apparent theft of morphine from an ambulance, and the investigation quickly focused on a particular paramedic, describe which officials outside the fire department you might consult prior to ordering the paramedic to answer your questions.

2. In Case Study 16-2, *Dennis v. Berne Twp. Board of Trustees,* the court reviewed

the issue of termination of a fire chief for failure to turn in fire reports to the state fire marshal. Describe the requirements in your state concerning fire reports, EMS run reports to your state officials, and their audit procedures to confirm there is full compliance with state law and regulations.

3. In Case Study 16-3, *City of San Diego v. John Roe,* the court addressed the issue of off-duty conduct and sexually explicit videos. Describe how "opinion letters" written by a firefighter to the local newspaper concerning fire department staffing levels might reflect poorly on a fire department, and what steps the department might take to prevent this activity without violating the First Amendment rights of an individual.

■■

Expand Your Learning

Read and complete the individual student or group assignment, as directed by your instructor.

1. One year after the Garrity decision, the U.S. Supreme Court in *Gardner v. Broderick,* 392 U.S. 273 (1968), ruled that a police officer may be disciplined for invoking his Fifth Amendment right to silence. Give examples of investigations where it would be appropriate for fire department investigators to confer with a local prosecutor prior to issuing a "Garrity Warning" to one firefighter, to compel him to answer questions that might incriminate another firefighter who is the "target" of the investigation.

2. On June 9, 2004, the National Labor Relations Board reversed its earlier precedence (known as the "Weingarten Rule") and held in *IBM Corporation and Kenneth Paul Schult, Robert William Bannon and Steve Parsley,* Cases 11-CA-19324, 11-CA-19329 and 11-CA-19334 (341 NLRB 148), that employees who are not in a union have no right to have a coworker accompany them to a company internal investigation interview. Discuss what actions a fire department should take when they are conducting an investigation of a part-time firefighter who wishes to be present during an investigative interview by the president of IAFF Local, which represents the career firefighters.

See Appendix A for additional Expand Your Learning activities related to this chapter.

Arbitration/ Mediation

 **17** CHAPTER

Mediation can resolve disputes, including "minimum manning" for each shift on the Newark (OH) Fire Department. Fire Chief Jack Stickradt (right), Consultant Retired Fire Chief Bill Kramer (left), and the author used the mediation process to resolve a grievance. *Photo by author.*

Key Terms

mediation, p. 218
five-step grievance
  procedure, p. 223

exhaustion of administrative
  remedies doctrine, p. 223

Having Sex on the Job Not Reason to Fire Firefighter

The *Plain Dealer* newspaper in Cleveland (OH) reported that an arbitrator upheld the grievance of a Delaware (OH) firefighter who was fired for having sex with a nurse while on duty, in the storage facility behind the station. In re *City of Delaware, Ohio, and IAFF Local 606*, FMCS #00-03172. An arbitrator ruled that termination was too harsh a penalty, but the firefighter will not receive back pay for the 18 months that he has been suspended. The city filed an appeal to the Ohio Court of Common Pleas and to the Ohio Court of Appeals, but neither court would reverse a "binding arbitration" decision unless there is proof of fraud by the arbitrator.

Sex on Duty—IAFF Post-Hearing Brief Refers to President Clinton

Counsel for IAFF Local 606 filed a post-hearing brief for the Delaware (OH) firefighter, stating: "Just cause for termination does not exist in this matter. It is interesting to note that, at about the time the City was considering the discipline of [this firefighter], the Nation was considering what if anything should be done to its President, who also engaged in sexual activity on duty. The Nation concluded that, although the President's activities were wrong, they were not sufficient to justify removal from office. The same conclusion should be reached here."

◆ INTRODUCTION

Public and private employers throughout the nation are seeking to avoid litigation in court concerning internal personnel issues. Instead of going to court, employers are using informal mediation (in which a mediator attempts to find a common ground for settlement) and formal arbitration (in which an arbitration hearing leads to a final decision for one of the parties).

◆ UNIFORM MEDIATION ACT

mediation Defined under the new Ohio Mediation Act as "any process in which a mediator facilitates communication and negotiation between parties to assist them in reaching a voluntary agreement regarding their dispute." Ohio Rev. Code 2710.01, effective November 29, 2005.

States throughout the nation are adopting the Uniform Mediation Act to provide the legal structure for individuals to resolve disputes without filing lawsuits. In the fire service, many fire departments and locales are adopting **mediation** as a step in the grievance resolution process and enjoying great success.

◆ U.S. SUPREME COURT—UPHOLDS ARBITRATION

In the private sector, an increasing number of employers are requiring new hires, as a condition of employment, to sign an agreement to submit any future employment claims to binding arbitration instead of filing lawsuits. Understandably, there has been much litigation over this issue. The U.S. Supreme Court on February 21, 2006, strengthened the right of employers to enforce arbitration agreements, holding in *Buckeye Check Cashing, Inc. v. Cardegna* that a group of employees, suing in Florida court about the high interest rates charged by a check cashing company, must submit the matter to arbitration.

◆ FIRE SERVICE—MEDIATION TO RESOLVE PENDING GRIEVANCES

In the fire service, mediation provisions are being added to fire department collective bargaining agreements as an effective method to resolve grievances arising out of discipline and other matters, thereby avoiding the expense and delay of binding arbitrations. See Appendix G, which includes the Cincinnati Fire Department and IAFF Local 48 Labor-Management Agreement (June 5, 2005–June 2, 2007). This agreement includes a mediation process for resolving pending grievances informally.

◆ KEY STATUTES

UNIFORM MEDIATION ACT

Many states are implementing mediation statutes. For example, Ohio adopted the Uniform Mediation Act on October 29, 2005 (Ohio Rev. Code 2710.01), which encourages private mediation to resolve workplace and other disputes. The legislative history of 2004 House Bill 303, 125th General Assembly is available at *www.legislature.state.oh.us.*

◆ CASE STUDIES

CASE STUDY 17-1 Firefighter Fails to Obtain Paramedic Certificate within Two Years—Results in Termination

Mark A. Marusa v. City of Brunswick, 2005 WL 605440 (Ohio App. 8 Dist.; March 16, 2005), 2005-Ohio-1135.

FACTS

A firefighter filed a lawsuit action against the city, alleging handicap discrimination and violation of the city charter, after he was terminated for failing to complete paramedic training. The Court of Common Pleas, Medina County, No. 03CIV0805, granted summary judgment to the city, and the firefighter appealed to the Ohio Court of Appeals.

Mark Marusa was employed by the City of Brunswick, first as a part-time firefighter and eventually as a full-time firefighter. According to city policy, he had two years from the time he became a full-time firefighter to complete his paramedic training. He did not complete the training within the two-year limit and was discharged.

On June 18, 2003, Marusa (the Appellant) filed a lawsuit against the city, William Lebus and Patrick Beyer. The city, Lebus, and Beyer filed an answer to Marusa's lawsuit on July 18, 2003. On November 13, 2003, the defendants filed a notice of serving their first set of interrogatories, requests for production of documents, and request for admissions to the firefighter. The firefighter did not respond to any of the defendants' requests.

On February 2, 2004, the defendants filed a motion for summary judgment, and stated "as a result of the admissions of [plaintiff] * * * and as a result of [plaintiff's] failure to provide any evidence to support his claims, there [were] no longer any material issues of fact[.]" Defendants also argued that they were entitled to summary judgment because the firefighter failed to exhaust his administrative remedies.

On March 15, 2004, Marusa filed a response to the defendants' motion for summary judgment. He argued that viewing the evidence most strongly in his favor as the nonmoving party, he had clearly established a prima facie case of breach of an implied contract and promissory estoppel because he was led to believe that his employment was permanent. He felt his employment was secure because the city did not enforce the two-year limit on paramedic training and did not seek other employment. He argued that he had established a prima facie case for intentional infliction of emotional distress because, contrary to defendants' representations that he was a permanent employee and would only be terminated for good cause, he was terminated after two years of employment which was "extreme and outrageous and intended to cause [him] severe emotional distress." He claimed that being terminated less than one week before Christmas caused him serious emotional distress.

The firefighter argued that he had established a prima facie case for his handicap discrimination claim because he had "a learning disability which affected his ability to understand the material and successfully complete the test portion of the required Paramedic Training."

In regard to his violation of the city's charter claim, Marusa argued that viewing the evidence most strongly in his favor, he "clearly established a violation[.]" The firefighter based his argument on the allegation that he was wrongfully terminated but maintained his part-time status and that, even after he passed the paramedic training test, he was not given the right of first refusal to new full-time positions, which was in violation of the city charter.

On March 19, 2004, the magistrate issued his decision and granted summary judgment to the city. The magistrate stated,

> Upon review of the evidence submitted by the parties pursuant to *Civil Rule 56(C)*, including evidence properly before the Court pursuant to the [Appellees'] unanswered discovery requests, now deemed admitted, the Court finds no material issue of fact exists which would preclude summary judgment on the issues presented.
>
> 2. The [Appellant] was a member of The International Association of Firefighters Local 3568 (hereafter IAFF); pursuant to a Collective Bargaining Agreement between the City of Brunswick and IAFF, the [Appellant] was required to resort to a 5 step grievance procedure as his sole and exclusive remedy for disputes concerning any type of discipline or discharge from employment.
>
> 3. The [Appellant] has alleged he did not receive proper notice of an administrative hearing. The evidence presented reveals he took no action in response, administrative or otherwise, other than filing this suit.

The magistrate found that Beyer and Lebus were entitled to summary judgment as a matter of law because there was no evidence "at all" before the court suggesting that they were acting in any capacity other than within the scope of their employment with the city. The magistrate also found that Marusa had other remedies that he could have pursued after he allegedly failed to receive notice of the administrative hearing. The magistrate determined "The [Appellant] can not [sic] by-pass the mandatory exclusive provisions of the collective bargaining agreement to pursue a breach of contract action (Count 1) or an alleged violation of the Brunswick City Charter (Count VII) by simply alleging he did not receive proper notice."

The magistrate concluded that because the firefighter failed to exhaust all of his administrative remedies, the City of Brunswick was entitled to summary judgment on Counts I and VII of Marusa's complaint. The magistrate awarded summary judgment to the defendants on the remaining counts of the firefighter's complaint because:

> [Plaintiff's] evidence, including the unanswered discovery requests deemed admitted, demonstrates the absence of genuine issues of material fact as to the essential elements of [plaintiff's] remaining causes of action. [Plaintiff] has failed to submit any evidentiary material showing a genuine dispute of the facts at issue. The un-rebutted facts demonstrate a complete failure of the evidence required to support the necessary elements to establish the remaining causes of action as set forth in Counts II, III, IV, V, & VI of [plaintiff's] complaint.

On April 1, 2004, the firefighter filed objections to the magistrate's decision. He objected to the magistrate's decision "on the grounds that [the] decision [was] against the manifest weight of the evidence and result[ed] in an unfair and inequitable result." The firefighter specifically objected to the following:

> 1. To the finding that no material issue of fact exists which would preclude summary judgment on the issues presented.
> 2. To the finding that [plaintiff] was required to resort to a 5-step grievance procedure as his sole and exclusive remedy for disputes concerning any type of discipline or discharge from employment.
> 3. To the finding that [Beyer and Lebus] were acting in any capacity other than within the scope of their employment with the [City].
> 4. To the finding that [plaintiff] failed to exhaust all his administrative remedies.
> 5. To the finding that [Appellant] failed to submit any evidentiary material showing a genuine dispute of the facts at issue.

On April 23, 2004, the trial court affirmed the magistrate's decision in full, entering summary judgment in favor of the defendants. The trial court found that "[u]pon careful independent review of the file, the Magistrate's Decision, and upon considering the briefs and oral arguments of counsel, the Court finds the Magistrate's Decision contains no error of law or other defect and there are no issues of fact which would preclude summary judgment."

The firefighter has now filed an appeal to the Ohio Court of Appeals.

YOU BE THE JUDGE

Was the firefighter's lawsuit properly dismissed?

HOLDING

Yes.

The three-judge Court of Appeals wrote,

In his sole assignment of error, Appellant has argued that the trial court erred in granting summary judgment to Appellees. Specifically, Appellant has argued that a genuine issue of material fact existed as to whether or not he exhausted all of his administrative remedies, whether or not Appellees violated the Brunswick City Charter, and in regard to his claims for breach of contract, promissory estoppel, emotional distress, and handicap discrimination. We disagree.

Before discussing each of the counts in Appellant's complaint and whether summary judgment was proper on said counts, we must address *Civ.R. 36*. Appellant's failure to respond to Appellees' request for admissions resulted in said admissions being admitted into the record. Pursuant to *Civ.R. 36:*

(A) A party may serve upon any other party a written request of the admission, for purposes of the pending action only, of the truth of any matters within the scope of Rule 26(B) set forth in the request, that relate to statements or opinions of fact or of the application of law to fact, including the genuineness of any documents described in the request. * * * The matter is admitted unless, within a period designated in the request, not less than twenty-eight days after service thereof or within such shorter or longer time as the court may allow, the party to whom the request is directed serves upon the party requesting the admission a written answer or objection addressed to the matter[.]

As previously noted, Appellees filed a notice of serving their first set of interrogatories, requests for production of documents and request for admissions to Appellant on November 13, 2003.

By the explicit terms of *Civ.R. 36(A)*, a party's failure to timely respond to request for admissions results in default admissions. When a party fails to timely respond to the request for admissions, "the admissions [become] facts of record, which the court must recognize." From a practical standpoint, however, a party typically moves the trial court to "deem" the matters admitted to bring the issue to the trial court's attention and to make the default admissions, which may not have been filed previously with the court, part of the trial court record. *See Id.*

As such, the following requests for admissions became facts of the record:

Admit that [Appellant], through his union representative, denied, at his pre-termination hearing held on or about December 19, 2002, that he had a disability.

Admit that [Appellant] was required to successfully complete his paramedic training within 2 years of his hire date of December 18, 2000.

Admit that [Appellant] failed to successfully complete his paramedic training within 2 years of his hire date of December 18, 2000.

Admit that [the City] was engaged in the performance of a governmental function, set forth in *Ohio Revised Code § 2744.01,* in connection with the subject matter of [Appellant's] complaint.

Admit that [Appellant] never notified the City of any alleged disability prior to his termination.

Admit that neither [Appellant] nor someone closely related to [Appellant] faced and was cognizant of an actual physical peril in connection with [Appellant's] termination from employment.

Pursuant to *Civ.R. 36,* Appellant's default admissions conclusively established that: 1) on or about December 19, 2002, Appellant through his union representative denied he had a disability; 2) Appellant was required to successfully complete his paramedic training within two years of his hire date of December 18, 2000; 3) Appellant failed to complete said training within two years of December 18, 2000; 4) the City was engaged in the performance of a governmental function, as set forth in *R.C. 2744.01,* in connection with the subject matter of Appellant's complaint; 5) Appellant never notified the City of any alleged disability prior to his termination; and 6) neither Appellant nor someone closely related to him faced and was cognizant of an actual physical peril in connection with his termination.

Appellees moved for summary judgment against all of Appellant's claims, relying on evidence that included Appellant's default admissions, an affidavit from the Administrative Services Coordinator for the City, the collective bargaining agreement between the City and Appellant's firefighter union, an affidavit concerning when Appellant was served with the request for admissions and his failure to respond, an inter-office correspondence signed by Appellant confirming that he was a full time firefighter and had two years from December 18, 2000 to complete the required paramedic training, and the City charter.

In the instant matter, the collective bargaining agreement between Appellant, as a member of the firefighters union, and the City contained a grievance procedure for employment disputes. Pursuant to agreement, the grievance procedure "shall be the sole and exclusive procedure for disputes concerning any type of discipline or discharge actions." The agreement outlined a **five-step grievance procedure**. The first four steps required the aggrieved party to deal with his supervisor, then his department head and finally the City manager. If the issue remained unresolved and "no agreement [was] reached in fifteen (15) working days, in Step 4, [the] grievance may [have been] submitted to a Mediator or the Civil Service Commission, but not both." Rather than follow the required procedure, Appellant filed suit in the court of common pleas.

The **exhaustion of administrative remedies doctrine** is a well-established principle of Ohio law. Specifically, the doctrine requires that a party exhaust available administrative remedies prior to seeking court action in an administrative matter. Pursuant to Appellant's collective bargaining agreement, his administrative remedy was the five-step grievance procedure.

As a general rule, the grievance procedures of a collective bargaining agreement must be exhausted as a prerequisite to any civil action against the employer. Appellant has claimed that he filed a grievance and a hearing was held on the grievance without any notice being given to him. But a review of the record reveals that Appellant has not provided any evidence to support such a claim. Appellant has not provided any evidence beyond his self-serving statements that he filed a grievance, that a hearing was held, that he was not notified of the hearing or that he attempted to pursue the other steps of the grievance procedure.

Based on the foregoing, we find that Appellees are entitled to summary judgment on Appellant's claim of breach of contract because Appellant failed to exhaust his administrative remedies under the collective bargaining agreement.

five-step grievance procedure   Under collective bargaining agreements between fire departments and local unions, it is common to have a multi-step process to file and process grievances.

exhaustion of administrative remedies doctrine   A firefighter who is covered by a collective bargaining agreement must normally file a grievance and process it through arbitration, prior to seeking any relief in a court of law.

LEGAL LESSONS LEARNED

A firefighter covered by a collective bargaining agreement must file a grievance and exhaust his administrative remedies prior to seeking judicial relief. Consider adding a mediation step to an existing collective bargaining agreement to resolve grievances without arbitration.

CASE STUDY 17-2 Firefighter Fired for Having Sex On Duty— Arbitrator Reinstates with No Back Pay: City Seeks Court Order to Reverse Arbitrator

City of Delaware v. International Association of Firefighters, Case No. 00CV-H-10-454, Ohio Court of Common Pleas of Delaware County, OH (March 12, 2001).

[This opinion and the Ohio Court of Appeals decisions are unpublished so the name of the firefighter has been withheld.]

FACTS

A firefighter was employed full-time by the City of Delaware (OH) for eight years, prior to his termination on May 19, 1999. The IAFF Local filed a grievance. The matter went to arbitration, and the arbitrator on August 3, 2000, issued a decision in favor of the firefighter, returning him to work but without back pay. The city filed an appeal to the Court of Common Pleas to modify or vacate the arbitrator's decision.

According to the trial judge's written decision, the firefighter "was married when he met [a] nurse, who was also married. She worked in the emergency room of Grady Memorial Hospital. The two carried on an extra-marital affair. To ensure that their meetings were not discovered by spouses or co-workers, they would meet in a storage building behind the fire station to have sex when she would get off work. [The firefighter], who was on duty while engaging in the sexual activities, carried a two-way radio with him to the storage shed so he would know if there was an emergency call. Additionally, the fire station had speakers on the exterior of the building which would sound an alarm when there was a run."

The court wrote,

> After ending the affair, [the firefighter] spoke to the Fire Chief and informed him of his adultery but did not reveal that it occurred on City property while he was on duty. Upon learning of the full extent of his affair, the City decided to discharge [the firefighter]. [The firefighter] filed a grievance which was heard by an arbitrator. After reviewing the arguments and evidence presented by the parties, the Arbitrator awarded [the firefighter] reinstatement of his position. The Arbitrator further ordered that [the firefighter] not receive any back pay for the eighteen months between the time of his discharge and his reinstatement.

The city appealed to the Court of Common Pleas and presented three arguments on why the arbitrator's award should be either vacated, or modified: (1) the arbitrator exceeded his power; (2) the arbitrator's award should be modified because there was a material mistake in the description of the firefighter's adulterous conduct as being less than very serious; and (3) the court should not be bound by the arbitrator's decision because the firefighter's affair violates public policy.

YOU BE THE JUDGE

Should the trial judge conduct a hearing and vacate or modify the arbitrator's decision?

HOLDING

No.

The Common Pleas judge wrote,

> An arbitration award carries a presumption of validity. This presumption is so strong that it cannot be overturned even if the decision is legally or factually wrong.

[Ohio Revised Code] 2711.10 (D) provides that in order to vacate the award of an arbitrator, the court of common pleas shall determine if the arbitrator exceeded his powers so that a mutual, final and definite award upon the subject matter submitted was not made.... Here, the Arbitrator derived his power from the Agreement which grants authority to interpret the Union Contract:

The Agreement provides, "the arbitrator's decision shall be based solely upon his interpretation of the meaning or application of the express terms of the Agreement." Article 11 of the Agreement, page 20. Further, the "ruling and decision of the arbitrator, *** shall be binding." Article 11 of the Agreement, page 19. Thus, it was the job of the Arbitrator to interpret Section One of article 11, entitled "Just Cause." It provides, "no member shall be removed, reduced in pay or position, suspended except for just cause." Section Two, "Progressive Discipline," states, "the City agrees to follow the principles of progressive, corrective action. The Fire Chief may skip any step of progressive action if the violations are of a *very serious nature.* Further, the City agrees to fairly and equitably discipline members." [Emphasis added.]

The Arbitrator construed the meaning of the phrases "very serious nature" and "just cause." The arbitrator then applied the meaning of the words to the sexual activities engaged in by [the firefighter] on the City property on duty. In light of the firefighter's ending the affair and the absence of any harm, the Arbitrator found that the sexual misconduct did not rise to the level of severity that constitutes just cause to discharge [the firefighter]. Arbitrator's Award, pages 22–23. Accordingly, the Arbitrator performed his duties according to the terms of the Agreement.

While this Court may not agree with the Arbitrator's conclusions or even believe the Arbitrator had adequate evidence to reach these conclusions, it is not the role of the trial court to replace the arbitrator's decision with its own.

The Arbitrator decided that eighteen months suspension without pay served as adequate punishment for [the firefighter's] immoral and risky behavior. It is not for this Court to determine the appropriate sanction. The arbitrator's decision signals that sexual encounters by City employees while on the job will result in severe financial repercussions. Thus, the eighteen month suspension without pay does not violate the public policy.

LEGAL LESSONS LEARNED

Arbitrators' awards are seldom reversed by a court of law. It is wise for both parties to consider informal mediation prior to going to binding arbitration. The city filed an appeal to the Ohio Court of Appeals, which affirmed the trial judge's ruling.

CASE STUDY 17-3   Cincinnati Fire Department and IAFF Local 48 Mediation Process

See Appendix G for the mediation process used by the Cincinnati Fire Department and IAFF Local 48. Mediation is an excellent process to reach settlement without the expense and delay of a formal arbitration.

■ ■

Chapter Review Questions

1. In Case Study 17-1, *Mark A. Marusa v. City of Brunswick,* the firefighter was fired for not obtaining his paramedic certificate within two years. He claimed to have had a reading disability. Describe what "reasonable accommodations" are provided in your state for students enrolled in paramedic school.

2. In Case Study 17-2, *City of Delaware v. International Association of Firefighters,* the court addressed the issue of on-duty sexual activities. If you were in charge of the investigation, and the captain in charge of the shift advised you that he had no knowledge of the firefighter's activities, and you got the same story from all of his fellow crew members, would you have recommended discipline for any of these personnel?

3. In Case Study 17-3, under the Cincinnati Fire Department mediation provisions, grievances can be submitted to mediation upon the agreement of management and the union/firefighter. If you were president of a union local and a firefighter was facing discipline for repeatedly being late for work, what arguments would you make to convince him or her to have the matter submitted to mediation?

■■■

Expand Your Learning

Read and complete the individual student or group assignment, as directed by your instructor.

1. In the case of *Mark A. Marusa v. City of Brunswick,* 2005 WL 605440 (Ohio App. 8 Dist.; March 16, 2005), 2005-Ohio-1135, what steps might the firefighter have taken with the fire department to get an extension of time on the two-year requirement to get his paramedic certificate?

2. Assume that you were the lead paramedic instructor for Mark Marusa, and he advised you that he had a reading deficiency, wanted additional time in taking the written exams, and wanted additional after-class instruction. Discuss the steps that the paramedic school might authorize so you could reasonably accommodate him.

See Appendix A for additional Expand Your Learning activities related to this chapter.

Module V: Legislative and Political Foundations

Legislative Process
Getting Helpful Laws Passed for the Fire Service (Amending the Fireman's Rule)

18 CHAPTER

Two firefighters were killed and two were injured in New Knoxville (OH), October 1, 2003, when the silo at a lumber yard, used to store wood fiber for electrical generation, exploded when firefighters on an aerial platform opened the top and sprayed water. Read NIOSH Report F2003-32 (*http://www.cdc.gov/niosh/fire/*). *Photo by author.*

Key Terms

From the Headlines

Kentucky Passes the Brenda Cowan Act—Felony to Injure a Firefighter

Lieutenant Brenda Cowan, Lexington (KY) Fire Department was killed by a rifle shot on February 13, 2004, when she was approaching a downed victim outside of a home on a domestic violence call. Senate Bill 91, enacted in 2005, applies to fire, emergency medical service, and rescue personnel who are assaulted "while personnel are performing job-related duties." See NIOSH Fire Fighter Fatality Investigation Report F2004-11 (*www.cdc.gov/niosh/face200411.html*).

Florida's Safe Haven Law—Vero Beach (FL)—Baby Left at Fire Station

A baby girl with her umbilical cord still attached was dropped off at a fire station under Florida's Safe Haven Law, passed in July 2000. Many states have passed laws that allow a mother to leave a newborn at any medical facility or fire station, with no questions asked.

Numerous NIOSH and other studies advise owners of silos to place warning signs at the base and the top of silos warning of the danger of opening the top to fight fires. This silo contained no such warnings. *Photo by author.*

Legislative change occurs because special interest groups, lobbyists, and individuals "get involved" to bring a need for change to the attention of their state and federal elected officials. The fire service can help influence change. Organizations such as the International Association of Fire Chiefs, the International Association of Fire Fighters, state fire chiefs' associations, and state union organizations can influence legislative change.

Our nation is involved in a "War on Terrorism," and the fire service needs to speak loudly and clearly in support of our legislative and financial needs as key players in domestic preparedness. This chapter will focus on the legislative process, and in particular, getting state laws enacted that amend the Fireman's Rule.

◆ **MODIFYING THE FIREMAN'S RULE**

This chapter focuses on states that have enacted laws either modifying or abolishing the Fireman's Rule. The Fireman's Rule is based on state court decisions holding that firefighters and police officers could not sue property owners for their injuries. Originally, the Fireman's Rule developed under the English common law during the existence of the feudal system of land ownership. Under this system, a firefighter was labeled as a licensee to whom the owner or occupant owed no greater duty than to refrain from the infliction of willful or intentional injury. Several states have abolished or modified the rule, particularly if the injuries were caused by the property owner's violation of building or safety codes.

In states that have not amended the Fireman's Rule by statute, it is difficult for the survivors of firefighters killed in the line of duty to win a lawsuit against business owners or homeowners. The insurance company for the business owner or homeowner will file a motion to dismiss the lawsuit prior to trial.

WIDOW OF FIREFIGHTER SETTLES WITH INSURANCE COMPANIES

Fortunately, the widow of an Ohio firefighter was able to reach a settlement with numerous insurance companies. Please read NIOSH Report 2001-16, on the death of Firefighter Bill Ellison, Miami Township, Hamilton County (OH), *http://www.cdc.gov/niosh/fire/.* His wife filed a lawsuit against the homeowner, a heating and ventilation company, and a plumbing company based on violation of building codes. A second water heater had been installed in the basement. Both water heaters used the same vent pipe, which must be a "pipe with a pipe" construction to avoid overheating. The lawsuit was filed in 2001 and settled after extensive depositions in 2004, for an undisclosed amount prior to trial. Complaint and other documents can be read online, *http://www.courtclerk.org/,* in Victoria Ellison v. Anderson Heating & Cooling, Case No. A0106222.

Hopefully, this chapter will encourage firefighters to become involved in the legislative process and work to have statutes enacted that will abolish or amend the Fireman's Rule in their states.

FLORIDA HAS ABOLISHED THE FIREMAN'S RULE

Fla. Stat. 112.182 provides:

> 112.182 "Firefighter rule" abolished.—
>
> (1) A firefighter or properly identified law enforcement officer who lawfully enters upon the premises of another in the discharge of his or her duty occupies the status of an invitee. The common-law rule that such a firefighter or law enforcement officer occupies the status of a licensee is hereby abolished.
>
> (2) It is not the intent of this section to increase or diminish the **duty of care** owed by property owners to invitees. Property owners shall be liable to invitees pursuant to this section only when the property owner negligently fails to maintain the premises in a reasonably safe condition or negligently fails to correct a dangerous condition of which the property owner either knew or should have known by the use of reasonable care or negligently fails to warn the invitee of a dangerous condition about which the property owner had, or should have had, knowledge greater than that of the invitee.

duty of care For a homeowner or business owner to be liable for injuries to a third party, there must be a breach of a legal duty of care for this injured party.

MINNESOTA HAS ABOLISHED THE FIREMAN'S RULE

Minn. Stat. 604.06 provides:

> The common law doctrine known as the fireman's rule shall not operate to deny a peace officer, as defined in section 626.84, subdivision 1, clause (c), or public safety officer, as defined in section 299A.41, subdivision 4, a recovery in any action at law or authorized by statute.

CALIFORNIA HAS MODIFIED THE FIREMAN'S RULE

Civil Code 1714.9 imposes liability for willful or negligently inflicted injuries sustained by police officers, firefighters, or any emergency medical personnel employed by a public entity, where the injury-causing conduct occurs after the landowner or occupant knows or should have known of the presence of the public safety officer.

NEVADA HAS MODIFIED THE FIREMAN'S RULE

Nev. Res. Stat. Ann. 41.139 provides,

> [A] peace officer, fireman, or emergency medical attendant may bring and maintain an action for damages for personal injury caused by the willful act of another, or by another's lack of ordinary care or skill in the management of his property, if the conduct causing the injury:... [o]ccurred after the person who caused the injury knew or should have known of the presence of the peace officer, fireman, or emergency medical attendant;... violated a statute, ordinance or regulation;... intended to protect the peace officer, fireman or emergency medical attendant; or prohibiting resistance to or requiring compliance with an order of a peace officer or fire fighter; or.... was arson.

NEW YORK HAS MODIFIED THE FIREMAN'S RULE

N.Y. Gen. Oblig. Law 11-106 provides,

> [W]henever any police officer or firefighter suffers any injury... the police officer or firefighter suffering that injury... may seek recovery and damages from the person or

entity whose neglect, willful omission, or intentional willful or culpable conduct resulted in that injury.

See also NY Gen. Mun. Law 205-e.

STATE APPELLATE COURT OPINION

Some state appellate courts have issued opinions that have limited the Fireman's Rule. For example:

- Arizona's Supreme Court has now confirmed the Fireman's Rule does not apply to off-duty conduct: *Espinoza v. Schulenburg,* 129 P.3d 937 2006 (Ariz. March 15, 2006); Ariz. LEXIS 27. The Arizona Supreme Court held that the Fireman's Rule does not bar an off-duty firefighter, who stops to aid at a motor vehicle accident and is injured, from suing the person whose negligence caused the accident. This case is reviewed in Chapter 2.
- Michigan court has declined to apply the Fireman's Rule to EMS personnel: see *McKim v. Forward Lodging, Inc.,* 266 Mich. App. 373, 702 N.W.2d 181, 2005 Mich. App. LEXIS 1108 (Court of Appeals of Michigan, 2005). Paramedic Katrina McKim responded to a person who fell on ice on a parking lot outside a hotel and conference center. The assistant manager who came to the aid of the person also slipped and broke her ankle. After assessing the assistant manager, the paramedic walked back to her ambulance for equipment, and she also fell on the ice, hitting her head and injuring herself. The paramedic filed a lawsuit against the hotel, but the trial court dismissed the case finding that the hotel did not have a duty to warn the paramedic of the obvious icy conditions. The trial judge refused to consider later evidence that the day prior to the accident, the hotel maintenance staff had sprayed hot water on the roof to dislodge ice and the water flowed over the parking lot. The Court of Appeals reversed and remanded the case for trial, finding that the common-law Fireman's Rule had been abolished by the Michigan legislature, Mich. Comp. Laws 600.2967, and the paramedic was an "invitee" of the hotel, rendering aid for the hotel's benefit.
- South Carolina court has refused to follow the Fireman's Rule: *Minnich v. Med-Waste, Inc.,* 564 S.E.2d 98 (S.C. 2002). A public safety officer employed by the Medical University of South Carolina was injured when he was assisting in loading medical waste onto a tractor-trailer owned by Med-Waste, Inc. The unoccupied truck began to roll forward, and Jeffrey Minnich was injured as he jumped inside and stopped the truck.

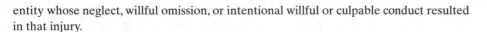

◆ **CASE STUDIES**

CASE STUDY 18-1 EMT Injured by Toxic Fumes—Files Suit against Chemical Company

Kapherr v. MFG Chemical, Inc., 625 S.E.2d 513 (9a. App. 2005).

FACTS

EMT Terrie Melissa Kapherr, an emergency medical technician, responded to a call to evaluate and treat personnel injured by a highly toxic gas and vapor cloud that had escaped from a local chemical plant owned by MFG Chemical, Inc. When she entered the toxic hazard zone, Kapherr developed chest pain, muscle tightness around her abdomen and back, and shortness of breath, symptoms that became a permanent asthmatic condition for her.

Kapherr and her husband filed suit: The trial judge dismissed it on the basis of the Fireman's Rule. They filed an appeal and argued that the Fireman's Rule (which precludes actions by public safety employees for injuries received as a result of the negligence causing the emergency to which they are responding) does not apply to emergency medical technicians generally and did not apply to her specific set of circumstances.

YOU BE THE JUDGE

Does the Fireman's Rule apply to EMTs?

HOLDING

Yes.

The three-judge Court of Appeals wrote,

Finding that the Fireman's Rule does apply, we affirm.

Kapherr sued MFG Chemical, alleging negligent release of the toxic chemicals. MFG Chemical moved for judgment on the pleadings on the ground that as an EMT responding to an emergency situation, Kapherr was precluded under the Fireman's Rule from recovering for injuries caused by the very circumstance that occasioned her presence at the scene. Kapherr responded by filing an affidavit that she had no firefighting duties and that under her employer's policy, she could only treat persons at secure scenes. The trial court granted the motion, entering judgment against Kapherr and her husband. Kapherr and her husband appeal, arguing that the Fireman's Rule does not apply to EMTs and in any case does not apply to her particular set of circumstances.

1. The Fireman's Rule in Georgia is a public policy of the State of Georgia that a public safety employee cannot recover for injuries caused by the very negligence that initially required his presence in an official capacity and subjected the public safety employee to harm; that public policy precludes recovery against an individual whose negligence created a need for the presence of the public safety employee at the scene in his professional capacity.

The Fireman's Rule provides that a public safety officer may not recover for the negligence that caused the situation to which he responded.

Here, the facts alleged in the complaint show that Kapherr was injured as a result of the negligence (accidental release of toxic fumes) that necessitated her presence at the scene. The primary question on appeal, therefore, is whether the Fireman's Rule applies to EMTs such as Kapherr. This hinges on whether an EMT is a public safety officer or employee as understood in this rule.

To answer this question, we look to the policy reasons behind this rule. First is the assumption of risk doctrine.

"[I]t is the nature of the job undertaken for the employee to be subjected to risks of injury created by people he or she is called upon to serve. By accepting that job the employee assumes a general or primary risk of injury.... The justification for imposing this general or primary risk is that the employee is paid to encounter it and trained to cope with it." EMTs, who like firefighters and police are the first responders to emergency situations where hazards are inherent, would appear to assume this general risk of injury.

(L)ike firefighters, EMTs "know that they will be expected to provide aid and protection to others in... hazardous circumstances"), (by the nature of his job, stadium security guard assumed risk of encountering an armed robber and being injured thereby). We see no reason not to apply this policy to Kapherr.

Second, it would be too burdensome to charge all who negligently cause a need for emergency services with the injuries suffered by the first responders trained to come and deal with the effects of those inevitable, although negligently created, occurrences. Indeed, "it offends public policy to say that a citizen invites private liability merely because he happens to create a need for those public services. Citizens should be encouraged and not in any way discouraged from relying on those public employees who have been specially trained and paid to deal with these hazards." [Citation and punctuation omitted.] (A)bsent the firefighter's rule, individuals who fear liability may hesitate to call for assistance in fighting a fire; "it is also well recognized public policy that the doctrine seeks to prevent a chilling effect that may occur if citizens in need of help were not free to solicit the assistance of professional rescuers for fear of tort liability." Here, when faced with an accidental release of toxic fumes, manufacturers like MFG Chemical should be encouraged to call on emergency services such as EMTs to handle the resulting injuries without fear of inviting additional liability.

Third, as stated by the California Supreme Court, "it is somehow unfair to permit a [public safety officer] to sue for injuries caused by the negligence that made his or her employment necessary." It is the emergency personnel's business to deal with emergency situations, (emergency medical personnel and other first responders "respond to public emergencies or crises for which they have been employed and specially trained"); thus, such a public safety employee or officer "cannot complain of negligence in the creation of the very occasion for his engagement." Here, Kapherr as an EMT was responding to a public emergency for which she had been employed and trained; she should not be allowed to complain of the negligence that created the very reason for her employment. "(P)ublic safety officers, whose occupation necessarily exposes them to certain risks of injury, cannot complain of negligent acts that create the very reason for their employment" [punctuation omitted].

Beyond the policy reasons that indicate that EMTs are "public safety employees" for purposes of the Fireman's Rule, other factors indicate that this phrase normally includes EMTs. OCGA § 31-22-9.1 (23), which concerns HIV tests, defines "public safety employee" as "an emergency medical technician, firefighter, law enforcement officer, or prison guard." The court referred to 28 CFR § 32.2 (j) (for purposes of death and disability benefits, "public safety officer" includes law enforcement officers, firefighters, rescue squad members and ambulance crew members) to determine to whom the Fireman's Rule applies. One state that codified the Fireman's Rule expressly included EMTs. N.H. Rev. Stat. Ann. § 507:8-h (I).

Accordingly, many other jurisdictions have concluded that the Fireman's Rule applies to EMTs. See *City of Oceanside,* supra at 629 (IV) (A) (California); *Melton,* supra at 875–877 (D.C.); *Randich v. Pirtano Constr. Company* (Illinois); *Sam v. Wesley* (Indiana); *Maggard v. Conagra Foods* (Kentucky); *Pinter,* supra at 111 (Wisconsin). But see *Kowalski v. Gratopp* (Michigan); *Krause v. U. S. Truck Co.* (Missouri). See generally Annot., "Application of 'Firemen's Rule' to Bar Recovery by Emergency Medical Personnel Injured in Responding to, or at Scene of, Emergency," 89 ALR4th 1079 (1991). Based on the policy reasons for the rule, we concur and hold that the Fireman's Rule applies to EMTs.

2. Kapherr maintains, however, that under her particular set of circumstances, the Fireman's Rule does not apply since under her employer's policy, she was to treat the injured at the scene only if the scene were secure. However, this argument ignores that as a first responder and professional rescuer, Kapherr necessarily assumed the risk of the hazardous circumstances at the scene, particularly at a scene involving the toxic release of chemicals. It also ignores the other policy reasons behind the Fireman's Rule, such as

the unfairness of allowing an EMT to recover for the very negligence that created the need for her employment.

The policies of Kapherr's particular employer do not diminish the dispositive nature of this inquiry.

Judgment affirmed.

LEGAL LESSONS LEARNED

EMTs and firefighters must understand the risks of their jobs. If the run is for individuals unconscious in a building, emergency personnel should always wear SCBA.

CASE STUDY 18-2   EMT Severely Injured in Gas Explosion—Sues Company Laying Underground Cable That Punctured Gas Line

Todd Randich v. Pirtano Construction Company, 804 N.E.2d 581, 589 (Ill. App. Ct. 2004).

FACTS

The EMT sued for negligence and willful and wanton misconduct. The trial court dismissed plaintiff's amended complaint under section 2—619(a)(9) of the Code of Civil Procedure, (735 ILCS 5/2—619(a)(9) (West 2000)).

Defendants are construction contractors. On April 29, 1999, they were working for Western Cable Communications (Western) installing underground television cable along a public utility easement granted to Western at the Wespark housing subdivision in Romeoville. Defendants' employees laid the cable underground through the use of a directional boring machine. In the process of boring into the ground, the defendants' employees punctured a natural gas main.

Northern Illinois Gas Company (NICOR) employees and members of the Lockport Fire Protection District (LFPD) were dispatched to the scene to contend with the gas leak and ensure safety. The plaintiff, an emergency medical technician, was one of the members of the LFPD sent to the scene.

The plaintiff was in the vicinity of the leaking gas when it was ignited by an unknown source within the housing development. A resulting explosion and fire completely destroyed two houses and damaged several others. More importantly, several NICOR employees and members of the LFPD were injured. The plaintiff was among the members of the LFPD who were injured in the explosion, sustaining "serious and permanent injuries."

On November 14, 2001, the plaintiff filed his amended complaint, alleging negligence and willful and wanton misconduct on the part of defendants. Specifically, the plaintiff's negligence claim alleges that the defendants failed to (1) investigate and ascertain the precise location of underground gas mains in the vicinity where they were operating the directional boring machine; (2) properly expose the gas main by hand digging before boring into the ground; and (3) arrange with NICOR in advance to turn off the gas prior to digging. The plaintiff's claim of willful and wanton misconduct basically sets forth the same factual allegations as the negligence claim, but it adds that the defendants acted with actual knowledge that a gas main was located within the utility easement where the defendants were conducting their drilling activities.

In response, the defendants filed a motion to dismiss under section 2—619(a)(9) of the Code [735 ILCS 5/2—619(a)(9) (West 2000)]. The trial judge granted the motion and dismissed the plaintiff's claims against the defendants on the ground that the Fireman's Rule prohibited the plaintiff's cause of action.

On appeal, the plaintiff asserts that the defendants cannot avail themselves of the Fireman's Rule because (1) the rule does not bar actions based on willful and wanton misconduct; (2) the defendants are not considered owners or occupiers of Western's utility easement; (3) public policy and the development of the deliberate encounter doctrine abrogate the Fireman's Rule; and (4) the application of the rule violates the plaintiff's equal protection rights.

YOU BE THE JUDGE

Should the EMT be permitted to have a jury trial to establish that the construction company was liable for willful and wanton misconduct?

HOLDING

Yes.

The three-judge Court of Appeals wrote,

> The fireman's rule limits the extent to which firefighters or other public officers are allowed to recover for injuries incurred when, in an emergency, they enter upon an owner's or occupier's property in discharge of their duty. *McShane v. Chicago Investment Corp.,* 235 Ill. App. 3d 860, 864 (1992). It provides that "an owner or occupier of land must exercise reasonable care to prevent injury to firemen that might result from a cause independent of the fire, but has no duty to prevent injury resulting from the fire itself." *Vroegh v. J&M Forklift,* 165 Ill. 2d 523, 527 (1995). Thus, the rule limits a fireman's right to recover for injuries arising out of the fire itself. *McShane,* 235 Ill. App. 3d at 865.
>
> The first issue in this case is whether the fireman's rule bars a cause of action based on willful and wanton misconduct. In Illinois, there is conflicting authority on this issue.
>
> The fireman's rule is a creature of case law, and thus the pertinent case law must be analyzed to understand and properly apply the rule. Originally, the fireman's rule developed under the English common law during the existence of the feudal system of landownership. Under this system, a firefighter was labeled as a licensee to whom the owner or occupant owed no greater duty than to refrain from the infliction of willful or intentional injury. In Illinois, the supreme court in *Gibson v. Leonard, 143 Ill. 182, 189(1892),* adopted this English common-law rule.
>
> Later, in *Dini,* our supreme court reshaped the fireman's rule to address what had been characterized as a barbaric formulation of the rule. In *Dini,* an inadequately constructed wooden staircase in the defendant's building collapsed and resulted in injury to several firefighters. *Dini,* 20 Ill. 2d at 412. Our supreme court rejected the common-law rule labeling firemen as licensees, finding it to be an illogical anachronism, originating in a vastly different social order, and pock-marked by judicial refinements. *Dini,* 20 Ill. 2d at 416. Instead, recognizing that firemen confer on landowners economic and other benefits that form a basis for imposing the common-law duty of reasonable care, the court held: "[A]n action should lie against a landowner for failure to exercise reasonable care in the maintenance of his property resulting in the injury or death of a fireman rightfully on the premises, fighting the fire at a place where he might reasonably be expected to be." *Dini,* 20 Ill. 2d at 416-17.
>
> Then, in *Washington v. Atlantic Richfield Co.* 66 Ill. 2d 103 (1976), the supreme court refined the reasonable care standard set forth in *Dini.* In *Washington,* firemen were

seriously injured in the process of extinguishing a car fire at a service station. *Washington,* 66 Ill. 2d at 104. The court was faced with the issue of whether the "liability of a possessor of land for injuries to a fireman extends to acts of negligence which cause a fire." *Washington,* 66 Ill. 2d at 105. In response, the court held that "while a landowner owes a duty of reasonable care to maintain his property so as to prevent injury occurring to a fireman from a cause independent of the fire he is not liable for negligence in causing the fire itself." *Washington,* 66 Ill. 2d at 108. Thus, from *Washington,* the current form of Illinois's version of the fireman's rule was born. It permits a firefighter to recover for injuries that result from an act unrelated to the specific reason he was summoned to the scene, but not for negligent acts that caused the emergency. This then leads to the issue of whether a fireman can recover for willful and wanton misconduct that causes an emergency.

In *Bandosz v. Daigger & Co.,* 255 Ill. App. 494 (1930), the court affirmed a judgment in favor of an administratrix of an estate of a fireman killed when a building exploded as a proximate result of the defendant's willful and wanton misconduct. The court recognized the principle that an owner or occupier of property is liable for willful and wanton misconduct to a firefighter in discharge of his duty. *Bandosz,* 255 Ill. App. 494.

In 1975, one year before the supreme court's decision in *Washington* was released, the Appellate Court, Third District, in *Marquart v. Toledo, Peoria & Western R.R. Co.,* 30 Ill. App. 3d 431, 432 (1975), addressed a situation in which a firefighter sued to recover for injuries incurred while fighting a fire that began with the explosion of a railroad car containing liquified propane gas. There, the court stated that "a fireman may recover where the injuries were caused by the willful and wanton misconduct of the owner or occupant of premises where the fire occurred." *Marquart,* 30 Ill. App. 3d at 434.

This expression of the fireman's rule comports with the purpose of the rule as articulated by the supreme court and recognizes that the current trend in the law favors amelioration of the harsh effects of the rule. In *Court v. Grzelinski,* 72 Ill. 2d 141 (1978), the supreme court held that the fireman's rule does not protect a defendant from a **products liability** action. The court observed the public policy considerations that shaped the fireman's rule in Illinois. It stated that "[s]ince most fires occur because of the *negligence* of the landowner or occupier, it was believed that the imposition of a duty to prevent fires from occurring or spreading on a person's premise would place an unreasonable burden upon the person who owned or occupied improved land." [Emphasis added.] However, this public policy consideration tended to undermine the duty of reasonable care placed upon a landowner or occupier. A compromise was reached with regard to firemen, recognizing that the risk of harm from fire is inherent in a fireman's occupation.

As the purpose of the fireman's rule was expressed in *Court,* the supreme court struck a balance by recognizing that the negligent acts of landowners or occupiers are often the cause of fires and that it is the public function of firemen to absorb the risk of injury stemming from those negligent acts. However, the supreme court has never stated that it is the duty of firemen to absorb the risk created by the willful and wanton misconduct of landowners or occupiers. In fact, as far back as *Gibson,* the supreme court stated that a fireman could recover for the infliction of willful or intentional injury. See *Dini,* 20 Ill. 2d at 416. We also note that, under the Local Governmental and Governmental Employees Tort Immunity Act (745 ILCS 10/5-106 (West 2000)), while a firefighter is insulated from liability for his or her negligent acts, a fireman is liable for his or her own acts of willful and wanton misconduct committed while acting within the scope of employment.

Furthermore, among other jurisdictions we have surveyed that have adopted the fireman's rule, most articulate it in the same fashion as our supreme court in *Washington.* See 62 Am. Jur. 2d *Premises Liability* §432 (1990). However, these jurisdictions favor

products liability The Fireman's Rule does not prohibit a firefighter in most states from suing a manufacturer of defective fire equipment or a defective appliance, including a gas water heater or other product causing a fire in a building.

recovery where the owner or occupier is liable for willful or wanton misconduct. See Annotation, L. Scheafer, *Liability of Owner or Occupant of Premises to Fireman Coming Thereon in Discharge of His Duty,* 11 A.L.R.4th 597 (1982); *Minnich v. Med-Waste, Inc., 349 S.C. 567, 564 S.E.2d 98 (2002).*

In the end, we reiterate the supreme court's holding from *Washington,* which provides: while an owner or occupier owes a duty of reasonable care to maintain his property so as to prevent injury occurring to a fireman from a cause independent of the fire, he is not liable for negligence in causing the fire itself. *Washington,* 66 Ill. 2d at 108. However, we decline to follow *Luetje.* Rather, we conclude that the immunity of the fireman's rule does not protect a defendant whose willful and wanton misconduct created the emergency or danger that caused injury to a fireman.

In accordance with our holding, we affirm the trial court's dismissal of plaintiff's negligence claim against defendants.

However, on the matter of plaintiff's claim of willful and wanton misconduct, plaintiff alleges that defendants failed to expose or locate the live gas mains, which they knew were in the same utility easement, before workers began boring into the ground. We find that plaintiff's claim of willful and wanton misconduct is not barred by the fireman's rule. Accordingly, we reverse the trial court's dismissal of this claim. We, however, express no opinion on the merits.

The second issue in this case is whether defendants' status as contractors working on behalf of the landowner/occupier, Western, precludes them from availing themselves of the protections of the fireman's rule. Plaintiff maintains that defendants do not meet this definition of occupation and, thus, cannot claim the benefits of the fireman's rule. We believe that plaintiff misses the point on this argument.

As the holder of an easement, Western was the owner of an interest in real estate entitled to the protections of the fireman's rule. The important determination to be made here is whether defendants' connection with Western allows them to avail themselves of the immunity provided by the fireman's rule. To make this determination, we look at the Restatement (Second) of Torts, section 383, which states:

"One who does an act or carries on an activity upon land on behalf of the possessor is subject to the same liability, and enjoys the same freedom from liability, for physical harm caused thereby to others upon and outside of the land as though he were the possessor of the land." Restatement (Second) of Torts §383 (1965).

Defendants' status as a servant or independent contractor in this matter is irrelevant. See Restatement (Second) of Torts §383, Comment a (1965).

Here, in accordance with section 383, defendants were carrying on their drilling activity to lay cable within the utility easement on behalf of the possessor, Western, and thus they share the same freedom from liability as Western. Accordingly, we conclude that defendants come within the scope of the protections of the fireman's rule.

Here, plaintiff, an EMT employed in public service with the LFPD, possessed specialized training and unique experience to anticipate and manage risks associated with emergency situations. We believe that placing plaintiff in a different class than NICOR employees who do not possess comparable training or experience is rationally related to the purpose underlying the fireman's rule. Accordingly, we reject plaintiff's equal protection challenge.

For the aforementioned reasons, we affirm the trial court's dismissal of plaintiff's negligence claim against defendants. However, we reverse the trial court's dismissal of plaintiff's claim of willful and wanton misconduct. Accordingly, the judgment of the circuit court of Du Page County is affirmed in part and reversed in part, and the cause is remanded.

Affirmed in part and reversed in part; cause remanded.

LEGAL LESSONS LEARNED

Courts in several states are now carving out exceptions to the Fireman's Rule and allowing firefighters and EMS a jury trial to prove willful and wanton misconduct by property owners and their subcontractors.

CASE STUDY 18-3 **State Trooper Responds to Accident on Snowy Highway and Is Struck by Second Motorist—Sues First Motorist**

Richard C. Fordham v. Ryan Oldroyd, 2006 UT App. 50, 2006 Utah App. LEXIS 16 (February 16, 2006).

FACTS

On December 28, 2003, Mr. Oldroyd was driving his vehicle on an off-ramp from the interstate highway in Salt Lake City when his car skidded on the snow and ice. State Troopers responded to the single-vehicle accident. Trooper Richard Fordham was getting out flares from his trunk when another motorist lost control and struck him. He was severely injured.

He filed a lawsuit in May 2004 against Mr. Oldroyd, alleging that his negligence created the chain of events leading to the injury. He did not sue the motorist who struck him.

After a limited period of discovery, Mr. Oldroyd filed a motion to dismiss the case based on the Fireman's Rule (sometimes called the "Professional Responder rule"). The trial judge granted the defense motion, and the trooper has filed an appeal.

YOU BE THE JUDGE

Was the lawsuit properly dismissed?

HOLDING

Yes. The three-judge court ruled that Utah follows the Fireman's Rule. The court noted that numerous states have narrowed or amended the rule by court decision, including Oregon, South Carolina, Connecticut, Illinois, and Nebraska. However, this is the first time that the Utah Court of Appeals has considered a case in which the injury did not occur on the premises of the home or business owner, and the court ruled that the trooper was barred under the facts of this case from suing the driver.

LEGAL LESSONS LEARNED

Firefighters and police officers need to turn to their state legislatures to get a law enacted allowing suits in which, if they can prove driver negligence, they can recover against the driver's automobile and, for premises liability, against the home owner or property owner's insurance.

■■

Chapter Review Questions

1. In Case Study 18-1, *Kapherr v. MFG Chemical, Inc.,* the court addressed the issue of an EMS response to employees overcome by chemicals in a chemical company. Describe the procedures to be used in a hazardous materials response

to ensure that EMS responders are safe while treating the patients.

2. In Case Study 18-2, *Todd Randich v. Pirtano Construction Company,* the court addressed the issue of an emergency responder injured in a gas line explosion. Discuss what procedures are followed by your local fire department when the smell of natural gas is detected in a home or office structure.

3. In Case Study 18-3, *Richard C. Fordham v. Ryan Oldroyd,* the court held that the state trooper could not sue the motorist in the single-vehicle accident. If your state senator were to introduce a bill in the state legislature that would authorize such a lawsuit, what do you expect would be the reaction of insurance companies that write automobile and homeowners' insurance policies?

Expand Your Learning

Read and complete the individual or student assignment, as directed by your instructor.

1. Florida enacted a statute that abolished the Fireman's Rule. Review the Florida Professional Firefighter's website, "FPF Legislative Successes and Issues Benefiting Firefighters and Paramedics" (*www.fpfp.org*) and identify proposed changes in law that might be helpful to firefighters in your state.

2. The Arizona Court of Appeals held in 2005 that the Fireman's Rule does not apply to an off-duty firefighter, who stopped to render assistance at a motor vehicle accident and was struck by another car. This injured firefighter may sue the driver of the car that ran into her. Discuss whether the firefighter has any basis to also sue the owners of the vehicle in the original accident, who had allowed their underage daughter to drive.

See Appendix A for additional Expand Your Learning activities related to this chapter.

Expand Your Learning Activities

Chapter 1: American Legal System: Search Warrants in Arson Investigations, Fire Code Enforcement, and Civil Litigation

1. Under the Foreign Intelligence Surveillance Act of 1978, (FISA) search warrants are required for domestic wiretaps (50 U.S. Code, Section 1802). Assume that a laptop computer has been seized from a terrorist in Pakistan, containing a listing of a dozen cell phones in New York City. Can the president of the United States direct the National Security Agency to place electronic intercepts on these cell phones immediately to prevent another 9-1-1 attack on New York City and Washington, DC?

2. The U.S. Supreme Court in *Michigan v. Clifford*, 486 U.S. 287 (1984) discussed delays in arson investigations. The Clifford residence was damaged by an early morning fire while they were out of town. The fire department had extinguished the fire by 7:04 a.m., at which time all fire officials and the police left the scene. Five hours later, an arson team arrived and found a work crew boarding up the house and pumping water from the basement. The homeowners had learned of the fire and called their insurance company, who sent the repair crew to the house. The arson investigators searched the house and found two Coleman fuel cans next to a crock pot and electric timer. Other evidence of arson was found on upper floors. Raymond and Emma Jean Clifford were charged with arson. The trial judge refused to suppress the evidence. The Cliffords appealed to the Michigan Court of Appeals, which reversed. The State of Michigan asked the U.S. Supreme Court to hear the case, and the court agreed. Justice Powell wrote the majority decision, finding that the search here was not a continuation of an earlier search, therefore the arson investigators needed a search warrant for the midday search. The evidence found in the house cannot be introduced in evidence, nor can the investigators testify about this evidence. However, "One of the fuel cans was discovered in plain view in the Clifford's driveway. This can was seen in plain view during the initial investigation by firefighters" and is therefore admissible.

   Develop a PowerPoint training presentation for fire inspectors, which discusses the "plain view" exception to the search warrant requirement.

3. In a case involving an EMT who checked for keys in a backpack [*State of Utah v. Rameen Rey Amirkhizi*, 2004 UT App. 324; 100 P.3d 225; 509 Utah Adv. Rep. 6; 2004 Utah App. LEXIS 339; Case No. 20030639-CA (Utah Court of Appeals, Third District, September 23, 2004)], the court held that an EMT's "search" of a patient's backpack was illegal and reversed the patient's conviction for possession of cocaine. The EMT had responded to a one-car motor vehicle accident on Interstate 80; the driver was standing outside his station wagon with minor injuries. The driver agreed to be transported to the hospital to be checked out, and he was strapped to a gurney and placed in the back of the ambulance. The EMT asked the patient if he would like to have the EMT put his keys in his backpack, and when the EMT unzipped the back pack, he found the cocaine.

   Develop a PowerPoint training presentation for EMS personnel, explaining when they may open and inspect a patient's personal property.

4. In *Paul Camiolo v. State Farm Fire and Casualty Company, et al.*, 334 F.3d 345 (3rd Cir. 2003); 2003 U.S. App. LEXIS 13258; Case No. 02-1603 (U.S. Court of Appeals for 3rd Circuit; June 30, 2003), the court addressed a civil suit against arson investigators. Paul

Camiolo was arrested for arson and murder of his parents for allegedly setting fire to the house where he lived with his parents. He was incarcerated for 10 months, until all charges were dropped when his expert, John Lentini, submitted a report that showed laboratory samples of the hardwood flooring contained gasoline and lead. He explained that lead was included in gasoline prior to the early 1980s and therefore the gasoline samples found by fire department arson investigator had not been used to start the fire. Camiolo filed a federal civil rights suit, 42 U.S. Code 1983, against the insurance company and fire department investigators. In pre-trial discovery, he sought transcripts of the investigators' testimony before the grand jury. The federal district judge dismissed the lawsuits based on the plaintiff's settlement with the insurance company and refused to disclose the secret grand jury transcripts. The 3rd Circuit Court of Appeals agreed.

Develop an SOP that describes when fire investigators must write a supplemental report and disclose to prosecutors newly discovered evidence that may exonerate a person arrested for arson.

5. In *James Gallo v. City of Philadelphia,* 161 F.3d 217 (3rd Cir. 1998); 1998 U.S. App. LEXIS 29807; Case Nos. 98–1071 and 98–1238 (U.S. Court of Appeals for 3rd Circuit, November 23, 1998), the court addressed a civil suit by a person arrested for arson. Mr. Gallo was acquitted by a jury for arson of his own business, Gallo's Cabinets. He filed a civil suit against the City of Philadelphia and several city and federal investigators, claiming that the city's fire marshal/fire lieutenant had originally concluded the fire was from a defective electrical appliance but changed it to arson based on pressure from the insurance company, and that the federal prosecutors did not disclose the original exculpatory report. The federal district judge dismissed the civil lawsuit. The U.S. Court of Appeals for the 3rd Circuit reversed, holding he is entitled to conduct pre-trial discovery in order to try and prove his allegations.

As a juror, discuss how much money you would award the plaintiff in compensatory damages to make him "whole" for his time in prison, lost income, and legal expenses, and in punitive damages against the city to "send a message" that grossly negligent investigations will not be tolerated.

6. In *State of Utah v. Troy Lynn Schultz,* 58 P.3d 879; 460 Utah Adv. Rep. 21; 2002 Utah App. LEXIS 112; Case No. 20010775-CA (Utah Court of Appeals; November 7, 2002), the defendant was convicted of felony arson to a vehicle after a canine trained in detection of accelerants "alerted" on the defendant's vehicle. The Salt Lake Fire Department dog named "Oscar" was allowed in the courtroom so the jury could see him perform. The court determined that the dog and his handler can provide "expert testimony" consistent with NFPA 921.

Prepare a PowerPoint presentation for a county arson team training on what type of training records should be kept on arson dogs in order to prove in court that they are reliable witnesses.

7. President George Bush has stated that Attorney General Gonzales confirmed he has the power under the U.S. Constitution to order National Security Agency intercepts involving communications between a domestic source and an overseas source, and he does not need a congressional statute to authorize this action to protect our nation.

However, there has been a lot of newspaper and other criticism of this position. For example, on December 25, 2005, the following headline appeared in the *San Francisco Chronicle,* "Why Bush Decided to Bypass Court in Ordering Wiretaps. Panel of Judges Modified His Requests," Hearst Newspapers. The newspaper's review of U.S. Department of Justice records of the 26-year-old Foreign Intelligence Surveillance Court show that since 1979, the 11-judge court authorized "FISA wiretaps" for at least 18,740 requests, under five presidents. The judges had modified or questioned only two wiretap requests from 1979 to 1999. "But since 2001, the judges have modified 179 of the 5,645 requests made under President Bush."

What provisions in the Constitution might the president and the U.S. attorney general rely on for this authority (please cite the actual language in the Constitution)?

Chapter 2: Line of Duty Death LODD and Safety: Litigation and Fireman's Rule

1. In Coos Bay (OR), on November 25, 2002, three firefighters were killed when the roof collapsed on an auto parts store. See NIOSH Fire Fighter Fatality Report, F-2002-50 (*http://www.cdc.gov/niosh/firehome.html*).

   Oregon State OSHA investigators issued 16 safety violations to the City of Coos Bay, carrying fines of $50,000, related to the fire. The citations alleged a breakdown in the incident command structure and communication problems, lack of a Rapid Intervention Team, and failure to maintain SCBA properly. The Oregon safety inspectors also cited the failure to follow incident management standards set by the National Fire Protection Agency (NFPA). See *http://www.iaff.org* (search "coos").

   The fire chief has advised the author that the citations were strongly contested by the fire department and were ultimately settled for $8,000 without any admission of wrongdoing. Discuss whether it is appropriate for state OSHA officials to issue citations based on NFPA "standards" when the NFPA is a private organization and the "standards" are not state statutes or regulations.

2. In *Hart v. Shastri Narayan Swaroop, Inc.,* 870 A.2d 157 (Md. Ct. App. 2005), a Maryland County fire lieutenant responded to a fire alarm and a 9-1-1 call at a motel. There was heavy smoke in the parking lot, so he activated his thermal imaging camera to find the stairway to the second floor and search for trapped occupants. He grabbed onto a railing but fell into the well of an open and unguarded stairwell pit, suffering severe injuries. He sued the motel, which filed a motion to dismiss citing the Fireman's Rule. The trial judge denied the motion and awarded $500,000 damages. The Maryland Court of Appeals reversed, and the firefighter appealed to the Maryland Supreme Court. The Maryland Supreme Court held that the Fireman's Rule prohibited the lawsuit and ruled in the favor of the motel, stating that the exposed stairwell was only "hidden" because of all the smoke and could clearly be seen by all under normal conditions. The motel owner therefore did not breach a duty to warn of hidden dangers.

   Maryland adopted the Fireman's Rule in 1925, on the basis of public policy that property owners should not be reluctant to call for emergency assistance because of potential liability to the firefighters and police officers who respond.

   Describe when the Fireman's Rule was adopted in your state, and provide examples of recent court decisions on the rule in your state.

3. In *Grahovac v. Munising Township,* 689 N.W.2d 498 (Mich. Ct. App. 2004), Grahovac was a volunteer EMT with the county, who responded to an accident. While assisting the motorist, the EMT was struck and killed by a township fire truck when the brakes failed. The estate of the deceased firefighter sued the township's fire chief for gross neglect in failing to ensure that the fire truck was properly inspected and maintained. When the trial judge declined to dismiss the fire chief, he appealed. The Court of Appeals held that the chief must stand trial, because under Michigan law, "absolute immunity" is granted only to the highest elected or appointed officials, who in a township is the township supervisor.

   Discuss whether fire chiefs and firefighters in your state should have the protection of "governmental immunity" statutes.

4. In *Giuffrida v. Citibank Corp.,* 760 N.E.2d 397 (N.Y. 2003), a New York City firefighter responded to a fire at a donut shop in a building owned by Citibank Corporation. He was fighting the fire when the low air bell sounded on his SCBA, advising he had six minutes of air left in the tank. He informed his lieutenant, who ordered him to stay until the rest of his crew exited. The firefighter backed out of the building with his crew. He was on the nozzle, sprayed water, and was the last to leave. As he turned to leave, his oxygen supply ran out. He was overcome by smoke and suffered severe burns and smoke inhalation. He filed suit under New York Code GML 205-a, which the court described as "a statute that creates a cause of action for firefighters who suffer line-of-duty injuries directly or indirectly caused by a defendant's violation of relevant statutes and regulations." He alleged that the donut shop had accumulated grease in the kitchen and no fire suppression system was operating above the deep-fat fryer, all in violation of law. The trial court dismissed

the lawsuit, because the firefighter failed to show that this grease violation was the proximate cause of his injuries. The Appellate Division agreed. The firefighter appealed to the New York Court of Appeals (the "Supreme Court" of New York State), which reversed and ordered the civil case back to the trial court for trial. The court referred to the affidavit of firefighter Vincent DiCicco, who stated he was one of the first to enter the donut shop and observed that the fire started in the kitchen area and that a properly functioning fire suppression system would have left "an easily detectable chemical residue." The court said, "The depletion of plaintiff's air was not a superseding cause of his injuries. It was the result of an act of courage that was part of plaintiff's efforts in battling the blaze."

Discuss whether a statute like New York's would serve as an encouragement of business and home owners to comply with building codes, including maintaining their fire suppression systems in good working order.

Chapter 3: Homeland Security: National Incident Management System, USA Patriot Act, and War on Terrorism

1. In *McNamara v. Hittner,* 767 N.Y.S.2d 800 (N.Y. Sup. Ct. App. Div. 2003), McNamara was a firefighter who was injured on the job. A civilian driver drove a car over the firefighter's ankle as he was stepping out of a fire truck. The firefighter sued the civilian, and a jury awarded him $300,000 for past pain and suffering and $2 million for future pain and suffering. The trial judge commented to the jury that it had been only four months since the 9/11 attacks on New York City and that "we should never forget what happened four months ago. We can't bring back the dead, but we should never forget our losses." The trial judge then asked for a moment of silence in remembrance. The jury awarded the firefighter $2 million for future pain and suffering. The Court of Appeals held that the trial judge made an error in his comments but did not require a mistrial. However, the court ruled that the jury's award of future damages was unreasonable and ordered it either reduced from $2 million to $400,000 or the firefighter may have a new trial.

   Discuss whether the injured firefighter, suing a civilian under an exception to the Fireman's Rule, should be able to appear in court wearing his Class A uniform, or whether this would unfairly influence the jury.

Chapter 4: Incident Command: Fire Scene Operations, Training, and Immunity

1. In 2003, the Ohio Bureau of Workers' Compensation, Division of Safety and Hygiene, after consultation with the Ohio Fire Chiefs' Association, the Ohio Professional Firefighters Association, and Ohio Rural Fire Council, issued new fire department safety and health regulations. The regulation on Incident Management, Ohio Administrative Code 4123:1-21-07, provides that, "An incident management system shall be established with written operating procedures applying to all members involved in emergency operations." 4123: 1-21-07 (A) (1). The regulations also provide that while operating at emergency incidents, "The employer shall provide an adequate number of personnel to safely conduct emergency operations." 4123: 1-21-07 (C) (1). (These regulations are posted on the Ohio Professional Fire Fighters Association website, *www.oapff.com.*)

   In 2005, Ohio amended the 2003 regulations to further clarify the "two-in/two-out" standard. The revised regulation (*italicized* words are new) provides: "*In interior* structure fires a minimum of four employees shall be required, consisting of two employees working as a team in the hazardous atmosphere, who shall remain in voice or visual contact with each other; and two members who are located outside the hazardous atmosphere, who shall be responsible for maintaining a constant awareness of the number and identity of those operating in the hazardous atmosphere and be prepared to perform rescue of those members if required. *Interior structure firefighting means the physical activity of fire suppression, rescue, or both, inside of buildings or enclosed structures, which are involved in a fire situation beyond the incipient stage.*"

Ohio is a federal OSHA state. There is no federal or state agency that routinely conducts "safety audits" of Ohio public fire departments. The "teeth" in these regulations are penalties that can be awarded to injured firefighters. In addition to normal workers' compensation coverage (medical costs, lost wages), an injured firefighter can file a Violation of Specific Safety Regulation (VSSR) claim and receive additional penalty compensation assessed against the employer.

Discuss whether there is any agency that conducts safety audits of fire departments in your state and whether there are workers' compensation regulations allowing an injured firefighter to receive additional "penalty" compensation for safety violations.

2. In *Downs v. Saperstein Associates Corp.,* 697 N.W.2d 190 (Mich. Ct. App. 2005), an apartment building in Detroit caught fire on April 1, 2000, and the Detroit Fire Department dispatched two ladder trucks as a "still alarm," for investigative purposes. One of these trucks was originally an aerial ladder and had been condemned on March 20, 2000, due to a leak in the hydraulic system. The aerial was being used only to transport firefighters and could not be raised to save victims who desperately needed rescue. The fire quickly grew to a fourth alarm. Smoke inhalation killed five people. Another person jumped from the eighth floor and lived, but is now a quadriplegic.

Lawsuits were filed by the families of the deceased and the quadriplegic against Chief Ronald Naumann, chief of the Detroit firefighting division, even though he was not even at the fire scene. The chief passed away while this case was pending, and his estate filed a motion for summary on the basis of qualified immunity. The trial judge denied this motion, and there was an immediate appeal taken to the Michigan Court of Appeals.

The three-judge Court of Appeals reversed and dismissed the lawsuit against the chief. The plaintiffs argued that the fire department was "grossly negligent" in sending the condemned aerial and that the chief owed a "statutory duty" to the residents of the apartment building because the city's Home Rule Charter requires the fire commissioner to appoint a director of the fire fighting division. The Court of Appeals disagreed. It said there was no common-law duty owed by Chief Naumann to the decedents or to the quadriplegic. He never met, spoke, or dealt with them and had nothing to do with this fire. There was no "special relation" with them and "absent a special relationship, a party has no duty to take actions to benefit a third party."

Discuss whether it is appropriate for a fire department to use an aerial or other apparatus for limited duty as a transport vehicle, after it is taken out of active firefighting service.

3. In *Clarke v. City of New York*, 796 N.Y.S.2d 689 (N.Y. App. Div. 2005), FDNY responded to a fire at a residence, conducted a primary search, and found no one. They were also informed that the residents, a mother and daughter, had left earlier that day. Relatives of Mr. Clarke, who suffered from cerebral palsy, arrived and told the IC that Clarke also resided in the house. They conducted a second search and found no one. After the fire was extinguished, Clark's body was found in a second-floor bedroom. He had died of smoke inhalation. His estate filed suit against the City of New York, alleging gross negligence by the fire department. The trial judge granted the city's motion to dismiss. The Court of Appeals agreed, finding the city did not have a "special relationship" with the victim.

Describe what type of investigation the fire department should conduct after the fire concerning why the first two search teams did not discover the victim. Discuss whether a written report should be prepared, given the high likelihood that the plaintiffs' counsel will attempt to use it in a lawsuit against the city.

4. In *Trinity Universal Insurance Company v. Lyons,* 896 So.2d 298 (LA. Ct. App. 2005), the fire department began a "junior volunteer firefighter" program in 1999. In March 2000, there were three suspicious fires, including a suspicious fire at the American Legion Hall. Lyons, a junior firefighter was charged with arson, after one of his friends told the assistant fire chief of his suspicions. This conversation was about three hours before the American Legion fire. The insurance company for the American Legion building paid their claim and then sued Lyons, his parents, and the city, seeking to recover their loss.

The trial judge granted the city's motion for summary judgment on the basis of governmental immunity. The insurance company appealed to the Louisiana Court of Claims, arguing that the city had a duty to conduct psychological testing of their junior firefighters, that the assistant chief had a duty to act immediately on the report that Lyons may be setting fires, and finally that the city is vicariously liable for the conduct of its junior firefighters.

The Court of Appeals reversed and remanded the case to the trial judge. The court rejected two of the insurance company's arguments: (1) there is no duty under Louisiana law to conduct psychological exams, and even if the fire department decided to start conducting these tests after Lyons' arrest, that does not impose a legal duty to conduct these tests; (2) because Lyons was not under the city's control at the time of these arson fires, they are not legally responsible for his off-duty conduct. However, the Court of Appeals said the trial judge should conduct a hearing to determine if the assistant chief was negligent in not promptly investigating the report that Lyons was an arsonist.

Discuss the advantages of psychological testing of firefighters, including junior firefighters, and describe the type of testing used by a fire department in your state.

Chapter 5: Emergency Vehicle Operations

1. Most states do not require emergency vehicles to come to a complete stop at red lights, stop signs, or other controlled intersections. Some fire departments, particularly in larger cities, have adopted a "full stop" policy.

   Discuss the advantages and disadvantages of such a policy, such as the Newark Fire Department policy (provided to the author by Chief Jack Stickradt):

2. In *Richard A. Conti et al. v. City of Columbus,* Case No. 04CVC07-7312, Franklin County Court of Common Pleas (June 29, 2005; final order dismissing case December 2, 2005), four Columbus firefighters were seriously injured on July 15, 2002, when they were dispatched on a ladder truck to an alarm drop at Ohio State University. The driver lost all air brakes going down a hill, and the ladder rolled and crashed through the wall of Patrick J's restaurant, injuring several patrons. The four firefighters were all covered by workers compensation, and they all eventually returned to active duty as firefighters. They filed this "intentional tort" lawsuit seeking punitive damages against the city, claiming that the city's maintenance garage, a fleet-wide repair shop that services city non-emergency vehicles, was grossly negligent in a long record of failure to repair the leaky air brakes adequately on this old aerial. This aerial was used as a "loaner" when front-line ladder trucks were in the shop for maintenance.

   Post-accident investigations by Columbus police, an expert retained by the IAFF, and two engineering firms retained by the city reached different conclusions. For example, the city's outside engineers concluded that the driver of the aerial knowingly left the station with "low air" warnings.

   The city filed a motion to dismiss, claiming governmental immunity. The trial court agreed, stating: "Every constitutional challenge to individual sections of O.R.C. 2744 [Ohio Political Subdivision Tort immunity statute] that has been decided by the Ohio Supreme Court has been denied." The trial judge then cited several cases in the various Ohio courts of appeal, all holding that Ohio public employees cannot sue their employer for intentional tort.

   Discuss whether a state legislature should pass a statute authorizing public employees, including firefighters, to sue their public employer and seek punitive damages from a jury when they have been injured from the apparent gross negligence of mechanics who failed to repair the brakes on their aerial.

3. In 2003, the Ohio Bureau of Workers' Compensation, Division of Safety and Hygiene, issued Ohio Administrative Code 4123: 1-21 (2003), which provides: "Employees shall be required to be seated and belted when apparatus is in motion, except while loading hose. Note: in wild land fire fighting, employees shall be required to ride inside the compartment, unless they are provided with seats, seat belts and rollover protection." To read the entire regulation, go to *www.legislature.state.oh.us/,* and click on "Laws, Acts, and Legislation."

Newark Fire Department
Standard Operating Procedure

OCCUPATIONAL SAFETY AND HEALTH

Subject:	***INCIDENT RESPONSE***	SOP# 403.03 IAFF Notice: 6/8/05
Written By:	CAPT. SPURGEON	**Initiated:** 7/1/05
Approved:		**Revised:**
Fire Chief:	JACK STICKRADT	Date: 5/19/05
Safety Director:	KATHLEEN BARCH	Date: 6/6/05

I. **Scope**

  A. This standard applies to the driver of a vehicle owned or operated by the department while responding to an incident. It was developed to establish safety guidelines during *Priority One* and *Priority Two* responses.

  B. This standard was developed to ensure that the department's response to both *Priority One* and *Priority Two* incidents are safe, appropriate, and efficient.

II. **Categories of Response**

  A. *Priority One:* Those incidents that pose a significant risk to life or property. *Priority One* response requires the use of all audio (siren and air horns) and visual (lights) warning devices in accordance with ORC 4511.041. These devices must be in use during the entire duration of the response unless an incident commander downgrades the response to a *Priority Two.* Units responding on a *Priority One* status shall identify such by the radio transmission "*Priority One*" (i.e. "Rescue One is responding *Priority One* to 123 Your St.") The initial response to the following types of incidents shall be considered *Priority One* emergencies:

   1. *A reported fire in a structure.* This will receive a full first-alarm assignment.
      1. A full first-alarm assignment shall normally consist of all stations with a minimum of 2 engines, 1 ladder company, and 1 medic unit.
         a. A "company" is defined as a firefighting apparatus staffed by at least three personnel.
      2. If "nothing showing" is reported by the incident commander, all companies excluding the first due shall downgrade their response to *Priority Two.*
   2. *A reported fire outside of a structure that involves the potential destruction of property or poses a risk to human or animal life (i.e. dumpster, refuse, brush, etc.)* This will receive 1 engine or ladder company assignment.
   3. *A reported car fire.* This will receive 1 engine or ladder company assignment.
   4. *Carbon Monoxide checks with a reported illness.* This will receive 1 engine or ladder company and 1 medic unit assignment.
   5. *Automatic fire alarms* shall receive a full assignment but are emergencies for the first due engine company only. All other units shall initially respond *Priority Two* and can be upgraded if needed.
   6. *Inside gas leaks* shall receive the same initial assignment as an automatic fire alarm.

(continued)

Newark Fire Department		
Standard Operating Procedure		
Subject:	*INCIDENT RESPONSE*	SOP# 403.03

7. *Outside gas leaks* shall receive 1 engine or ladder company as the initial assignment.

8. *Mutual aid requests for fire support* shall receive 1 engine or ladder company as the initial assignment.

9. *Life-threatening Priority One medical incidents* (reports of unresponsiveness or unconsciousness) shall receive 1 medic unit and 1 engine or ladder company as the initial assignment.

10. *On all other Priority One medical incidents where an assist is required* the first due unit shall respond on a *Priority One* status and the additional unit shall respond *Priority Two.*

11. *On all Priority One medical incidents where an assist is not required* the response shall consist of one medic unit.

12. *Violent scenes* (stabbings, shootings, or fights in which a weapon is involved) shall receive the same assignment as a report of unresponsive or unconscious with the addition of 1 command unit. No units shall respond on a *Priority One* basis unless law enforcement personnel have secured the scene.

   1. In the event that a unit arrives on an incident scene that has not been secured, those unit(s) shall stage.

B. *Priority Two:* Those incidents that do not pose a significant risk to life or property. Audio and visual warning devices shall not be used during *Priority Two* responses unless ordered by an incident commander to upgrade the response to *Priority One* status. Units responding to such incidents shall only identify themselves as responding (i.e. "Rescue 1 is responding 123 Your St.") The initial response to the following types of incidents shall be considered *Priority Two:*

1. Carbon Monoxide checks without a report of illness.

2. Public service calls to assist the public when there is no immediate threat to life or property.

III. Response Guidelines

A. Apparatus and vehicles engaged in *Priority Two* responses shall obey all applicable traffic safety rules and regulations and shall not exceed the posted speed limit.

B. Any and all traffic preemptive systems shall be utilized before exiting a division building prior to a *Priority One* response.

C. Apparatus and vehicles engaged in a *Priority One* response shall at all times govern their response by the traffic, weather, and road conditions present at the time of response.

1. The maximum speed of travel shall *not* exceed posted limits by more than 10 mph.

D. During a *Priority One* response, drivers shall bring their vehicles to a complete stop for any of the following:

1. When directed by a law enforcement officer.

2. Stop signs.

| **Newark Fire Department** |
| **Standard Operating Procedure** |

| Subject: | *INCIDENT RESPONSE* | SOP# | 403.03 |

---

    3. Red traffic signals.

    4. Blind intersections.

    5. When the driver cannot account for all lanes of traffic in an intersection.

    6. When other intersection hazards are present.

    7. When encountering a stopped school bus with flashing warning lights.

  E. Drivers shall proceed through an intersection only when the driver can account for all lanes of traffic in the intersection.

  F. Drivers shall bring their vehicles to a complete stop at all railroad grade crossings and shall not cross the tracks until determining that it is safe to do so.

IV. Responsibilities

  A. Drivers shall be directly responsible for the safe and prudent operation of their vehicles in all situations.

  B. When a driver is under the direct supervision of an officer, the officer shall assume responsibility for the actions of the driver and shall be responsible for immediately correcting any unsafe condition.

As previously discussed, in Ohio, an injured employee can file a Violation of Specific Safety Regulations (VSSR) claim, and the Ohio Bureau of Workers' Compensation will send out investigators to determine whether the employer has an effective program to comply with the safety regulations. The Ohio Industrial Commission can, after an evidentiary hearing, find a VSSR violation and award the injured employee additional workers compensation. This penalty is assessed against the employer "dollar for dollar."

Discuss what regulations exist in your state that mandate the use of seatbelts and what steps can be taken by a fire department to help ensure compliance.

Chapter 6: Employment Litigation: Age, Beards, Free Speech, and Promotions

1. Beards and religious practices have been the subject of litigation in the fire service. On February 2, 2005, a Philadelphia firefighter was suspended for refusing to shave his beard for religious and medical reasons. On May 5, 2005, in *Curtis Deveaux v. City of Philadelphia,* Case Control No. 021818, Court of Common Pleas of Philadelphia County, a trial judge issued a temporary restraining order preventing the termination of a Philadelphia firefighter, who is of the Muslim faith and assigned to Engine 44. The lawsuit filed on his behalf by the American Civil Liberties Union alleges that he has been a Muslim for about five years, and he is "obligated to follow the Qur'an and the Sunnah (the way ordained by the Prophet Mohammed). These authorities teach that to be Muslim means to live entirely as a Muslim, and to follow the commands of the Qur'an and the Sunnah in every part of one's life." The complaint also alleges that the firefighter "grew his beard because he has a medical condition called Pseudofolliculitis Barbae. This condition, which is common among African American men because of the natural curl of their facial hair, is caused by shaving."

Discuss whether you, as fire chief, would terminate this firefighter or seek some accommodation.

[Note: On September 22, 2005, the Philadelphia court ruled that facial hair is a safety hazard, and the firefighter must shave or be terminated. In September 2005, U.S. District Court Judge James Robertson in Washington (DC) came to the same conclusion in the case of firefighter Hassan A. Umrani, a Muslim firefighter who has worn a full beard since his first day on the job 16 years ago.]

2. In *Cassidy v. Scoppetta,* 365 F.Supp.2d 283 (E.D.N.Y. February 4, 2005); U.S. District Court in New York City, the court addressed the August 25, 2003, memo of the FDNY chief of operations, which read, in part, "The following members have a history of service connected medical leaves that indicates a detail to a less active unit would be in the members best interest." The memo contained the names of 22 firefighters who were being transferred to less active fire stations, or from ladder companies to less active engine companies. The union filed a grievance, sought a temporary restraining order in state court, and filed this lawsuit in federal court under 42 U.S.C. 1983, claiming a deprivation of their Fourteenth Amendment rights to procedural due process. The U.S. District Court held for the city and dismissed the lawsuit. The judge wrote, "The Court will not discuss in detail the numerous past instances where employees of the state have challenged [in federal court] personnel decisions under the Due Process and Equal Protection Clauses [of the 14th Amendment]. See, e.g, *Kelley v. Johnson,* 425 U.S. 238 (1976) ... upholding the validity of the county's hair growing regulation for male members of its police force."

   Discuss how a fire department might address the issue of transferring firefighters with a history of workers compensation claims to slower stations and less strenuous activities.

3. In *Firefighters United For Fairness, Inc. v. City of Memphis, Tennessee,* 362 F.Supp.2d 963 (W.D. Tenn. March 28, 2005); U.S. District Court, a group of ten African-American lieutenants, who are members of an unincorporated association of African-American firefighters, filed suit alleging the promotion process followed in 2000 for battalion chief violated their constitutional rights under 42 U.S.C. 1983.

   The City of Memphis, at the urging of the U.S. Department of Justice, retained an industrial organizational psychology consulting firm, Performance Associates, to develop and administer a promotion test for fire lieutenants and battalion chiefs. The lieutenants' test included a written job knowledge test (22.5 percent), a practical video test (70 percent), and a seniority credit based on years of service with the Memphis Fire Department (7.5 percent). The battalion chiefs' test included a practical video (48 percent), an in-basket exam (27 percent), a group interpersonal skills exercise (17.5 percent), and seniority credit (7.5 percent).

   Candidates could review their scores and submit written appeals ("red lines"). The consulting firm received 264 appeals. The consultants corrected about two dozen scores and issued a revised promotion list. The U.S. District Court judge granted the city's motion to dismiss the lawsuit, finding that the consulting firm did a fair job. "Test administrators openly acknowledged that undertaking such a large and complex task as administering the promotion process naturally would entail the risk that some candidates might be slighted by arbitrary graders, faulty technology, or human error. In all instances, the administrators attempted both to minimize the effect of these variables and to allow affected candidates to challenge the instances where they felt those variables had rendered their scores incorrect or unreliable."

   Discuss the internal appeal process used by a fire department in your state to challenge a promotion exam.

Chapter 7: Sexual Harassment: Hostile Work Atmosphere, Pregnancy Discrimination, and Gender Discrimination

1. The EEOC lists numerous lawsuits and settlements on their website, *www.eeoc. gov/litigation/settlements/index.html,* "EEOC Litigation Settlements Monthly Reports." Read *EEOC v. Tom Lange Co., Inc.,* No. 05-0955 (W.D. Pa. September 22, 2005), and

EEOC v. Carmike Cinemas, Inc., No. 5:04-CV-673-BO(1) (E.D. N.C. September 26, 2005). Discuss why these cases may be relevant to the fire service.

2. Discuss whether it is appropriate for a fire department investigating a recent incident of alleged misconduct to terminate a firefighter for prior acts of misconduct that have just been discovered in the course of the investigation. See the following case.

   In *Ogden City v. Harmon,* 2005 UT App 274 (June 16, 2005), 116 P.3d 973, Captain Daniel Harmon, Ogden City Fire Department, was fired on December 15, 2000, after an investigation into a complaint received on September 8, 2000, revealed several prior incidents of misconduct. The Court of Appeals upheld the firefighter's termination and held that the fire department had a right to set the level of discipline based on the captain's prior conduct. (1) In the fall of 1996, as firefighter's union official, he coordinated a fund-raising event for the Muscular Dystrophy Association, in which he permitted female entertainers to pose topless with firefighters. (2) In the summer of 1999, while he operated a lawn fertilizing business on his off-duty time, he provided a retired battalion chief with a bottle of Round-Up weed killer: The bottle was filled instead with urine from the captain and two other firefighters. (3) During a summer 2000 training exercise, he urinated into a drafting pit, or water reservoir, being used by his and another fire crew. (4) About eight to ten years earlier, he had urinated into a shower stall occupied by another firefighter. (5) In November 1998, the captain failed to caution a female probationary firefighter after she held up a cucumber or zucchini and said to him, "Do you know what they call these where I'm from? Home wreckers." (6) He tolerated male firefighters "humping" one another to imitate sexual intercourse.

   The Utah Court of Appeals wrote, "Although an incident's remoteness in time may be relevant in mitigating the degree of discipline imposed, it does not erase the fact that Harmon committed a violation of department rules which merited discipline."

3. Describe whether public employers such as a police department or a fire department, should prohibit employees from dating lesser-ranking employees. See the following case.

   In *Anderson v. City of LaVergne,* 371 F.3d 879 (6th Cir. 2004), the U.S. Court of Appeals upheld a city's no-dating policy and set aside the jury verdict for the plaintiff police officer who was forced to resign or be fired. He will not collect any of the $10,283.86 in back pay and $5,000 in intangible damages. The court upheld the actions of the police chief of the City of LaVergne, TN.

   Michael Anderson began a dating relationship in 1999 with Lisa Lewis, an administrative assistant for the police department. The police chief learned about it after three months and ordered him to "cease all contact with each other" outside of the workplace. Lewis outranked Anderson. The chief issued this order out of concern that intra-office dating between employees of different ranks might lead to sexual harassment claims against the department. The trial judge allowed the case to go to a jury, finding that the policy had nothing to do with the work of the police department. The 6th Circuit Court disagreed, stating: "Contrary to the district court's conclusion, the City's policy is rationally related to a legitimate interest. The City barred dating relationships between police department employees of different ranks to promote its interest in avoiding sexual harassment suits. Such preventative policies are common among government employers. For example, this court has upheld policies prohibiting marriage among municipal employees, *Vaughn v. Lawrenceburg Power Sys.,* 269 F.3d 703, 712 (6th Cir. 2001), and requiring the transfer of one spouse if two employees of the same school marry. *Montgomery v. Carr,* 101 F.3d 1117, 1130-31 (6th Cir. 1996)."

4. Read the following article written specifically for this textbook by Deputy Fire Chief Mike Cardwell, Deerfield Township (OH) Fire and Rescue. Describe how NFPA standards can help fire and EMS departments avoid sexual harassment charges.

Equality of Performance and NFPA, by Michael Cardwell

Removing the potential for sexual harassment issues in the firehouse must continue past the hiring and initial training phases. It is an issue that must constantly be

Deputy Chief Mike Cardwell, Deerfield Township (OH) Fire and Rescue, conducted live burn training. He invited Cincinnati Channel 5 News to videotape from inside the house (the videographer was a volunteer firefighter wearing full turnout and SCBA), showing how quickly the light wood trusses failed and the roof collapsed. The International Association of Fire Chiefs obtained a copy of the video and has shared it nationwide. *Photo by author.*

addressed at all levels of the organization. Nowhere is this more important than emergency scene operations. And laying the ground work for sexual equity on the emergency scene is ensuring equity of performance.

The fire service is steeped in tradition, a team of professionals (paid or volunteer) working together to accomplish the difficult job of fighting fire. One of the hallmarks of any successful team is trust. Each member of the team trusts each other to complete their part of the operation. If everyone does his or her part, the team succeeds. The failure of one member of the team often leads to failure to complete the job.

Take a football team. In order for the team to succeed, the offense has to score points while the defense has to keep the other team out of the end zone. Breaking it down further, the offense can have the best quarterback, but if the receivers can't catch the ball, the offense breaks down. In short, everyone has a task to perform for the job to get done.

A fire company operates in much the same way. Each member of the company has a task to perform in order for the job to get done. Every firefighter on the company trusts the others to complete their tasks, and is entrusted to complete their own. If one task can't be completed, the job doesn't get done.

An effective training program can go a long way towards ensuring every firefighter has the capability of completing all tasks required on the fire ground. Starting in recruit school or basic firefighter training, potential firefighters are exposed to the rigors of the job and the expectations they will face when they become part of a fire company. An effective basic training program is based in the reality of the firefighting conditions a rookie firefighter may be expected to meet.

NFPA 1001, Standard for Fire Fighter Professional Qualifications (2002 ed.) clearly identifies what is expected of firefighters on the emergency scene. The standard uses Job Performance Requirements (JPR's) to illustrate operations the firefighter is expected to perform on the fire ground. For each operation, specific knowledge and skills required to complete the operation are identified. The bottom line, in order to meet the requirements of the job, the firefighter must have this knowledge, and be able to perform these skills.

Let's use ventilation as an example. The operational expectation is for a team of firefighters, given the necessary equipment, to successfully create a vertical ventilation opening in a structure. In order to complete this job, a firefighter would be expected to have knowledge of fire behavior, building construction, and ventilation theory. In order to complete this job, a firefighter would have to have the skills to place and climb a ground or aerial ladder, operate a power saw, and use the saw to create a vent opening.

Creating sexual equity begins at the recruit level. Regardless of how personnel are selected for entry into the organization, every firefighter will have the same expectations on them when they get on the job. Therefore every recruit must meet the same standards of performance. The key is to make sure those performance expectations are realistic. Many departments have long-standing drill books for new recruits, outlining all the skills a new recruit is expected to meet. How many of these drill books have been updated to meet current operations? How many recruits are expected to show mastery of equipment which is no longer even carried on front-line apparatus?

While many argue that this is an important part of "passing on" the tradition and pride of an organization, it also creates the opportunity for unfair or discriminatory practices. For example, how fair is it to wash out a recruit who cannot lift the end of a 50′ pole ladder when the department doesn't carry 50′ ladders on their trucks? Using JPR's outlined in NFPA 1001 is an excellent way for a department to ensure their training practices are based on current FD operations.

More and more it is becoming standard practice for recruit schools and academies to utilize job or skill sheets when evaluating students. These sheets objectively identify who will participate in completing the objective (an individual, pair, or team), what behavior or job must be performed, and conditions it must be completed under, and to what degree it must be done. Creating job or skills sheets which match national performance standards is an effective way of creating equity in the evaluation of recruits.

But no matter how effective a recruit is in school, they still have to pass muster in "the real world." Back to our football team analogy, how often has a prized recruit come onto a team only to fail miserably when faced with the competition at the next level. It's no different in the fire service. It's almost impossible to recreate actual fire ground conditions on the training ground. So once again, the newly appointed "probie" firefighter must pass another test—this one just as important as any test in recruit school.

Sexual harassment is often based on a belief system that says women can't do the job as well as men. To defeat this bias, it is imperative that every new recruit come to the company with a solid knowledge and skill set to draw on. But knowledge and skills are only part of what a new firefighter needs to be successful; they need the opportunity to succeed as part of a team.

Here job performance standards for both individuals and companies can be very effective. If everyone on a company is required to regularly meet department-wide performance standards, the department can ensure everyone is up to the job. But the added bonus comes from allowing everyone on a company the opportunity to see each other meeting the same standard. Additionally, company-based performance standards require the company, the team, to work together to meet a standard. Again, everyone sees the benefit of working together to achieve a goal. The rookie is more quickly assimilated into the company, and accepted not as a male or female, but as a teammate.

The final step in ensuring gender equity when it comes to emergency operations has long been viewed as the glass ceiling of the fire service—promotion. Many women who were promoted faced discrimination in promotional process. If they did succeed, their success was attributed to external politics, internal favoritism, or worse. Again, old prejudices die hard, and because there are so few women in top command positions across the country, the lack of positive female role models is often used (incorrectly) as proof by some that only men can hold positions of rank.

The solution for overcoming this final operational bias and achieving gender equity can again be found in national standards. *NFPA 1021 Standard for Fire Officer Professional Qualifications* (2003 ed.) identifies JPR's for each of four levels of Fire Officer. For each, knowledge and skills required to successfully meet these performance requirements are outlined. Each of the four levels is an ascending skills set for aspiring officers to meet.

Unlike many internal promotional processes in place, NFPA 1021 doesn't list specific courses or educational requirements for each level. Instead, the focus of the standard remains on jobs or operations the officer would be expected to meet in the performance of their job. It allows an organization to craft a promotional process that is performance-based.

Any officer candidate who is able to meet the requirements in a performance-based system will start building trust. Trust is critical for any newly promoted officer. In order to perform proficiently, the team must have trust in the person in charge. That trust comes, in part, from the belief that the person in charge deserved the promotion they received.

Gender equity can be found when everyone has to meet the same standards, and those standards are based on an expected level or performance. If a group of officers is promoted at the same time, each met the same standard, each should have the opportunity to start with an equal footing. In a smaller organization, where only one person is selected for promotion, a solid and equitable promotional process will ensure the best candidate is selected.

Chapter 8: Race Discrimination

1. In *Jones v. R.R. Donnelley & Sons Company*, 541 U.S. 369 (2004), the U.S. Supreme Court held that plaintiffs have four years to file suit for race discrimination in federal courts, and plaintiffs do not have to be concerned about the state statutes of limitation (such as the two-year statute of limitations in Illinois). "Statutes of limitation" define how long a plaintiff has in which to file suit in state courts and exist in every state. Describe how long a plaintiff has to file a race discrimination suit in your state courts.

2. Describe the EEOC's "Four-Fifths Rule" concerning when hiring or promotion exams are considered to have had an adverse impact on minority candidates. See the following case.

   In *Antonelli v. State of New Jersey,* _____ F.3d _____, Case No. 04-2573 (3rd Cir. August 17, 2005), the U.S. Court of Appeals upheld the dismissal of a lawsuit filed by 27 Caucasian applicants to become firefighters. They filed suit under 42 U.S.C. 1983, alleging that New Jersey officials violated their constitutional rights to equal protection under the Fourteenth Amendment by using an improper entrance exam that benefited minority candidates.

   The court stated that the Fourteenth Amendment's "due process" clause prohibits states from intentionally discriminating among individuals on the basis of race. Discriminatory intent, in the words of the Supreme Court, "implies that the decision maker ... selected or reaffirmed a particular course of action at least in part 'because of,' not merely 'in spite of' its adverse effects upon an identifiable group." *Personnel Ad'r of Mass. v. Feeney,* 442 U.S. 256, 279 (1979).

In 1977, the U.S. Department of Justice, Civil Rights Division, filed a lawsuit against the State of New Jersey and 12 cities, alleging they were engaged in employment discrimination denying entry-level positions to African-American and Hispanic firefighters. In 1980, the defendants agreed to a consent degree in U.S. District Court, requiring affirmative action programs. In 1999, the New Jersey Department of Personnel designed an entrance exam that had three components: (1) a multi-choice cognitive test, to assess ability to read and perform basic math; (2) a biographical questionnaire; and (3) a physical performance test. All candidates were required to achieve the minimum cut-off scores.

The New Jersey department followed the Four-Fifths Rule of the EEOC, 28 CFR 5014 (2004), which says that there is adverse racial impact if the passing rate of minorities is less than four-fifths that of the rate of Caucasian candidates who took the test.

The overall passage rate for the exam being challenged in this case is 61 percent for Caucasians; 53 percent for African-Americans; and 52 percent for Hispanics. The mean scores were as follows: Caucasians (49.92); African-American (50.21); and Hispanic (49.19).

3. Discuss what steps a fire department can take to help ensure that similar discipline is imposed on Caucasian and African-American firefighters for similar misdeeds. See the following case.

In *Williams v. City of Akron,* 107 Ohio St.3d 203, 837 N.E. 2d 1169 (Ohio Supreme Court, December 14, 2005), Gerald Williams is an African-American police officer who won a jury verdict of $1.7 million for disparate discipline. The Ohio Supreme Court ruled that he was not entitled to this verdict.

Williams, while off duty, slapped his wife in the head, and she dialed 9-1-1. He struck her again, this time breaking her jaw. He was indicted for domestic violence and pleaded guilty to aggravated menacing. He was fired for this conduct and for lying in an internal affairs investigation.

His discharge was upheld by the city's Civil Service Commission. He filed a civil suit in Common Pleas Court, alleging his discharge was motivated by racial discrimination. The city submitted evidence of its reasons for terminating the officer and asked the trial judge to dismiss the lawsuit. The court refused. The case was tried by a jury, and at the close of the evidence, the city again asked the judge to dismiss the lawsuit. The court again refused and instructed the jury that, in order for the plaintiff to establish a prima facie case, the plaintiff must prove by a preponderance of the evidence that (1) he is a member of a protected class, (2) he suffered an adverse employment action, (3) he was qualified for the position, and (4) he was treated differently from a similarly situated Caucasian police officer. The jury returned a verdict of $1.72 million. After the jury verdict, the city again asked the trial court to dismiss the suit, and the court again declined.

The Ohio Supreme Court held that the trial judge was erroneous in not dismissing the case, and said, "We hold that in Ohio, an appellate court, in determining whether the trial court erred in denying a motion for a directed verdict at the close of the plaintiff's case, can review a plaintiff's case-in-chief, including, in a discrimination case, plaintiff's prima facie evidence of discrimination, if the defendant has properly preserved the issue for appeal by renewing the motion for a directed verdict at the close of all evidence."

4. Describe how African-American firefighters might establish proof of discriminatory intent in a promotion process. See the following case.

In *Firefighters United For Fairness, Inc. v. City of Memphis, Tennessee,* 362 F.Supp.2d 963 (W.D. Tenn. March 28, 2005); a federal district court judge dismissed a lawsuit by ten African-American lieutenants, finding no racial discrimination in the promotion exam, and declining to find racial discrimination by the comments of a senior African-American officer. The plaintiffs are members of an unincorporated association of African-American firefighters, who filed suit alleging the promotion process followed in 2000 for battalion chief violated their constitutional rights under 42 U.S.C. 1983. The trial judge held a bench trial (no jury) on July 26–27, 2004, and held the plaintiffs failed to prove race discrimination. The trial judge pointed to the trial testimony of Plaintiff Lieutenant Ulysses Jones, who testified he had complained about the test procedures to Memphis Fire Department Director Chester Anderson (also African-American), and asked Anderson why he had appointed a Caucasian lieutenant to serve as liaison to the consulting firm retained to administer the test. The trial judge wrote in his opinion dismissing the lawsuit, "At trial, Plaintiff Jones testified that when he asked MFD Director Anderson, also an African-American, why Lt. Stevens was chosen to serve as liaison, Anderson replied 'you know we can't trust you bros.'"

The trial judge wrote: "Though this statement, despite being attributed to an African-American, might be offered to suggest impermissible racial discrimination, standing alone it is not sufficient to show the City discriminated against Plaintiffs. In assessing the relevancy of a discriminatory remark, we look first to the identity of the speaker. *Ercegovich v. Goodyear Tire & Rubber Company,* 154 F.3d 344, 354 (6th Cir. 1998). Though the Court does not suggest that African-Americans may not impermissibly discriminate against other African-Americans in hiring or promotional processes, the fact that Director Anderson and other key decision makers in the MFD [Memphis Fire Department] are African American mitigates the claim that African Americans were targeted for discriminatory treatment in the promotion process. Further, the remark, standing as it does as the only evidence of any discrimination, seems more an isolated, off-the-cuff remark in a conversation between two African American MFD employees."

Chapter 9: Americans with Disabilities Act (ADA)

1. Discuss the safety measures that EMS workers should take with HIV patients. Share with the class a well-written department SOP or procedure from a fire or EMS department in your state concerning HIV patients. Read the following case:

   On July 30, 2004, the U.S. Department of Justice filed a motion to join a lawsuit by a citizen with HIV in *John Gill Smith v. City of Philadelphia,* Case No. 03-6494, U.S. District Court for Eastern District of Pennsylvania (go to *www.usdoj.gov* and search under "EMTs"). The U.S. Department of Justice memorandum in support of their Motion to Intervene states, "This action stems from the alleged failure of defendant City of Philadelphia ("City") to ensure that its emergency medical system does not discriminate against individuals with disabilities."

   The case stems from a prior settlement on March 18, 1994, in which the city's Emergency Medical Services Unit responded to a 9-1-1 call regarding a young man who collapsed on the floor of a school building. "As soon as the EMS workers learned that the complainant had HIV disease, they stepped back from him and refused to place him on a stretcher. A nearby teacher had to lift the complainant onto the stretcher so that he could be transported to the hospital." The city entered into a settlement agreement with the U.S. Department of Justice, including payment of damages, agreement not to discriminate against patients with HIV, and to train EMS personnel about methods to prevent transmission of HIV during treatment.

   On February 20, 2001, a call was placed to 9-1-1 by John Gill Smith, the plaintiff in this case. He was experiencing severe chest pain while at his residence in the city. Two EMTs responded. The Department of Justice memorandum states that when they learned he had HIV, "One EMT abruptly exited plaintiff's home and remained outside, providing no assistance, throughout the emergency visit. The other EMT refused to touch plaintiff, speaking to him only from across the room. Although the EMTs transported plaintiff by ambulance to the hospital, plaintiff alleges that the EMTs failed to provide him with emergency medical assistance, and additionally, utilized harassing and threatening language with him."

Chapter 10: Family Medical Leave Act (FMLA)

1. The FMLA regulations of the U.S. Department of Labor make it clear that "serious health condition" includes pregnancy (see 29 CFR 825.114(a)(2)(ii), *www.gpoaccess.gov/cfr/ index.html*). Discuss how a fire department should manage postpartum depression, in

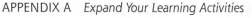

which a firefighter/paramedic is suffering from depression after the delivery of her baby, and needs intermittent time off to visit with her psychologist.

Chapter 11: Fair Labor Standards Act

1. Numerous fire departments use part-time "on-call" personnel. When a firefighter is "on-call" for a particular evening shift, he or she must remain alcohol- and drug-free, and be available for rapid response. Review the FLSA and the Department of Labor regulations concerning on-call personnel. Describe when personnel are entitled to be paid for their on-call time.

2. Robert Beck, president of the Cleveland Police Patrolmen's Association filed a lawsuit claiming the city cannot refuse requests for comp time leave simply because the city cannot afford to pay overtime for substitute police officers. The city won before the U.S. District judge in Cleveland on motion for summary judgment; the decision was reversed and the case ordered back for further hearing by the three-judge 6th Circuit Court of Appeals, *City of Cleveland v. Robert Beck,* 390 F.3d 912 (6th Cir. November 12, 2004), holding that the city cannot deny a timely compensatory leave request solely for financial reasons. On June 20, 2005, the U.S. Supreme Court denied the city's petition to hear their appeal, 125 S.Ct. 2930 (Mem), 162 L.Ed.2d 867, 73 USLW 3732. Now that the U.S. Supreme Court has refused to hear the city's appeal, discuss how this decision might influence mayors, police chiefs, and fire chiefs in Michigan, Ohio, Kentucky, and Tennessee (6th Circuit states) and elsewhere.

3. When Congress passed the amendments to the FLSA [Sec. 203(y)], and President Clinton signed them into law on December 9, 1999, there were numerous lawsuits by paramedics already filed throughout the country. In 2004, Houston Mayor Bill White announced a $72 million overtime settlement. Louisville settled its case worth as much as $35 million. Discuss how fire departments can seek an opinion letter from the U.S. Department of Labor concerning their overtime payments (*www.dol.gov*) and thereby have a court decide they at least were acting in "good faith" in relying on the opinion, thereby avoiding "liquidated damages" of double back pay awards.

Chapter 12: Drug-Free Workplace: Random Drug Testing and Firefighter DUIs

1. On October 13, 2004, a new Ohio law became effective, providing that employees who test positive in a post-accident drug test will not receive any workers' compensation for the injury, unless they can overcome the "rebutable presumption" that the drugs in their body caused the accident. Ohio House Bill 223 amended Ohio Revised Code 4123.35 (see www.legislature.state.oh.us). The Ohio Supreme Court declared a prior Ohio statute unconstitutional, and undoubtedly there will be legal challenges filed against this law.

   In 2005, The Arizona Supreme Court in *David Grammatico v. The Industrial Commission*, Case No. CV-04-0197 (AR 2005) struck down a 1999 statute that denied workers' compensation coverage to injured workers who tested positive in a post-accident drug test or refused to take the test.

   Discuss whether a firefighter in your state who has an accident driving an emergency vehicle and tests positive is denied workers' compensation coverage for his or her medical expenses and lost wages. Also discuss whether the firefighter is subject to discipline, including termination.

2. On June 6, 2005, the U.S. Supreme Court in *Gonzales v. Raich,* —— U.S. ——, 125 S. Ct. 2195, 162 L. Ed. 2d 1 (2005) decided an important case about a California law that had authorized residents to grow marijuana for their personal medical needs. Read the decision and discuss that holding of the U.S. Supreme Court and its implication for other states that may wish to authorize use of marijuana.

Chapter 13: Emergency Medical Service (EMS) and the Health Insurance Portability and Accountability Act (HIPAA)

1. In *Wigand v. Spadt,* 317 F.Supp. 1129 (Dist. Neb. 2004), a female firefighter/paramedic's paramedic certificate was cancelled by the medical director after she failed to improve her patient care skills in several areas. He suggested she take a 48-hour refresher course and

reapply for her paramedic certificate. She was demoted to the rank of firefighter and had her pay reduced. She filed a lawsuit in federal district court, claiming she was the victim of sexual discrimination and suffered a violation of her constitutional rights. The federal judge granted the motion of the fire chief to be dismissed from the lawsuit, because he was a public official and enjoyed qualified immunity. The judge refused to dismiss the medical director on the basis of qualified immunity, but he found that the preliminary evidence indicated that male paramedics were treated in the same manner when questions arose about their skills.

Discuss the quality review procedures that EMS officers and medical directors should implement to improve the level of service. Describe appropriate methods of retesting paramedics and EMTs when their skills are in question.

2. The U.S. Department of Health and Human Services reports federal Medicare/Medicaid fraud to the Department of Justice for criminal prosecution. Several fire departments considering "insurance only" billing of residents (and thus waiving collection of fees for residents without insurance) have sought written opinions confirming this waiver is lawful. If a private ambulance company told a nursing home that they would not bill those without insurance, the ambulance company might be prosecuted for fraud. Explain why fire and EMS departments can practice "insurance only" billing of residents and those who work and pay taxes in the fire district. Read the U.S. Department of Health and Human Services, Office of Inspector General, Advisory Opinion 05-10, issued June 16, 2005, and Opinion 04-14, issued November 4, 2004 (go to *http://oig.hhs.gov/fraud/advisoryopinions.html* or *www.hhs.gov/oig* under "Fraud Prevention and Detection").

3. Many states have a "privilege" that protects from disclosure the hospital quality review records of physicians after surgeries that do not have good outcomes. In some states, there is a similar protection for quality review records for EMS runs. Describe the advantages of these statutes and whether fire and EMS departments not having a quality assurance program could face civil liability.

Chapter 14: Physical Fitness

1. A 2005 study of 950 San Diego firefighters revealed that 31 percent had elevated blood pressures, 28 percent were overweight or even obese, 51 percent had less than average flexibility, 55 percent were at risk for back problems, 39 percent had various degrees of hearing loss, and 17 percent were at risk for heart disease. Describe how stretching equipment could be used to increase flexibility and strengthen back muscles. Describe some of the "best practices" followed by fire departments in your state.

2. Effective December 15, 2003, public safety officers are covered for line of duty deaths that are a direct and proximate result of a heart attack or stroke, as defined in the *Hometown Heroes Survivors Benefits Act of 2003*. Under this statute, the heart attack or stroke must have occurred within 24 hours of "non-routine stressful or strenuous activity." Congress has amended the line of duty death benefits for firefighters, which now extends to firefighters who have died within 24 hours of "strenuous" on-duty activity. (Go to *www.ojp.gov/BJA/grant/psob/psob_death.html,* and click on the Hometown Heroes Survivors Benefits Act of 2003.)

   If a firefighter dies of a heart attack within 24 hours of duty, but his only strenuous activity on duty that day was when he and the rest of the crew worked out for one hour together, using a treadmill and a stair climber, explain whether his family is entitled to the federal death benefits.

3. A firefighter is a member of a county Urban Search and Rescue Team and also a federal Task Force, both of which require an annual physical. He fails the physical because of a cardiac problem. He advises the teams that he has decided to resign from both teams, but he does *not* want his medical condition to be shared with his home fire department. Discuss whether the county USAR or the federal Task Force team should disclose the physical problems to the firefighter's home department.

Chapter 15: Critical Incident Stress Management

1. In many states, the actual voice recordings of citizens calling 9-1-1 for emergency assistance are "public records," and local TV and radio stations will use these calls when they are covering the story. Describe what types of calls you would not allow your reporters to broadcast if you were the director of a TV station. Also describe whether you would support any state law that would restrict use of these 9-1-1 calls.

2. Assume you are a member of a CISM team. In the course of a debriefing, a firefighter/paramedic expresses concern that a fellow firefighter "really needs to be here" because he has expressed some suicidal thoughts. Describe what action the CISM team member can take in light of the new Ohio law on testimonial privilege.

3. The county 9-1-1 dispatcher receives a call from a fire chief requesting the activation of the CISM team. They had been to a difficult motor vehicle accident run, involving a head-on crash that killed a family in their van, including three children, and the extrication of a drunk driver in the other vehicle who was only slightly injured. During the debriefing, a firefighter advises that the family killed in the accident were his neighbors and good friends, and he now knows the address of the drunk driver. He tearfully blurts out, "Don't be surprised if this guy disappears from the face of the earth." Describe your options as a CISM team member. What actions, if any, should the fellow firefighters in the debriefing take?

Chapter 16: Discipline: Misconduct On Duty and Off Duty and Ethical Decision Making

1. In the Key Statutes section of this chapter are the Ohio Rev. Code provisions that provide career (full-time, paid) firefighters with protection against suspension and termination. Discuss what protections exist in your state concerning career firefighters and what protections protect part-time and volunteer firefighters.

2. The inappropriate use of fire department computers to view illicit websites is becoming a problem in many fire departments. To reduce misuse of fire department computers, discuss the advantage of advising employees that there will be audits conducted that trace the use of fire station computers. Describe how such audits can be conducted.

Chapter 17: Arbitration/ Mediation

1. Under the new Ohio mediation statute, Ohio Rev. Code 2710.03 provides that a mediation communication is generally "not subject to discovery or admissible in evidence in a proceeding." Describe why such a provision helps the fire department, the union, and the firefighter reach agreement about a grievance over discipline imposed on a firefighter.

2. The new Ohio mediation statute provides in Ohio Rev. Code 2710.05 that there is no "mediation privilege" if there is "an imminent threat or statement of plan to inflict bodily injury or commit a crime of violence." Describe why the law has such an exception. Also discuss why the law has a similar exception for attorney-client communications, minister/rabbi communications, and testimonial privilege under the new Ohio Critical Incident Stress Management.

Chapter 18: Legislative Process: Getting Helpful Laws Passed for the Fire Service (Amending the Fireman's Rule)

1. Identify the names and addresses of your local representatives to your state legislature, and contact them about how to gain their support for legislation helpful to the fire service in your state.

Real-Life Legal Lessons Learned Scenario

Captain Tim Keene of the Delhi (OH) Fire Department and incident commander of the Ohio Task Force Team for Hurricane Katrina, in a Blackhawk helicopter over St. Bernard Parish (LA) October, 2005. *Photo supplied by Captain Tim Keene.*

Lessons Learned from St. Bernard Parish, Louisiana
by Timothy M. Keene, Commander, Ohio Task Force, and Captain, Delhi Township (OH) Fire Department, with 23 years in the fire service.

On October 8, 2005, I was deployed to St. Bernard, Louisiana, as the incident commander of an Ohio Task Force. This task force consisted of 6 engine companies, 2 truck companies, and one medic unit. The purpose of the task force was to help restore fire protection to the devastated area.

One of the most important lessons or realizations that must be foremost in the mind of the command staff is the job you are being sent to do. It became clear that we were sent to support the St. Bernard Fire Department in a recovery effort, and that any effort to "take over" would not be tolerated well. Our approach was "Anything we can do to help." With this attitude in mind, we were able to communicate with the St. Bernard Fire Department openly. Every meeting that we attended ended with the question, "What can we do for you today?"

This attitude was then used toward the members of the Ohio Task Force. These members were highly motivated and anxious to do whatever they could to improve the situation. This proved evident in the very first operation meeting. The task force wanted to get out into the community and help the residents. It was the job of the Incident Management Team (IMT) to promote this spirit without losing sight of the mission.

Another area that an IMT must be aware of—Politics. The IMT must understand that we are a part of a larger picture and the IMT is not the final answer in this area. In this case there were many elected folks that were trying to recover from a major disaster. The IMT must be patient. The IMT must find the way to work through all of the bureaucratic hoops.

As far as the IMT members and the members of the Ohio Task Force, I would recommend that CISM (Critical Incident Stress Management) be added as part of the response. Our team members did things and saw things that have changed their lives. CISM would have been a great benefit to our team.

The amount of personnel sent is questionable. These personnel were "on duty" 24 hours a day for the entire time they were in St. Bernard. This is very taxing and could lead to injuries from accidents. If there would have been a way to work 24 hours and then be off 24 hours, it may have been a better approach.

From the IMT side of things, even though it was difficult to be responsible for the operations every day, it did not seem to be overwhelming.

Equipment was a problem during this deployment. Office equipment, such as laptops and copiers and printers were hard to come by. Satellite hook up for Internet would have been a great help. Cellular (Satellite) phones are almost a must. We must remember in this situation, the entire cell sites system were damaged because of the storm. Getting a signal was difficult.

It would have been helpful to have only one contact person at home. Many times we would contact several different people to get answers. If one person had been the contact, taken the question, researched the problem or made some calls, then returned the information to the IC, it would have been a great help.

I offer this information only as suggestions to improve the operation. I feel the team we put on the field during my deployment was outstanding! Tasks these men accomplished during their time were way beyond anything I could imagine.

As for me, I like a challenge. I am very anxious to get the call to go again.

Example of Fire Chief Promotion Process

Fire Chief Nathan Bromen was promoted chief of Deerfield Township (OH) Fire and Rescue Department in January 2006. Chief Bromen holds a Masters degree from Grand Canyon University, a Bachelor of Science degree from the University of Cincinnati, and is a graduate of the Nation Fire Academy's Executive Fire Officer program.

Choosing a fire chief candidate to lead a fire organization is a substantial decision for any community. There are numerous expectations and complexities with managing a modern-day, multi-disciplined fire service delivery model and organization. The community leadership must select a competent candidate who will both instill public confidence, as well as have organizational support from within. When asked to recall the steps taken in our local fire chief replacement process, I reflected back on several philosophies that have had foundational significance for our community

First, as a community that was rapidly expanding with explosive commercial and residential growth, our department needed to continue progressing, because we were a new organization. After a decision had been made by the community seven short years ago to separate a previous joint fire district, we started a new fire department from scratch. Although most people may consider these circumstances to be a move backward, the community needed identity in the wake of continuous annexation by the neighboring city. Remodeling one station, renovating another fire station with a substantial addition, and constructing a third were just a few projects, in addition to replacing half of our apparatus fleet. Being progressive was not only a philosophy, but it was also truly a mandate for our survival. This grass-roots start-up provided opportunity, with virtually no or very few boundaries, and an ability to build accordingly to meet our needs. We have had tremendous community support from the beginning.

Second, the organization would need a fire chief who had balanced experience in the multiple disciplines with which we provide service. The candidate did not need to be an expert in every area but having sufficient knowledge and experience would help to gain respect among those charged with those responsibilities. As a typical, fast-growing suburban community, we provide a wide array of services. These include fire prevention and education, fire suppression, full EMS service and transport at the ALS level, specialized rescue services, hazardous materials response (with county team supplementation), and response to terrorism/WMD-related incidents (we are contiguous to a very large amusement park facility with supporting complexes, which is a specific target hazard). In addition to the basic career fire and EMS certifications, our organization requires all full-time employees to be paramedics, become certified fire safety inspectors, and develop skills in specialized rescue or support functionality. Both candidates were paramedics who had extensive background and certifications in each of these areas and who actually exceeded minimum qualifications significantly.

Third, the community had certain educational and professional expectations. In our somewhat upscale, suburban community, our government leadership felt that the fire chief should hold certain professional qualifications and educational degrees. The fire service is clearly in a transition phase from that of a trade to a profession, and the Board of Trustees felt that formal education was important. Again, both candidates were college graduates within fire-service-related disciplines and were either a National Fire Academy executive fire officer or a current EFO program student. In the future, having a chief fire officer designation may become the desired standard.

Fourth and finally, we embraced a philosophy and desire to promote from within the current ranks. Although the fire chief selection process first required the verification of an adequate number of competent, internal candidates, we were blessed to have two qualified and experienced internal candidates. All full-time hiring and promotions since our original inception seven years ago has come from within the full-time and part-time ranks of the department. Obviously, continued motivation from existing staff is at least partly dependent on future hiring or promotional opportunity. Caution may normally be needed because organizations can become stagnant and develop thinking that is limited to traditional actions within an existing department history. However, our department was still new, with continuing dynamic input from all levels and ranks within the organization.

In summary, each community should identify what is important to them and select who can best meet their needs. There are often basic philosophies that provide a foundation of expectations. Succession planning is a key component to the long-term success of every organization. We had focused on this in the past, so we had an adequate selection pool from which to choose, as opposed to going outside for someone who may not be a long-term match. The community was the ultimate benefactor, because the second fire chief candidate was appointed to the deputy fire chief position, and together we serve our community as a team.

Example of Position Posting
Fire Chief—Deerfield Township Position Posting
08/10/2005

Deerfield Township Fire Rescue Department is accepting applications for the position of:

Fire Chief

This is a career position with a starting salary in the range of *$70,000–75,000;* the standard employer PFDPF pension pickup of 24%, additional employer pickup of employee's 10% pension contribution, and full benefit package including comprehensive health insurance, vacation time, holiday time, and sick time.

Applicants must meet the following minimum certification requirements:
   Current employee of Deerfield Township
   10 years minimum service as a Fire Department Officer in a career or combination Fire Department serving a population of at least 20,000
   5 years experience as a Chief Officer in a career or combination Fire Department serving a population of at least 20,000
   State of Ohio certified FF I, or FFIB, or equivalent
   State of Ohio certified EMT-Basic
   Valid Ohio Driver's License

Applicants must meet the following minimum education requirements:
   Bachelor's Degree
   Preferences for an EFO (Executive Fire Officer) Graduate or Enrollee from National Fire Academy
   Equivalencies to the above, which can be documented by a candidate

Applicants will be subject to a post-hire physical, a complete background investigation and a psychological evaluation.

Applications are available from Deerfield Township Administrative Offices, 3378 Townsley Drive, Deerfield Township, between 08:00–16:30 hours, Monday thru Friday. In order to be considered for this position, a completed application with accompanying letter of intent for this position must be received by the Assistant to the Administrator no later than August 19, 2005 at 16:30 hours. Deerfield Township is an equal opportunity employer.

POST: All Deerfield Township locations, to be removed at 16:30 hours 08/19/2005

Deerfield Township Fire Department Job Description
August, 2005

Identification:
POSITION TITLE: Fire Chief
DEPARTMENT: Fire
IMMEDIATE SUPERVISOR: Township Trustees via Township Administrator

Job Summary
Performs duties of department head and senior administrator under direct supervision from the Township Administrator; responsible for fire protection, emergency medical services, and overall operations of the fire department; responsible for personnel, equipment, and facilities; responsible for fire code enforcement, fire prevention, and public fire education; performs related work as required.

Supervisory Responsibilities
Manages 4–7 subordinate supervisors who supervise a total of 90 employees in the Fire and EMS Department; is responsible for the overall direction, coordination, and evaluation of this unit; directly supervises three non-supervisory employees; carries out supervisory responsibilities in accordance with the organization's policies and applicable laws; assures effective organizational structure and utilization of chain-of-command; responsibilities include interviewing, hiring, and training employees; planning, assigning, and directing work; appraising performance; rewarding and disciplining employees; addressing complaints and resolving problems.

Essential Job Functions
Manages daily operations of the fire department; manages all full-time and part-time department employees.

Forecasts, plans, proposes, and administers departmental budget; approves expenditures.

Prepares specifications for apparatus and equipment purchases.

Supervises emergency management operations, including but not limited to, fire protection, EMS, hazardous material mitigation, emergency preparedness, fire prevention/education and code enforcement.

Develops and implements operating procedures for all functions of fire department.

Functions as Township representative to other Fire/EMS agencies; negotiates contracts and/or agreements for mutual aid assistance.

Reviews and approves building plans for construction relating to fire code compliance.

Plans and implements programs for anticipated growth of department to meet needs.

Maintains statistical data pertaining to operations; analyzes and reports fire protection and EMS status on a continuing basis; analyzes and projects fire loss statistics relative to insurability.

Develops and implements fire/EMS related resolutions.

Other Job Functions

Departmental representative to local fire chiefs' association; acts as liaison between fire and other township departments; may actively participate in fireground command operations and/or fire suppression and prevention activities; may investigate and/or document fire cause and origin investigations; may investigate fire code violations; supervise pre-fire planning, fire inspections, and acts as official spokesperson for all departmental functions.

Qualifications

To perform this job successfully, an individual must be able to perform each essential duty satisfactorily. The requirements listed below are representative of the knowledge, skill, and/or ability required. Reasonable accommodations may be made to enable individuals with disabilities to perform the essential functions.

Education and/or Experience

Minimum Bachelor's Degree in Fire Science or related field (Masters Degree preferred). Related experience and supplemental training may offset degree requirement. Five years experience as a high ranking officer of a department of similar size and complexity.

Language Skills

Ability to read, analyze, and interpret the most complex documents; ability to respond effectively to the most sensitive inquiries or complaints; ability to write speeches and articles using original or innovative techniques or style; ability to make effective and persuasive speeches and presentations on controversial or complex topics to top management, public groups, and/or boards of directors.

Mathematical Skills

Ability to work with mathematical concepts such as probability and statistical inference, and fundamentals of plane and solid geometry and trigonometry; ability to apply concepts such as fractions, percentages, ratios, and proportions to practical situations.

Reasoning Ability

Ability to define problems, collect data, establish facts, and draw valid conclusions. Ability to interpret an extensive variety of technical instructions in mathematical or diagram form and deal with several abstract and concrete variables.

Other Knowledge, Skills and Abilities

Must possess all skills and training required of firefighter; must possess high degree of managerial skills; must possess superior communication skills; must have knowledge of budgetary process and local and state laws relating to Fire and EMS policy; knowledge of local and national fire code and plan review process; must be familiar with state and federal employment rules; must possess negotiating skills; must have thorough knowledge of inspection and investigative procedures.

Required Certificates, and/or Licenses

Valid Ohio Driver's License; career level Fire Certification; EMT Certification (Paramedic Certification Preferred)

Physical Demands

The physical demands described here are representative of those that must be met by an employee to successfully perform essential job functions. Reasonable accommodations may be made to enable individuals with disabilities to perform the essential functions.

While performing the duties of this job, the employee is regularly required to talk or hear. The employee frequently is required to sit and use hands to finger, handle, or feel. The employee is occasionally required to stand; walk; reach with hands and arms; climb or balance; stoop, kneel, crouch, or crawl; and taste or smell. The employee must frequently lift and/or move up to 10 pounds and occasionally lift and/or move more than 100 pounds. Specific vision abilities required by this job include close vision, distance vision, color vision, peripheral vision, depth perception, and ability to adjust focus.

Work Environment

The work environment characteristics described here are representative of those an employee encounters while performing the essential functions of this job. Reasonable accommodations may be made to enable individuals with disabilities to perform the essential functions.

While performing the duties of this job, the employee is occasionally exposed to wet and/or humid conditions; moving mechanical parts; high, precarious places; fumes or airborne particles; toxic or caustic chemicals; outside weather conditions; extreme cold; extreme heat; risk of electrical shock; explosives; risk of radiation; and vibration. The noise level in the work environment is usually moderate.

Selection Guidelines

Formal application, rating of education and experience; oral interviews and reference check; job related tests may be required.

The duties listed above are intended only as illustrations of the various types of work that may be performed. The omission of specific statements of duties does not exclude them from the position if the work is similar, related or a logical assignment to the position.

The job description does not constitute an employment agreement between the employer and employee and is subject to change by the employer as the needs of the employer and requirements of the job change.

Diversity Goals

The City of Cincinnati and Cincinnati Fire Fighters Union Local 48 have the following provision in their 2005–2007 Labor-Management Agreement:

ARTICLE 40

DIVERSITY GOALS

The City of Cincinnati and the Union support racial and ethnic diversity and acceptance throughout the City. The Union supports the Cincinnati Fire Department's efforts to increase racial and ethnic diversity within each fire station.

The Cincinnati Fire Department and the Union have undertaken efforts to achieve diversity, such as incentive transfers. The Cincinnati Fire Department and the Union remain committed to ensuring that diversity is promoted, valued, and supported within the Fire Department, as there is a direct relationship between the human composition of these institutions and the attitudes and image regarding the City of Cincinnati.

The City of Cincinnati and the Union recognize the intrinsic value and social benefit of racially balanced public service institutions and urges the continued commitment to pursuing this balance whenever possible.

Sample Labor-Management Agreement

The City of Cincinnati and Cincinnati Fire Fighters Union Local 48 have the following provision in their 2005–2007 Labor-Management Agreement:

ARTICLE 33

DRIVERS LICENSE

No employee may operate any City owned vehicle or private vehicle on City business without an unrestricted operator's license required for the particular type of apparatus operated. Employees who have court-granted waivers to drive to, from or at work while they are on suspension may not operate City equipment regardless of any court exemption. Restrictions for medical reasons (e.g. eyeglasses) are not subject to this policy.

Employees must notify their Immediate Supervisor of any driving restriction not later than the next business day after the restriction is imposed. The supervisor is required to inform the Fire Chief through the proper chain of command.

The Fire Department will accommodate employees on restricted or suspended driving license for a period of time not exceeding 180 days for the first serious vehicular offense by placing the employee in a job assignment not requiring the operation of a vehicle for work. Employees with their first serious vehicular offense shall be referred to the Public Employees Assistance Program (PEAP) to be evaluated by a Substance Abuse Professional (SAP), and shall follow recommendations of the SAP or be subject to a written reprimand. Employees with restrictions of more than 180 days and/or more than one conviction of serious vehicular offenses are subject to disciplinary action.

Sample Labor-Management Agreement: Letter of Agreement—Health and Wellness

The City of Cincinnati and Cincinnati Fire Fighters Union Local 48, Labor-Management Agreement for 2005–2007 have the following provision in an Appendix:

LETTER OF AGREEMENT

HEALTH AND WELLNESS

CITY OF CINCINNATI (CITY) AND IAFF LOCAL 48 (UNION)

Whereas, the City and the Union agree that the implementation of a Joint Health and Wellness Program for fire fighters and the medical guidelines of NFPA 1582 may be beneficial to both parties, the Parties agree as follows:

Section 1. The City shall provide $200,000 per year towards a Joint Health and Wellness Program.

Section 2. The City and the Union agree that during the duration of a Joint Health and Wellness Program and in the interest of the health and wellness of all Cincinnati fire fighters, all members shall have a physical on a bi-annual basis or as recommended by the Health and Wellness Physician. Participation in the program is mandatory for all members. Failure to participate may result in disciplinary action.

Section 3. Physicals conducted by the Joint Health and Wellness Physician shall be done on duty. Members may elect to participate in physicals or other evaluations through the Joint Health and Wellness Program on their off duty time; however, if a member elects to participate while in an off duty status, said off duty time spent shall not be considered time worked and shall not be compensable. Additionally, referrals of members by the Joint Health and Wellness Physician to the member's personal treating physician shall be conducted in an off duty status; said off duty time spent shall not be considered time worked and shall not be compensable.

Section 4. The confidentiality of all personal medical information obtained shall be in accordance with federal and state law, and no personal medical information may be disclosed to any person without written permission of the member.

Section 5. Parties have further agreed that the Joint Health and Wellness Physician and the member's personal treating physician retain exclusive control over any recommendation for all duty status determinations through the Health and Wellness Program, and such duty status determinations shall be entirely independent of and without control by EHS. However, if for any reason, the Fire Administration has concerns over a member's fitness for duty, this section does not prohibit an independent medical evaluation by the Employee Health Services physician or another licensed physician chosen by the City.

Section 6. A Health and Wellness Program Steering Committee shall be formed and consist of one (1) representative from Fire Administration, one (1) representative from the Union, and one (1) representative from City Administration. The Steering Committee shall recommend by unanimous consent a Health and Wellness Physician, who shall be subject to approval by the City. The Steering Committee shall also determine by unanimous consent the requirements of the Health and Wellness Program.

Section 7. Any study or report regarding the health and wellness of fire fighters utilizing information submitted to or gathered through the Health and Wellness Program shall be submitted to Local 48 sixty (60) days prior to publication or release. Any such study or report shall conform to all federal, state and local laws related to patient confidentiality.

Section 8. This Letter of Agreement supersedes the Health and Wellness Side Letter, dated July 1, 2004, and shall expire on June 2, 2007.

Sample Labor-Management Agreement: Grievance Mediation Procedures

The City of Cincinnati and Cincinnati Fire Fighters Union Local 48,
Labor-Management Agreement for 2005–2007 have the following provision.
(Labor-Management Agreement kindly provided by Assistant Chief Mike Kroger,
Cincinnati Fire Department.)

GRIEVANCE MEDIATION PROCEDURES

In recognition of the desirability of resolving disputes by mutual agreement, the Union and the City, hereinafter referred to as the "parties," mutually agree to the following policies and procedures for the mediation of grievances pending arbitration, pursuant to the provisions of Article 6 of the collective bargaining agreement between the parties.

1. Grievance mediation is available to the parties at Step Three of the grievance procedure.
2. A grievance will be referred to mediation at the request of the Union unless the parties mutually agree to not mediate a particular grievance.
3. The parties shall mutually agree to a mediator to serve in the capacity of grievance mediator. The mediator must be an experienced mediator and/or arbitrator with mediation skills. The mediator may serve as an arbitrator for the same issue for which he or she is a mediator only with the agreement of both parties.
4. The mediator will be asked to provide a schedule of available dates. Cases will be scheduled in a manner which assures that the mediator will be able to handle multiple cases on each date, unless otherwise mutually agreed.
5. The grievant shall be represented by the Union and shall have the right to be present at the mediation conference. The City and the Union may each have no more than three (3) representatives (not counting the grievant) as participants in the mediation effort. Neither party shall be represented by an attorney. On the day of the mediation, persons representing the parties shall be vested that day with full authority to resolve the issues being considered.
6. The mediator may employ all of the techniques commonly associated with mediation, including private caucuses with the parties. The taking of oaths and the examination of witnesses shall not be permitted, and no verbatim record of the proceeding shall be taken. The purpose of the mediation effort is to reach a mutually agreeable resolution of the dispute. There will be no procedural constraints regarding the review of facts and arguments. There shall be no formal evidence rules. Written materials presented to the mediator will be returned to the party presenting them at the conclusion of the mediation conference.
7. Mediation efforts will be informal in nature and shall not include written opinions or recommendations from the mediator unless mutually agreed to by the parties and the mediator. In the event that a grievance which has been mediated is appealed to arbitration, there shall be no reference in the arbitration proceeding to the fact that a mediation conference was or was not held. Nothing said or done by the mediator or the parties may be referenced or introduced into evidence at the arbitration hearing.

8. At the mediation conference the mediator shall first seek to assist the parties in reaching a mutually satisfactory settlement of the grievance which is within the parameters of the collective bargaining agreement. If a settlement is reached, a settlement agreement will be entered into in writing at the mediation conference. The mediator shall not have the authority to compel the resolution of a grievance.

9. If a grievance remains unresolved at the end of the mediation conference the mediator may, if requested by either party, render a verbal opinion as to how the grievance is likely to be decided if it is presented at arbitration. This opinion is non-binding and inadmissible in any subsequent arbitration proceeding.

10. If a settlement is not reached, the matter may go forward to Step Four Arbitration Level. All applicable time limits for appealing a grievance to arbitration contained in the collective bargaining agreement shall commence on the day of the mediation conference.

11. The dates, times and places of mediation conferences will be determined by mutual agreement of the parties. Each party shall designate a representative responsible for scheduling mediation conferences.

12. The parties agree to timely schedule grievance mediations when a grievance is advanced to mediation.

13. The fees and expenses to be charged by the mediator shall be negotiated between him or her and the parties. Fees and expenses for grievance mediation shall be paid equally by the City and the Union.

14. The parties agree to mutually examine and review the grievance mediation process and procedures adopted herein twelve (12) months from the date of execution of the collective bargaining agreement. The purpose of said examination and review is to revise, alter, correct or otherwise improve the grievance mediation process and procedures if such is deemed necessary.

Glossary

affirmative action programs Federal court orders that require fire departments and other public employers to adopt programs to hire, retain, and promote minority employees.

amicus curiae In cases before the U.S. Supreme Court and other courts, the court may allow others with an interest in a case, such as the U.S. Solicitor General or the American Civil Liberties Union (ACLU), to file "friend of the court" briefs.

assumption of duty A fire department assumes a duty when it responds to assist a resident, such as when power lines are down, but is not liable for injury or death, unless its conduct increases the danger to the resident.

balancing test Employees such as firefighters and police officers have a restricted right of free speech, balanced against the government employer's right to avoid undue public concern.

bifurcated trial In lawsuits in federal court under the FLSA and other statutes, the trial judge may ask the jury in the first phase of the trial to determine if the employer violated the statute. In the second phase, the jury will be asked to determine back pay and other damages for each plaintiff.

bona fide occupational qualification The Age Discrimination in Employment Act (ADEA) prohibits employment discrimination against individuals age 40 or older unless the employer can show that age is a legitimate occupational qualification given the duties and risks of the job.

class action lawsuit This is a lawsuit filed on behalf of all similarly situated employees, such as a lawsuit on behalf of all paramedics on a fire department who, prior to the 1999 change in the FLSA, did not receive overtime after 40 hours in a work week.

class of jobs ADA and U.S. Supreme Court decisions require employees to prove that their disability prevents them from performing not only their current job but also a broad range of work.

coerced Statements obtained from an employee, who is subject to being fired for refusal to answer, may be considered inadmissible in a criminal trial unless the employee is first granted immunity from criminal prosecution.

constructive discharge Describes a work atmosphere that is sufficiently severe or pervasive to alter the conditions of employment. A plaintiff must show that the abusive working environment became so intolerable that resignation qualified as a fitting response.

deliberate interference In lawsuits filed against state governments or political subdivisions alleging deprivation of constitutional rights, plaintiff must prove deliberate misconduct on interference with an individual's rights.

disability Under the ADA, an individual with a physical or mental impairment that substantially limits one or more major life activities.

dismissed without prejudice When a lawsuit has been dismissed by the trial judge "without prejudice," the plaintiff may later file a new complaint if additional facts are discovered.

disparate impact Physical agility tests in which the pass rate of women is less than 80 percent of the pass rate of men are deemed by the Equal Employment Opportunity Commission (EEOC) as evidence of their inappropriate impact on women. In age discrimination and race discrimination lawsuits, statistical evidence may be admissable to prove adverse impact.

duty of care For one party to be liable for injuries to another party, there must be a legal duty of care owed to this injured party.

equal protection Employees facing discipline can assert under the Fourteenth Amendment that the employer is treating them more harshly than others for similar misdeeds.

equivalent position Under FMLA, when an employee returns from leave, the employee is entitled to his or her old job back immediately or a job that is equivalent.

estoppel Under FMLA, if the employer has granted leave, even though the medical condition did not qualify as a serious health condition, the employer may have waived its right to challenge the basis for the leave in court.

exhaustion of administrative remedies doctrine A firefighter who is covered by a collective bargaining agreement must file a grievance and process it through arbitration, prior to seeking any relief in a court of law.

exigent circumstances Emergency circumstances, including extinguishing fire, will authorize warrantless entry of private property.

failure to train Lawsuits may be filed by injured civilians under 42 U.S.C. 1983, alleging failure to adequately train emergency responders and dispatchers.

failure-to-accommodate Under the ADA, an employer must reasonably accommodate an employee with a disability if this will assist the employee in performing the essential functions of the job.

Fireman's Rule A rule prohibiting public safety officers, including firefighters, EMS, and police officers, from filing suit against a property owner for their death or injury, unless there has been an extraordinary hazard created by the property owner.

FISA Court Congress established this court under the Foreign Intelligence Surveillance Act (FISA) of 1978, in which selected U.S. District Court judges review requests for foreign intelligence electronic surveillance wiretaps.

five-step grievance procedure Under collective bargaining agreements between fire departments and local unions, it is common to have a multi-step process to file and process grievances.

front pay In race and sex discrimination lawsuits, damages can be awarded not only for lost back pay, but also for equitable relief in the form of expected lost future earnings.

good faith Under the FLSA, if the trial judge determines that the employer did not carefully review the federal regulations and U.S. Department of Labor opinion letters on overtime pay, plaintiffs may also be awarded a penalty payment of "liquidated damages" (the back pay award can be doubled).

governmental immunity A political subdivision, such as a city, township, or village, may enjoy immunity from liability under state statutes as long as its employee was adequately trained and was not acting with willful or wanton misconduct.

hostile work environment Prohibited by Title VII of the Civil Rights Act of 1964, it includes practices ranging from direct requests for sexual favors to workplace conditions that create a hostile environment for persons of either gender, including same sex harassment.

intermittent leave Under the FMLA, employees may take up to 12 weeks of leave, including leave in which the employee takes time off intermittently.

irreparable harm For plaintiffs seeking an injunction, such as firefighters asking the court to stop the launch of a wellness and physical fitness program, there must be evidence that the plaintiff will be permanently harmed unless the program is enjoined.

last-chance agreement Occurs when an employee seeks treatment for alcohol or drug addiction and signs an agreement that if he or she tests positive upon return to work, employment will be terminated.

major life activity ADA requires persons claiming to be disabled to show that their condition prevents them from performing major life activities.

malicious purpose In many states, the political subdivision enjoys "governmental immunity" and the public employee enjoys "qualified immunity" unless the employee acted with malicious purpose, bad faith, or in a wanton or reckless manner. Ohio Rev. Code 2744.03 (A) (6) (b).

mediation Defined under the new Ohio Mediation Act as "any process in which a mediator facilitates communication and negotiation between parties to assist them in reaching a voluntary agreement regarding their dispute." Ohio Rev. Code 2710.01, effective November 29, 2005.

medical review officer A physician or other medical expert licensed to confirm a "positive" drug test.

mixed motive Employers who discipline a female employee more harshly than male employees can be sued for sexual harassment and will be liable if the jury concludes the employer's conduct was motivated in part by the plaintiff's gender, even if the employer's conduct was also motivated by a lawful reason.

petition for a writ of habeas corpus A petition filed by a person held in a jail or prison seeking a judge to review the lawfulness of his or her detention.

pregnancy discrimination Title VII was amended by the Pregnancy Discrimination Act of 1978, which prohibits discrimination on the basis of pregnancy, childbirth, and related medical conditions.

prejudgment interest A plaintiff who proves discrimination can be awarded interest on the money that he or she should have been paid if the employee had not been unlawfully terminated or forced to resign, or had not been passed over for promotion.

prima facie case A plaintiff must present sufficient evidence of race or sex discrimination, and then the burden shifts to the employer to prove race or sex was not a factor in the employment decision.

products liability The Fireman's Rule does not prohibit a firefighter in most states from suing a manufacturer of defective fire equipment or a defective appliance, including a gas water heater or other product causing a fire in a building.

public records 9-1-1 audio tapes of callers may be considered records open to the public and TV and radio stations for public broadcast.

qualified immunity Fire, EMS, and other governmental officials will normally not be held personally liable for their negligence, as long as they did not act with willful or wanton misconduct.

random drug testing Employees are selected at random, often by computer search, for a drug test.

rescue doctrine Protects individuals acting as "Good Samaritans" when trying to rescue or render aid to others in an emergency from liability for their negligence, as long as they acted in good faith, did not exceed their scope of authority, and were not grossly negligent.

restored to prior position An employee who takes leave under FMLA is "entitled, on return from such leave to be restored by the employer to the position of employment held by the employee when the leave commenced; or to be restored to an equivalent position."

retaliation Employee who files an EEOC complaint is protected from discipline, if he or she can prove there is a causal connection between the filing of the EEOC complaint and the subsequent discipline.

reverse discrimination A city or other political subdivision can defend affirmative action for minorities if there has been a history of prior race discrimination and the corrective actions taken were reasonable to overcome that history.

self-incrimination The Fifth Amendment protects firefighters and others from being compelled to answer questions that can incriminate them in a crime; if an employee is granted written immunity from projection, then the employees must answer questions concerning his/her conduct.

sex discrimination The EEOC definition is as follows: "It is unlawful to discriminate against any employee or applicant for employment because of his/her sex in regard to hiring, termination, promotion, compensation, job training, or any other term, condition, or privilege of employment. Title VII also prohibits employment decisions based on stereotypes and assumptions about abilities, traits, or the performance of individuals on the basis of sex."

special need showing When a fire department seeks to justify blood draws and other invasions of privacy, a public employer must demonstrate a special need to protect the public.

state-created danger EMS personnel may be liable if they unnecessarily increase the risk of danger to a patient by acting with willful or wanton misconduct.

subterfuge Congress has authorized mandatory retirement for firefighters age 55 or older, as long as there is a state or local bona fide retirement plan, and it is not used in a discriminatory manner to force only certain employees to retire.

summary judgment A trial judge dismisses the lawsuit prior to trial, based on his or her conclusion that, as a matter of law, the facts in the case support one party's position.

Title VII Refers to Title VII of the Civil Rights Act of 1964. Title VII protects individuals against employment discrimination on the basis of sex as well as race, color, national origin, and religion. Applies to employers with 15 or more employees, including state and local governments. It also applies to employment agencies and to labor organizations as well as to the federal government.

unreasonable searches and seizures Under the Fourth Amendment to the U.S. Constitution, evidence seized without a court-ordered search warrant may be determined to be inadmissible in evidence, unless the seizure is deemed by a judge to be reasonable under the circumstances.

waiver of right to sue After the 9/11 attacks, Congress established a no-fault Victim Compensation Fund, which includes a provision that all claimants waive their right to sue the airlines or others for damages arising out of the terrorist attacks.

willful or wanton misconduct Operating an emergency vehicle or performing other duties with gross negligence or failure to exercise any care towards the public.

Index